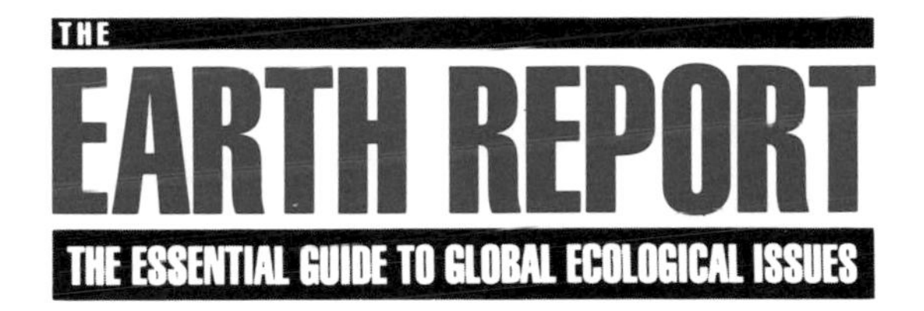
THE
EARTH REPORT
THE ESSENTIAL GUIDE TO GLOBAL ECOLOGICAL ISSUES

GENERAL EDITORS

EDWARD GOLDSMITH AND NICHOLAS HILDYARD

PRICE STERN SLOAN, INC.

THE EARTH REPORT was edited and designed by Mitchell Beazley International Limited, Artists House, 14–15 Manette Street, London W1V 5LB

Editor Frank Wallis
Art Editor Roger Walton
Assistant Editor Simon Ryder
Assistant Art Editor Paul Drayson
Designer Susie Hooper
Editorial Assistant Lesley Crabb
Picture research Brigitte Arora
Production Ted Timberlake

Published in the United States and Canada by
Price Stern Sloan, Inc.
360 N. La Cienega Blvd.
Los Angeles, California 90048

Library of Congress Cataloging-in-Publication Data
The Earth Report
Includes index.
Contents: Man & the natural order/Donald Worster -- Politics of food aid/Lloyd Timberlake -- Man and Gaia/James Lovelock -- Nuclear energy after Chernobyl/Peter Bunyard -- [etc.]
1. Environmental protection. 2. Ecology. 3. Man -- influence on nature. I. Goldsmith, Edward, 1928–. II. Hildyard, Nicholas.
TD170.E17 1988 304.2 87–21451

ISBN 0–89596–673–0
ISBN 0–89586–678–1 (pbk.)

Note: Although all reasonable care has been taken in the preparation of this book, neither the publishers nor the contributors or editors can accept liability for any consequences arising from the use thereof or from the information contained herein. The Publishers will be grateful for any information that will assist them in keeping future editions up to date.

Photoset in Great Britain by Bookworm Typesetting, Manchester
Reproduction by David Bruce Graphics Ltd., London
Printed in Spain by Graficas Estella, S.A., Navarra

CONTENTS

How to use this book

THE EARTH REPORT is divided into two sections, the first containing signed essays on six topics of importance and the second some 400 shorter articles arranged alphabetically. The A-Z entries are in strictly alphabetical order; thus **2,4,5–T** follows **Tropical forests** and precedes **2,4–D.**

The A-Z entries carry both internal cross-references to other, relevant entries (which are indicated by *qv* in parentheses) and “”see also” cross-references, set at the end of each entry, that refer to entries containing additional material. The “”see also” references carry the following symbols:

ⓐ agriculture
ⓒ conservation
ⓔ energy
ⓕ foundations of ecology and basic concepts
ⓛ lifescan — entries relating to health and nutrition
ⓝ nuclear
ⓟ pollution
③ Third World

The symbols enable the reader to follow a specific theme. Thus the “”see also” from **Acid rain** to **Coal** carries the symbols ⓔ and ⓟ, which means that the entry on **Coal** discusses its energy and pollution aspects. If the reader wishes to pursue the theme of pollution, the entry will yield information; however, the “see also” from the same entry to **Deforestation** carries the symbols ⓒ and ③, which mean that the entry has no information on pollution, but does deal with conservation and the Third World.

There are no cross-references in the essays, but the reader will find unfamiliar terms defined in the A-Z section.

Measurements Both metric and imperial measurements are given, with the exception of measurements for siltation, which are given in acre-feet. See below for conversion factors.

Currency Is given in United States dollars ($), Australian dollars ($A) or pounds sterling (£). The conversion rate used is $1.60 = £1.

Billions Because of differing usage between the United Kingdom and the United States, billions are represented as thousand millions: £15,000 million = £15 billion.

Abbreviations and conversions

Bq	becquerel
cm	centimetre (0.393701 inches)
cu m	cubic metre (35.3146 cubic feet)
ft	feet (30.48cm)
gm	gramme (0.035273 ounces)
kg	kilogramme (1,000 grammes 2.204 pounds)
km	kilometre (0.62137 miles)
kph	kilometres per hour
lb	pound (0.4536kg)
mph	miles per hour
MW	Megawatts (a million watts)
microgramme	a millionth of a gramme
milligramme	a thousandth of a gramme
mol	mole (the SI unit of amount of substance)
mSv	milliSieverts
tonne	1,000kg (2,204lb)
oz	ounce (28.35gm)
sq	square

1 hectare (ha) = 2.47105 acres
There are 100ha in a sq km

Foreword

This book documents relentless, seemingly unending environmental destruction – the drainage of wetlands, the destruction of coral reefs, the extinction of species, the erosion, desertification and salinization of agricultural land, and its paving over to accommodate roads, housing estates and factories. It deals with the contamination of groundwater and surface waters by agricultural and industrial chemicals, the acidification of lakes by acid rain (which is also involved in the degradation and death of forests), the apparent depletion of the ozone layer by fluorocarbons and other industrial chemicals, and the destabilization of climate as a result of the accumulation of carbon dioxide and other greenhouse gases in the atmosphere.

It is also about escalating human problems such as the population explosion, massive urbanization with an ever-growing proportion of humanity living in squalid shantytowns, the impoverishment, the malnutrition and the famine which, for the first time in the human experience, has become chronic over a entire continent and is rapidly spreading to others, the inexorable growth in the incidence of the "diseases of civilization," in particular cardiovascular disease and cancer, and the equally inexorable spread of infectious diseases to areas where they were previously unknown.

The book shows that people's reactions to all this is influenced by one or other of two different and opposed views of the world. At one extreme the natural world or "biosphere" we inherited is seen as fundamentally hostile – "red in tooth and claw"` to use Tennyson's famous phrase – and the lives of those who derive their sustenance directly from it as "nasty, brutish and short" to use Hobbes's equally famous phrase. Those who hold that view see it as our duty to transform the world as radically as possible to make it more hospitable and hence to ensure that our lives become more pleasant and civilized, and also longer. This transformation they call "progress", and equate it with economic growth, measured in gross national product (GNP).

The opposite view is that we ourselves have created that hostility; that our needs are not so much material as biological, social, cultural, aesthetic and spiritual, and were admirably satisfied in the natural world we inherited. Most people see some merit in both views, and try to reconcile them in their daily lives.

Unfortunately, the first view dominates our modern world. All benefits are seen as derived from the man-made world or "technosphere," to which economic growth gives rise. Human welfare is measured by per capita GNP. Support comes from modern economics, which attributes no economic value to those benefits provided free by the normal functioning of the biosphere, such as fresh air to breathe, clean water to drink, fertile soil, regular rainfall and a predictable climate, without which we could not eat. It must follow that in destroying our environment, as we are doing, and hence in depriving ourselves of such biospheric non-benefits, we are seen as incurring no costs. This, partly at least, explains why, in terms of the dominant world-view, all this destruction is viewed with such indifference. Those who nevertheless concede that such destruction cannot occur with total impunity, deny that it is in any way connected with economic development. Indeed, how can a process that has been identified with progress give rise to such undesirable side-effects? The destruction, it is maintained, can only, on the contrary, be the work of the poor, whose life has remained "nasty, brutish and short," and who are so preoccupied with day-to-day survival that they cannot consider the side-effects of their activities on the environment.

The population explosion is also seen as the product of poverty, the poor being necessarily insecure and hence requiring as many children as possible to look after them in their old age. Malnutrition and famine are also attributed to poverty, the poor being unable to invest in modern agricultural technology, and thereby being forced to make use of archaic and unproductive agricultural methods, while, for want of cash, they are unable, when the harvest fails, to buy the food they require to feed themselves and their families.

In other words, economic development is not only exonerated of any blame for causing the environmental and human problems described in this book, but is actually made out to provide their only solution.

Such an interpretation is accepted by many people today, since most of us are employed by, and have thereby become dependent for our livelihood on, the continued expansion of the international agencies, government departments, institutions and business corporations which have come to replace the original social groupings in which man has spent the greater part of his experience on this planet.

Such an interpretation is also convenient for those who direct such surrogate social groupings, and whose income, professional status, power and influence are clearly at stake. Thus, it is not surprising that this interpretation is clearly reflected in all the literature published or sponsored by the World Bank, the Food and Agricultural Organization of the United Nations (FAO) and the other institutions sponsored by the development industry, rationalizing and thereby legitimizing the development enterprise to which they are so totally committed.

Today, however, an increasing number of people are coming to question the dominant world-view, and one can see emerging among them the rudiments at least of what we might refer to as the "world-view of ecology." In terms of this world view, it is only in the aberrant man-made world in which we live that we require the material goods made available to us by economic development. Their role is largely to provide some sort of compensation for real benefits, those that satisfy our real needs, and of which economic development is depriving us. As economic development systematically transforms the biosphere, however, so those real needs are ever-less well satisfied, and so is our requirement for material compensations correspondingly increased.

That economic development is the cause of all the problems considered in this book seems clear. How else can one explain that destruction is occurring in both temperate and tropical countries, whether their economy is centrally planned, as in the communist block, or under a regime of free enterprise, as in the West?

How else can one explain that more destruction has been wrought to the fragile fabric of the biosphere during the last 40 years, since global development has really got under way, than during the preceding two or three million years of the human experience on this planet?

Indeed, we can no longer afford to delude ourselves, the situation is too critical. Our choice is a simple one: either we put our society on a radically different course so as to reduce, rather than continue to increase, its destructive impact on the biosphere, or we delegate this task to the four horsemen of the Apocalypse.

Edward Goldsmith

Donald Worster

"Never before in our history has the organic world around us been in so much trouble . . . We are creating an environment of gashes, wounds, disorganization and death"

A few years ago I came down a backcountry road in Wisconsin looking for a place where a man had given his life. The road had once been the route of pioneers moving west, then a farm road running through dry, sandy, marginal fields. In the days of Prohibition it had carried illegal whiskey distilled hereabouts, some of the last trees having been cut down to cook the bootlegger's brew. Then in 1935 another sort of settler came along. It was the time of the Great Depression, and he could buy a lot of land, 120 acres in all, land abandoned by its owners, for a little money in back taxes. The land had no economic value left in it. The man, whose name was Aldo Leopold, knew that but did not mind; he was not after gain or even subsistence. He began coming out regularly from the city of Madison, where he taught at the university, to plant trees. For 13 years he planted and nurtured. Then, in 1948, he died fighting a forest fire on a neighbour's land. Knowing those few details, I came wanting to know what manner of man he was and what he had died for.

There was no publicity, no tour guide provided, but the dense forest of pines was a sufficient announcement that here was Leopold's place, now all grown up again to natural splendour. I walked through an open field rich in wild grasses and forbs to a small, grey, weathered shack where he had stayed on those weekends, regaled by the smell of his new pines coming up and the sound of birdsong and wind in their branches. From the shack, I found my way down a short path to the Wisconsin River, rolling silently between its pungent banks, the warm summer sun glinting on its ripples. One August years ago Leopold, as recalled in a sketch he wrote and collected in *A Sand County Almanac*, found the river "in a painting mood," laying down a brief carpet of moss on its silty edges, spangling it with blue and white and pink flowers, attracting deer and meadow mice, then abruptly scouring its palette down to austere sand. For me, that painting had long disappeared but not the memory of it, which had been lastingly captured in the words of the man who had seen it, and who, in its presence, must have stood for a breathless moment or two, intensely sure that he had nature on his side. The land, he realized, could come back from its degradation. It had an inexhaustible capacity to create harmony and grace out of the most ordinary, valueless materials, even out of silt, minute spores, and the trackings of herons along a bar. And no matter how abused its history had been, it could regenerate itself. In the thirties, a time of national despair brought on by severe economic and ecological collapse, that must have been a reassuring fact to discover. Today, it is still a needed fact. And Leopold's homeplace is where the world can come to see demonstrated those processes of natural regeneration, aided by human commitment and intelligence. There is, after all, a way back to the Garden.

I am increasingly an admirer of Leopold's wisdom and, though aware of the need for collective effort, find the most hopeful message in his individual, unaffected dedication to a humble dream. He did not turn the job of restoration over to someone who had more money or authority or free time. Though a busy professional, active in research and teaching, with another home and a family in the city to care for, he put his own shovel and his own back to work rehabilitating his piece of land.

Leopold does not appear at any point to have had the notion that what he was doing on those cutover, depleted acres was building an investment for himself or others, that one day the property would again be worth something on the market, for lumber or recreation. Though trained in the modern school of natural resource management – he was a game specialist in his other life – he was not here to "manage" the wildlife or forests or water, at least not as profitable commodities that we manipulate for our own instrumental ends. What brought him out on weekends was, first, a desire to know the place intimately and, then, to apply his knowledge and love toward its healing. He came as a kind of doctor, a "country doctor" we might say, who had found an ailing, neglected patient who could use his care. In accepting moral responsibility for that patient, he disregarded all chance of remuneration except for whatever satisfaction he could get simply in watching the recovery progress.

"Finding damage done to a thing of beauty, he felt an urge to bring back its glory"

Alternatively, though it comes to the same thing, we could say that he came as an artist and, finding damage done to a thing of beauty, he felt an urge to bring back its glory. Health for him was beauty, beauty was health. Leopold had the kind of aesthetic temperament that seeks not to find an outlet for its own subjective impulses, but to learn what there is of wonder in the world, latent or achieved, and to become its appreciator, its caretaker, accepting that beauty has an objective existence outside oneself, that beauty is a quality that can be discovered as well as invented. No one who had settled Leopold's place before him had paid much attention to its aesthetic qualities; perhaps, given the pressures of survival, they could not. They saw only the surface of things. "The incredible intricacies of the plant and animal community," he wrote, "the intrinsic beauty of the organism called America," had disappeared under the heavy tread of settlement without much thought or perception.

What draws visitors like myself to Leopold's wild, blooming garden today is a shared idea that the world of nature has a pattern of order that we are bound to respect and care for, perhaps even risk our lives to save.

I don't know where the idea of such an order originates. Probably it has been there in the human mind all along, like the ability to count and make things and raise children. Certainly the earliest cultures, pagan and animistic, had a strong, lively view of it. So too, though with varying degrees of commitment, have such modern religious traditions as Judaism, Christianity, Taoism, and Hinduism. Christians, for example, sometimes speak rapturously of the "Creation," having in mind some ordered arrangement of natural things that is at once rational and beyond comprehension. Its harmony is functional, but not narrowly utilitarian for them. It demands that they admire as well as use it. Christians along with Jews go on to insist that there must be Someone who contrives that beauty out of nothing and constantly holds it intact in time and space. The heavens declare the glory of God, they say with the psalmist, and the Earth shows His handiwork. But surely the power conjured up as explanation came later, the awareness of beauty in the world was there first. The Taoists of China, in contrast, believe that the order of nature has not been contrived by any outside force but inheres in nature itself – that there is a Way, and all things move together harmoniously in it and under their own power. Whatever these differences in concept, every religionist can agree that, because there is luminous beauty in the world, there is some obligation on the part of humans to respect it. Although we did not design or organize that order, we are capable of being, and therefore are obliged to be, its stewards and guardians. That is one of the fundamental truths that men and women of many faiths have held to be self-evident.

I grew up believing that religion had a permanent enemy in science, that the victory of one must mean the defeat of the other. It still seems clear that religion has lost considerable ground in modern times to science as a source of authority. But how astonishing to find out that science, whatever its competitive effect on any particular creed, itself rests finally on an awareness of natural beauty that is very like that in religion. Take away the assumption that the world is an orderly

Aldo Leopold working on his land in Wisconsin in the late 1930s

whole whose parts all work together toward a self-regulated stability, that there is an arrangement and coherence to things that can be understood, and science would cease to exist. I see now that science, and every branch of it, begins with some aesthetic ideal. It is a bedrock principle. And quite possibly, like its religious analogue, it cannot be proved once and for all by appealing to facts or texts, but instead is derived from some deeper process of insight – from an intuition that comes to almost everyone living in close, observing relationship with nature. Scientists commonly do so; it is not surprising, therefore, to find that many of them have spoken with awe and delight about the exquisite order they find in the world.

Aldo Leopold was such a scientist, an early ecologist proficient in the study of food chains, population cycles, and energy flow; and it was ecological science, he maintained, that had opened his eyes to a nature the pioneers never saw.

When it comes down to reducing the perception of natural order to some experimentally testable theory, however, scientists have always run into disagreements. Their process of analyzing and describing inevitably leaves something significant out. Struggle then ensues over who has the most nearly adequate explanation. Perhaps, generation improving on generation, we will arrive one day at a complete account of the world that will satisfy everyone's data, but it seems unlikely. Science has an impressive method for explaining the coherence of nature; it may be a better one than religion has. But nonetheless it is imperfect and incomplete, narrowing as it does the ineffable whole to some manageable (or measurable) portion, inevitably setting in motion the counterforces of scepticism and challenge.

Scientists, moreover, are people em-

Leopold's land today: "an open field rich in wild grasses and forbs . . ."

bedded, like the rest of us, in their societies. The order they find in nature, and the relation of order to disorder, is bound to be influenced by their social and historical circumstances. They must to some extent see what their times allow them to see. From the mid-nineteenth century on, the times have become more and more disorderly, and that condition has been reflected in the attitudes of many scientists.

Beginning with Charles Darwin and his theory of evolution, ecologists have insisted that nature is not to be understood as a fixed or permanent order but is constantly undergoing change, much of it violent and destructive. Did that understanding destroy the age-old confidence in natural coherence? For a long while it did not. Darwin, for example, did not let the new way of seeing shake his conviction that nature manages to contrive a remarkable degree of order, that despite all the ragged opportunism of individual organisms striving for success, all the catastrophic upheavals of geology and climate, all the evidences of imperfect adaptation, there is still a pattern to be found in the sum of things. Wherever he looked, even in the most tumultuous settings, he discerned a condition of beauty, though it was more the beauty of process than of fixed relationships. Nature in his view remained a system tending toward balance and, as such, offered a model for humanity – not a model frozen in time but one that was through and through historical, dynamic, and innovative.

Such was the understanding that the scientist Leopold had, too, and his work on the land was meant to restore that process of growth and movement to its former vigour. The beauty of the organic world for him lay in its continuous creativity rather than in any rigidly prescribed table of organization.

Now some 40 or 50 years after Leopold, and more than a century after Darwin, some ecologists have carried the idea of continuous change in nature so far that they have begun to lose sight of the order and pattern that is also there. Anarchic competition is all that many of them find operating. There is no discernible direction in nature, they say, no clear, predictable pattern of events over time, no overall direction discernable, no point of "climax" or "balance" or "mature state" that nature ever reaches, and therefore no comprehensive standard by which we can evaluate the effect of our own human actions on the whole. There is in fact no whole, they insist, there are only fragments. Nature appears in this view as a multitude of limited, specific processes going on, all of them grinding against one another and never merging into some unified flow or outcome. It would seem that these ecologists, having completely discredited older integrative concepts like "the balance of nature" as being too riddled with exceptions to be true or meaningful, too imprecise to test mathematically, are now unable to find new ones to put in their place. Even the idea of the ecosystem, which over the last half century has been the ecologist's favourite model for describing the unity and organization in nature, has come under attack as a mere fiction, impossible to verify, and therefore worthless. The sound notion that science has always depended on fictions or metaphors to conceive of nature has been taken by some to the extreme position of insisting that all order exists only in the human mind, that nature is nothing but disorganized raw material on which we are free to impose our ideas and desires.

"The essential ingredient in aesthetic apprehension is the ability to look beyond the level of isolated details and perceive their underlying cohesion"

I believe we are going through a period of some uncertainty among ecologists about the basic coherence of the organic world. It will not last. Ecology will come back, and must if it is to continue as a human enterprise, to some consensus on the whole. But in this unsettled interim the lay public, who have come to rely on the authority of ecology to reveal what good stewardship of the land is and what it is not, must often feel befuddled and uncertain. And feel so at a most critical time, when the forces of environmental change are accelerating everywhere and we want to know as fully as we can what it means to "damage" nature. Leopold did not find science an unpredictable ally in his mission, but suddenly we do. Possibly too, we have expected more guidance from it than it can ever give, and we have not trusted enough our own direct, intuitive perceptions.

The philosopher Alfred North Whitehead argued that we need to rely less on the authority of reductive empirical research and more on "the habit of aesthetic apprehension." He meant an ability to see wholes instead of pieces. It develops through exercising that faculty in our minds that we associate with the arts, like landscape painting or music or poetry, though it is broader than any such specialized activity. Different people may exercise their faculty in different ways: through science or religion, as I have said, or through bird-watching, photography, or simply walking through nature with senses all alert. Whatever the activity, the essential ingredient in aesthetic apprehension is the ability to look beyond the level of isolated details and perceive their underlying cohesion. The details remain important, but the habit of looking at them too closely for too long can atrophy the aesthetic faculty. One loses the awareness of how things are joined together, how they form patterns with one another, how fitness is achieved. Nature then ceases to please the eye, and in the

Aldo Leopold at the University of Wisconsin, 1938

saddest cases the eye does not even know it should be pleased. When the aesthetic awareness is well developed, on the other hand, one sees easily and surely the deeper harmony within, and the pleasure it affords is intense. Words like "beauty" and "integrity" come readily to mind. Indeed, such qualities become the most significant realities that exist, and their perception and enjoyment is the highest form of living.

The beauty discovered in nature through aesthetic apprehension has inspired people repeatedly to try to construct harmonies of their own, in the landscape as well as in song and picture. All human art, I am sure, has its primal impulse in the deep observation of nature. We see its close relation of parts, its adaptation of means to ends, its wonderful suitability, and impressed by what we have seen or heard, we set out to express it in our own limited way. In every period of history and every corner of the Earth people have done so, from the gardens of Japan to the hedgerows of rural England, from the aquarium in a child's bedroom to the miniature rainforest recreated in the foyer of a modern office building. None of these creations is "nature" in some final or complete sense of the term, but all are efforts by the human mind to find some part of the whole they can grasp, imitate, and make their own.

But these days we feel more and more compelled, as Aldo Leopold was, to try to repair the Earth's beauty rather than merely select from it. Before, it did not seem to need our help; it was our shining exemplar. Now, however, for those whose aesthetic faculty is vigorous and searching for inspiration, much of that larger glory of nature has departed, not to be encountered again in our lifetime. We sense that we live in a vandalized world, It is as though someone had burned down many of our great museums with their priceless treasures, and we were left with only a few plaster casts turned out by unskilled copyists.

This degradation of natural beauty was what Leopold lamented on his abandoned farm. It is not merely stupid or unfortunate; it is wrong. Every person or nation has a right, he allowed, to derive a living from the Earth and to participate in the processes of natural creativity; but no one no matter how desperate his condition or elevated his sense of self, has any right to diminish the complexity, diversity, stability, fruitfulness, wholeness – in short, the beauty – of the natural world. Everyone has the responsibility, whether acknowledged or not, to get his living in such a way as to preserve that beauty. In so many words, that was what Leopold called the idea of a "land ethic". He worked it out in his head as he worked his shovel in the soil. I submit that it is one of the most important ideas anyone has put forward in the twentieth century. If observed consistently, at least as consistently as we have ever observed any ethical idea, not one of our institutions, philosophies, systems of knowledge, or modes of life would remain the same. Something bigger than pines would come up out of the Wisconsin sand.

Never before in our history has the organic world around us been in so much trouble. We seem to be thriving – at least as a species we are replenishing ourselves everywhere – but not much else is. Whole forests in both the tropical and temperate zones are dying from acid rain, radiation, air pollution, timber harvesting, slash-and-burn agriculture. Harbours, estuaries, seas as broad as the Mediterranean, the River Rhine are all in decline from toxic wastes. Between now and the end of the century the extinction rate among higher vertebrates will be as much as 400 times higher than the average rate that has prevailed over the history of evolution. As these creatures that have evolved with us now disappear by our hand, many of the Earth's native ecosystems will go with them. We are creating an environment of gashes, wounds, disorganization, and death. The order of nature may be a difficult ideal to define precisely and follow, but we are simply disregarding it. Each minute of each day a rainforest that would have covered a large part of Leopold's farm is chewed up by bulldozers and fires. I say that not to belittle his work, of course, but to suggest what we are up against these days. The present onslaught on the global environment far exceeds any impact of westward-moving pioneers in North America, or ancient European farmers draining marshes, or paleolithic hunters killing game. It is the result, I believe, of a radical change in human culture, one that has left large numbers of us wholly indifferent to ideas of natural beauty, health, and order.

For the last two centuries or so, our ideas of worth have been largely determined, especially in Western civilization, by the forces of industrialism. Those forces have generally led us to think that it is necessary and acceptable to ravage the landscape in the pursuit of maximum economic production. There can be no doubt about this outcome; it is clearly written in the historical record of England, the United States, and every nation that has been brought under the industrial system. By industrialism I mean the extensive mechanization of productive processes in clothing, food, transportation, and the like, typified by the large, centralized factory. Goods in this economy are provided to consumers, not directly by their own effort, but through elaborately organized commercial markets. But that is only the external aspect. It has also an inner dimension, vast, complex, and effective: the habits of

"Harbours, estuaries, seas . . . , the River Rhine are

thought and perception that are needed to make the system and its demands appear reasonable. All economic systems are, after all, first mental systems. The revolution in modern production started in the mind. Or more accurately, it started in a few people's minds, and from there it spread to others, until eventually a full-blown culture of industrialism could be said to exist, more or less unified in the pursuit of certain goals.

Modern industrial culture first sprang from the minds of a rising class of capitalists or entrepreneurs during the seventeenth and eighteenth centuries. Even now, it is that group which dominates overwhelmingly the institutions, politics, media, and thinking patterns of industrial life. We might as well say then that we live within the culture of industrial capitalism. There are, to be sure, some important variations on that culture – industrial socialism, for example – but all preach much the same notion of what has worth and what does not. Where they diverge is in their degree of concern for a just distribution of the products of manufacturing, their skill in the efficient management of industrial systems, and their willingness or reluctance to use the power of the state to impress their ideas on people. Globally, however, it has been the industrial capitalists who have been the most decisive voice in modern times; almost all the world is now their factory, and the fate of the Earth is largely in their hands.

all in decline from toxic wastes."

Until recently, any suggestion that nature has an intrinsic value or beauty that must be preserved has been viewed by many industrial leaders as a serious threat. They have had another, rival set of values to pursue. Industrialism has sought not the preservation but the total domination of the natural order and its radical transformation into consumer goods. The environment has been seen to exist mainly for the purpose of supplying an endless line of those goods and absorbing the by-products of waste and pollution. Whatever has not been produced by some industry and placed on the market for sale has had little value. It has been viewed, in the most negative word that industrial culture knows, as "useless." Since the only way industrialists can use nature is to disorganize it, in order to extract the specific commodities they value, typically they have regarded as most useless of all those very qualities of stability, harmony, symbiosis, and integration that characterize the living world in the composite. They have tended to devalue both the services that natural systems provide people, like a forest regulating stream flow, and the aesthetic satisfaction that contemplating such order affords. As Leopold conceded in a letter to the American conservationist William Vogt, the idea of a land ethic has been incompatible with the drives of industrial civilization; one insists that we respect the order of nature, the other urges us to triumph over it.

What is more, industrialism has made a large number of people (including, as I have indicated, some scientists) sceptical of the very idea of order in nature. Constant innovation, constant change, constant adjustment has for a long time now been the normal experience in this culture. We have so far forgotten that life can be otherwise that we have come to accept as natural much of the chaos, uncertainty, and disintegration we find in our institutions, families, and communities. I think it is accurate to describe modern industrial societies as, on the whole, actively seeking disequilibrium. We have so learned to associate that condition with the possibilities of personal satisfaction, with full self-realization, with the more abundant life, even with justice and liberation, that we have even felt threatened by any talk of maintaining or restoring the natural order. We have been afraid of "stagnating" or "falling behind" or being kept "in our place" by repressive forces. By that way of thinking the notion of preserving nature, or trying to restore some semblance of its order, has been known to evoke fear and hostility.

This may be the greatest revolution in outlook that has ever taken place. Traditional societies tended to see and value the order in nature; we of the modern industrial era have tended to deny it. And therein lies the deepest source of our contemporary environmental destructiveness.

Such a revolution did not occur with ease or quickness; it took centuries to beat down all the old ways of thinking. Similarly, if industrialism is ever to give way to some other culture and society, then it will take a long, difficult struggle to revise its habits of thought. Billions of people on the planet are just being converted to the industrial way of life and its stance toward nature; while the rest of us, living in more economically affluent nations, show little willingness to abandon completely the system. In a thousand ways we acknowledge with regret the price we pay – that nature pays – for our productive wealth, but we are unwilling to go back to any pre-industrial state. Caught increasingly in a dilemma between the economic institutions that support us and our awareness of their chaotic impact, we often try to resolve it by acquiring all the goods and money we can and then, privately, make our escape to some rural retreat, some place in the woods, some island in the sun, as far away from the Great Necessary Evil and as close to the beauties of nature as we can get. Obviously, on a planet as small and overpopulated as ours, that is a strategy that cannot work for very many or for very long.

The only practical way out of the dilemma is to begin using the opportunity that industrialism affords us, in the form of wealth, comfort, leisure, and receptivity to new ideas, to transcend it. We must set about building the new in the shell of the old. And we must keep at that building until the old shell cracks and falls away, revealing some new post-industrial culture growing within. What will the life and thought of that new culture look like? What kind of economic base will it involve – a return of technology and production to the locus of the small community, a communal rather than privatized pattern of ownership, a decline in the power of nation states and international corporations, or a more extended planetary consciousness and economy? How could we feed ourselves other than the way we do now? Where might we turn for

"We try to resolve it by . . . escape to some rural retrea

a new set of values and perceptions? What changes will we need in our scientific models of nature? No one can give precise answers to those questions until the old shell has cracked enough to reveal what is developing inside, until the reform is actually made. This much seems clear: the next stage of our relation with nature will not involve some archaic hunting or farming revival, for there are far too many of us now for that solution to be realistic. Nor will it involve making exact restoration of native ecosystems to their former mix of plants and animals, a task well beyond our abilities. In economic terms, the stage beyond industrialism is likely to involve our learning how to satisfy a clearly defined but limited set of human needs without diminishing the numbers and variety of other living things anywhere on Earth. What that will require from us will take a while to imagine. We must first decide that it is worth the effort, that we have finally had enough production and consumption (at least those of us in affluent countries) and can turn our energies to other ends. Then we must begin where industrialism did in its rise to authority, by influencing the minds of many people toward a different hierarchy of value.

some place in the woods, some island in the sun . . ."

It is too much to expect that the richest among us, who have profited most from industrial expansion, will furnish many leaders in this reformation. Once they were the vanguard of change, rising out of obscurity to challenge established authority; now, with notable exceptions, they have too much invested in the present system to welcome its demise. Nor ought we look for much leadership from the army of industrial workers, for they will tend to follow out of a sense of desperation and necessity whoever is presently giving them jobs. When the post-industrial future comes into being, it will be the achievement primarily of those who have been the least dependent on the old productive processes and who have had the freest minds, the greatest discontent, and the most compelling alternative vision. They will be, it follows from all I have said so far, men and women of the most fully developed "habits of aesthetic apprehension." Call them artists if you like, but they will not be a brigade of specialists in some fine arts discipline. Instead, they will be in the tradition of all those who, from the early years of the industrial revolution onward, shocked and outraged by the ugly, ruthless new ways, have constituted "the opposition." I think of them as the party of Henry David Thoreau and William Morris, John Muir and Richard Jeffries, H.M. Tomlinson and Rachel Carson, and Leopold of course – a party of writers, painters, and scientists. But many in the party have left no novels, paintings, poems, or naturalist essays, have left no record of their thinking except for the mountains, forests, and marshlands they have saved or restored. Famous or obscure, these are the people who are likely to lead us to transcend the present industrial culture.

I come now to the core of what I want to say. Industrialism will be reformed substantially only when there is a powerful desire among enough people to experience, preserve, and nurture more natural beauty, not only in the remoteness of Alaska or the Himalayas but in the everyday environments where they dwell.

Evidence suggests that such a desire may be gathering force all around us. Travel posters try endlessly to stir it up. All the conservation organizations around the world make it their leading theme, filling their magazines with colour photographs of unspoiled coral reefs, spectacular butterflies in green jungles, even swamps, deserts, and cold sterile icefields that travellers once fiercely cursed and hated. Indeed, all the media – movies, television, music, radio – participate endlessly in this largely unorganized, subversive education in aesthetic awareness. So do the schools, in art and biology classes, in student outings, in films and recordings, which release young minds from an exclusive training in abstract and bookish theories into the realm of concrete nature, complete and whole and mysterious. These experiences, I am hopeful, are having an effect. The sensibilities of vast masses of people are being opened, perhaps as never before, to the green world, and there is no telling where that opening will lead.

Ideas and feelings must, of course, find effective political expression if they are to produce fundamental change. They need leaders who can translate sensibility into power. They need people who will act on their feelings. Already, in every industrial nation, such activists have begun to appear by the millions, demonstrating and voting in the name of beauty, morality, health, and ecological integrity. I think of the Greens in Germany, the Green Party in the United Kingdom, the Sierra Club and Friends of the Earth in the United States. They still constitute a minority in the midst of a vast, chaotic, international mass society, but increasingly they are effectively mobilized to win important battles. When one recalls how little of such political activity there was anywhere a mere fifty years ago, the prominence of these groups is simply astonishing. They have changed the tone of political discourse not only in their own countries but across the planet. How much power they will eventually acquire, how effective they will be in transforming the industrial economy and its institutions, are questions that no one can answer with assurance; but we can be certain that this is a new kind of politics whose potential will only grow.

The way to reform, Aldo Leopold wrote, lies not in building roads into lovely country, but in building "receptivity into the still unlovely human mind." Whitehead made much the same point when he complained of "the stone-blind eye with which even the best men of [the past] regarded the importance of aesthetics in a nation's life." Both were aware that often it is people of considerable intellect and ability who have been most indifferent to the natural order and who stand in greatest need of learning to see it whole again. Both Leopold and Whitehead understood that enlightenment will not occur simply through the accumulation of more and more scientific knowledge. Mere facts do not of themselves correct that blindness they thought was most consequential. In fact, it sometimes seems that the more facts we gather, the more knowledge we disseminate, say, in science or economics, the less able we are to see into the heart of things. Of course we can use facts to improve our relationship with nature; they have been known to open a few minds. But the blindness from which we suffer most grievously is one that only art, in the larger and looser sense I have been employing the term here, as a way of experiencing and knowing the organic order around us, can remedy.

In the industrial era we have made the mistake of disregarding beauty altogether or of assuming it is something that we humans must impose on the chaotic realm of nature. A more complete and humble view would be to see that nature constitutes a different and greater kind of art than anything we have created. It is not a project that gets finished at a certain hour, is wrapped up and sold, and later hung on a wall or put in a bookcase.

Nature is a creative work that has been going on for billions of years, was going on before there was any mind to translate it into human understanding. It is a ceaseless and infinitely inventive process of creating. It has no purpose behind it that we have been able to find and agree on, but it does display at every moment an order far more complicated and marveleous, and I am tempted to say wise, than any we have been able to design. It is the grandest beauty there is, speaking to us of the most comprehensive order and value. In restoring some of that primordial beauty to a ravaged Earth, we may, with Aldo Leopold, learn to see again much we have forgotten.

Lloyd Timberlake

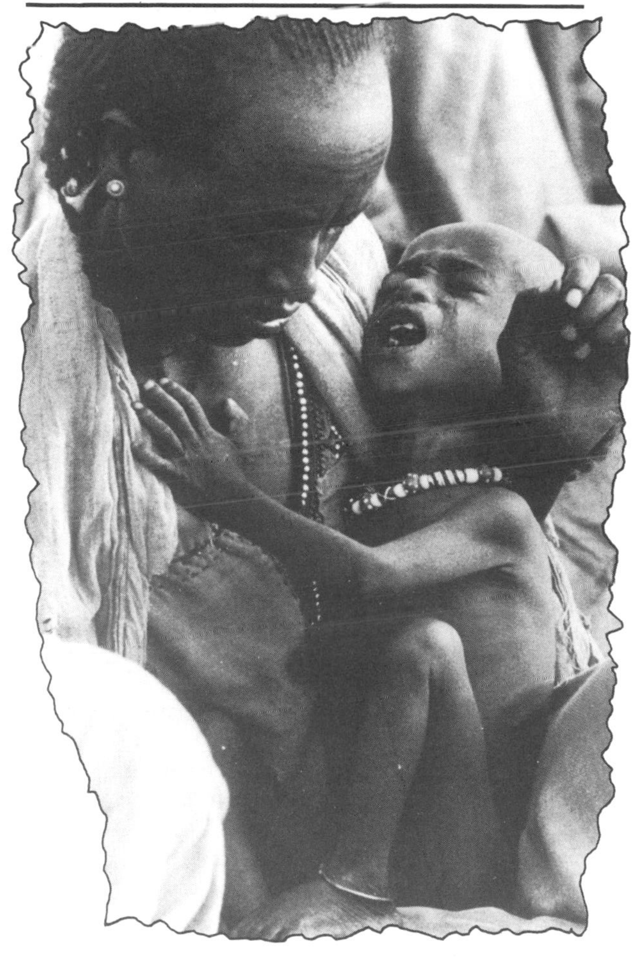

"Most of the issues associated with international food aid point to one overwhelming conclusion – moving free or cheap food from country to country is a very inefficient way of seeing to it that the hungry eat"

The giving of food by one person to another is an act deeply embedded in worldwide cultural and religious traditions. Jesus, the Buddha, Muhammed, Jewish laws – all enjoin us to feed the hungry. In most societies, rich and poor, giving food to guests has a cultural and social significance beyond mere entertaining.

I grew up in the southern part of the United States, where once you accepted another man's "bread" you were not expected ever to gossip about him or his family. I gained weight while conducting a survey of the poorest and hungriest families in the slums of Dhaka, Bangladesh, because it would have been a great dishonour to them if they had not offered me food and drink once I crossed their thresholds; and there was of course no way such hospitality could be refused. I had similar experiences in the Ethiopian highlands during the famine in 1984, where guests were fed, though local people died of hunger. In neither case did I *need* the food; and the givers could not really *afford* to give it to me. The exchange did not do any of us any logical good, but it was the way things were done, and we all honoured the tradition.

It is hardly surprising then, that the ways in which nations pass food one to another are even less logical. And these transfers too often fail to do anyone any real good. The mechanics of food aid are in fact bound up in a confused tangle of motives, mixing human charity with various political and financial incentives of both the givers and the recipients.

In fact, the giving of large amounts of food by a rich and well-fed nation to a poor nation with a high proportion of hungry people can even do more harm than good; though this defies the belief of ordinary citizens in both the donor and recipient nations. If there are mountains of unwanted food in the North and masses of hungry people in the South, what could be more reasonable, more charitable and more satisfying than shifting those mountains south? Toward the end of the most recent African crisis in 1985, there was even a movement in Europe called "Move the Mountain", bent on getting European grain surpluses to Africa. But, as we shall see, if the mountain had been moved then, Africa's long-term crisis would have been compounded. How can giving food to "hungry" nations be a bad thing?

One of the most obvious answers is that a surprisingly small proportion of food aid reaches people who are either chronically hungry or people who are temporarily hungry because of an emergency. Food assistance (worth $US 2,600 million annually in recent years – some ten per cent of all official development assistance) can be divided roughly into three different categories. Only about five per cent of the total in 1984-85 went in the form of emergency food aid. In a given year, something like one quarter is for "project" food aid – that is, food used to get various things done in the recipient nations, often as "food-for-work" schemes, in which people are paid in food for working on projects. The remainder goes in the form of "programme" aid, food donated outright or sold at subsidized rates to help with recipient governments' balance of payments or to support their budgets.

All three forms of food aid present their own problems and are subject to intense debate. In the early 1980s, before the African drought/famine crisis, the general debate over food aid seemed to be getting somewhere. A number of groups in and out of northern governments were going as far as to call for a halt to all food aid except emergency relief aid. These groups were joined by eminent Africans such as biologist Edward Ayensu of Ghana. Then when more than 20 African nations slid toward famine, this debate dried up; the participants thought it tactless at best to talk during such a tragedy of limiting any kind of food transfers. There is a pressing need to revive this debate.

Before going into the political debate surrounding programme and even project food aid, it is worth pausing to look at the surprisingly severe problems raised by the apparently apolitical act of getting food to starving people in a crisis. Even discus-

Types of aid and the problems they create

10% of all official development assistance is food aid – $2,600 million a year – of which...

10%

5%
25%
70%

5% is Emergency aid
25% is Project aid
70% is Programme aid

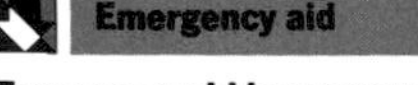

Emergency aid

Emergency aid is a response by donor countries to a famine or a natural disaster such as an earthquake. It is supposed to be a one-time donation, but may become semi-continuous.

Problems...
- It is sometimes sent when it is not needed – aid supplied to Guatemala after the 1976 earthquake arrived at a time of good harvest
- Because of bureaucratic time-wasting it often arrives too late – 400 days elapsed between the promised EEC delivery date to Ethiopia in 1984-85 and the actual delivery date
- When it arrives too late, people are often in the wrong place – e.g., in camps instead of on their farms
- It is often not fully distributed
- It is often the wrong kind of food in that it may convert tastes to food that cannot be produced locally, or that will compete with local food and thus impoverish farmers
- It may destroy traditional ways of coping with emergencies

Project aid

Project aid is most commonly supplied as "food for work", but may also be food for mothers and pre-school children, or food for schools, youth camps and hospitals.

Problems...
- It is sometimes switched from food-for-work to emergency aid, thus throwing the poor out of work
- It leads to a neglect of conservation as farmers leave their holdings to work on projects
- The work done is sometimes inappropriate – farmers may be asked to build roads used by the rich, or even to improve rich men's farms
- Publicity without explanation sometimes misleads recipients into misjudging the food's overall nutritional value – mothers may feed babies only grain and butterfat, which were intended only as supplements to traditional fare
- Food may miss its target – e.g., food for schools will not reach those children too poor to go to school

Programme aid

Programme aid is usually supplied on the understanding that the food will be sold, the money raised being used for specified developments. Or it may be used to create a food reserve.

Problems...
- It is often designed to help the donors more than the recipients
- It favours city dwellers and privileged classes such as civil servants, police and soldiers at the expense of the poor
- The food is often not what the local farmers produce
- The donors have no incentive to rationalize their production and cut back on surpluses
- The recipients have no incentive to get their emergency storage and distributions systems right
- The recipients are at the mercy of fluctuating demand or prices elsewhere in the world, which may divert expected supplies to more profitable markets
- The recipients become addicted to annual "fixes" of imported food

sing these issues can make one appear churlishly opposed to the whole exercise. Indeed, during the height of the African famine, many journalists and politicians appeared more concerned with documenting the amounts of food *not* getting to the hungry than the amounts that were.

One of the main failures of emergency relief efforts involves timing. The Food and Agriculture Organization warned in 1982 that Ethiopia would need a lot of food aid in 1983 to avoid widespread hunger. There was little response from donors. Such warnings became more strident during 1983. In May 1984 I took about 25 journalists to look at the situation in the Ethiopian highlands, and at the same time the UN High Commissioner for Refugees also took journalists there. All wrote that millions faced starvation. Again there was little response.

It was only after televised footage of babies dying appeared in the world's living rooms in the autumn of 1984 that governments responded. But they responded to a three-year-old drought as if they were reacting to a sudden hurricane the day before. Food aid poured in, overwhelming port and storage facilities, transport and distribution systems, and the data base on the identities and locations of the most vulnerable. Thus only about three-quarters of the food delivered to Ethiopia was ever distributed. In the Sudan it was an even lower 64 per cent.

That which was delivered was handed out only after people had been driven off their own land and been assembled in famine camps. Had it arrived earlier, farmers could have stayed on their land and rehabilitated it so they would have been ready for the return of the rains.

As the crisis deepened, affecting some 35 million people in more than 20 nations, more and more donors swung into action – but most swung very slowly because of the laborious bureaucratic processes used to expedite such donations. An average 400 days, well over a year, elapsed between the date upon which the European Community agreed to deliver food and the date upon which it actually delivered it. This meant that food promised in the dry year of 1984 arrived in many African countries during 1985. In some places it was off-loaded in the ports during the rainy season and could not be moved inland over flooded roads. In other cases it arrived as local farmers were bringing in their first good harvests in years. Farmers found not only no market for their own grain, but no place for it in warehouses full of European wheat. The record 1985 harvests in Kenya, not one of the countries hardest hit by shortages, were too big for Kenyan warehouses. Yet "emergency" food aid continued to pour into that country.

Donors not only send emergency food *when* it is not needed; they often sent it *where* it is not needed. A drought or a hurricane can obviously destroy crops; earthquakes rarely do. Yet donors often respond to earthquakes with piles of food. In their seminal book *Against the Grain*, Tony Jackson and Deborah Eade of OXFAM detail the many ways in which disaster relief aid often goes wrong, and tell how the Guatemalan earthquake of 1976 coincided with huge domestic harvests. Farmers and agricultural officials pleaded with donors and relief agencies not to send food. But food came, virtually destroying the agricultural economy for the following year. Many complained that the food aid was a bigger disaster than the original earthquake.

Also, once food is given for a specific disaster, it tends to keep on coming needed or not. Guatemala has continued to receive much larger quantities of food aid ever since that quake. Haiti first received large amounts of food aid in 1954 after Hurricane Hazel, and according to one report from a local missionary, "it never stopped coming." Three leading researchers into international food issues, Frances Moore Lappé, Kevin Danaher and Rachel Schurman, describe in their book *Betraying the National Interest* how in 1985 Haitian peasants brandished machetes to stop US food aid from being landed by helicopter because of the disruptive effects that the food would have on local markets.

It is more often the much more plentiful programme aid, rather than emergency aid, that disrupts markets and changes the eating and coping habits of local people. But in smaller societies, relief food can also have a permanent negative effect. In 1972 there was a four-month drought and severe frosts in New Guinea, then an Australian colony about to gain independence. Colonial administrators organized imports of Australian rice and Japanese canned fish to feed 150,000 people for 18 months. The operation had the long-term effect of destroying traditional mechanisms for coping with emergencies, such as stricken villages temporarily migrating to less affected areas. All began to rely on outside help, and many people were converted to a taste for rice and canned fish, foods they could not produce themselves.

Project food aid, the second most common style of food assistance, appears on the surface eminently worthy and worthwhile. In its most usual form, "food-for-work" projects, people in need of food are paid in food to do work which, in theory, improves their ability to fend for themselves. Paul Harrison, in his recent book, *The Greening of Africa*, says that "food-for-work is far better than free relief food. It does not create a feeling of dependence or beggary. According to social surveys, participants feel that they are supporting their families by their labour". Harrison

400 days between a promise of food and the date it actually arrived . . .

visited the world's largest food-for-work scheme, in which the World Food Programme (WFP) has been paying thousands of peasants in the Ethiopian highlands to construct terraces to conserve soil and water and to replant trees on hillsides for the same purposes. Each worker gets an average of three kilos of wheat and 120 grammes of vegetable oil for a day's labour – about enough to keep a family of six alive. He found that three out of five participants said that this food had saved them from starvation.

But this massive effort also ran into problems. During the height of the famine, when the food was most needed, the Ethiopian government switched priorities to the movement of emergency food aid, and the food for work was delayed; 50,000 workers were laid off, which meant that 300,000 people (families of workers included) joined the relief roles. Harrison also found that the free food skewed the economics of the conservation work, encouraging methods that were not cost-effective. It also discouraged farmers from beginning needed soil and water preservation themselves, lest they be left out of the project. Finally, the Ethiopian government was apparently lulled into relying overmuch on the WFP project for all of its conservation work, with the result that after ten years of work only four per cent of the land suffering from serious degradation had been treated. By 1985, three times as much cropland was being degraded to poor pasture each year as was being reclaimed by the project.

The Ethiopian project is generally thought to be one of the best. There are other accounts of food-for-work projects in which the poor are virtually forced to build roads that primarily benefit the

In a food-for-work programme, Ethiopian peasants built 700,000km of terraces to hold rainwater

wealthy, or even to improve the farms of the rich while their own small-holdings are neglected.

But even the best efforts prove again and again how difficult it is to get the right foods to the right people in the right manner. This is almost always a more complex matter than project designers and administrators realize, and there are almost always unplanned side-effects and spin-offs. Another form of project food aid is mother-and-child nutrition programmes aimed at these obviously vulnerable members of society. One study in a part of Central America found that children's nutrition improved only when the programme ceased. When the grain and butter fat was available, mothers fed their children only that. They had, after all, been told how good this foreign food was for their children. When it was not available, they returned to giving their children traditional fare, local fruits and vegetables, which contained a lot of vitamins and minerals not found in the supplementary food.

Dr. Frances Stewart of Somerville College, Oxford, in a short study of "Food Aid: Pitfalls and Potential," lists many of the ways in which "targeted" food aid misses its target. For instance, trying to reach hungry children through school-feeding programmes obviously misses the children of the really poor households, who do not go to schools. And children who *do* receive the food at schools often take it home to share with their families, or they are simply given less food at home, so they do not get any more food overall. She finds that food aid can be used in food-for-work schemes to increase employment or to get important facilities

In the Sudan, only 64 per cent of the food delivered was ever distributed

installed, such as sanitation works. Or it can be used to support a food stamp programme. But these tend to be administratively cumbersome and require extra funding for moving and storing the food. She concludes that it is simpler to finance the needed interventions with cash, perhaps raising the cash by selling the food aid on the open market (which, as will be seen, donors often make difficult).

Programme food aid: the grain junkies

In today's world, despite the fact that 730 million peoople in 37 developing countries (excluding China) do not get enough calories to support an active working life, food is much more abundant than cash. This fact is at the very heart of most food aid issues. Nations with food surpluses tend to be much more interested in getting rid of that food in ways that benefit the producing countries rather than the recipient countries. This is a harsh, but hardly a radical conclusion. The normally conservative World Bank, prone to understatement, concludes that "the quantity of food aid is more closely related to the needs of donors than those of recipients." And Sir William Ryrie, former head of Britain's Overseas Development Administration, wrote in 1986 that the bulk of food aid "is frankly more a means of disposing of European agricultural surpluses than of helping the poor". This is little understood by northern taxpayers, who assume that the term "food aid" means "food transfers that help the hungry". It all too often means "food transfers that help the rich."

When the United States passed Public Law 480 in 1954 and began large-scale international food aid, its stated goals

"Farmers and agricultural officials pleaded with the donors and relief agencies not to send food. But food came, virtually destroying the agricultural economy for the following year"

were to "expand international trade among the United States and friendly nations . . . to make maximum efficient use of surplus agricultural commodities in furtherance of the foreign policy of the United States, and to stimulate and facilitate the expansion of foreign trade in agricultural commodities produced by the United States." Since then the law has been reformed slightly, and countries of the European Community, and others, have adopted policies to see that more food aid is used for the development of *developing* nations.

But basic motives have changed little, as witnessed by the anomalies of recent food aid figures. In 1984-85, 25 donor countries provided more than 100 developing countries with about 12 million tonnes of cereals, 430,000 tonnes of vegetable oil, 356,000 tonnes of skimmed milk powder, 98,000 tonnes of other vegetable products, and 21,000 tonnes of meat and fish products, according to the World Bank. The major donors are the United States, which accounts for about half of all food aid, the European Community, 30 per cent; and Australia, Canada and Japan, which together contribute about 14 per cent.

But who gets the food? "The distribution, quantity and nature of food aid sometimes bears little relation to dietary deficiency," note the World Bank. Egypt, a middle-income country where the average calorie intake is about 28 per cent higher than that needed for a healthy diet, gets 20 per cent of all cereal aid. For example Togo, a country poor in money and food, gets only six per cent of Egypt's per capita bonanza. Despite the fact that Egypt runs food subsidy schemes that make bread cheaper than chicken feed, it is a nation of great strategic and economic importance to the West. Lappé and her colleagues note that in 1983 six out of the top ten recipients of US food aid were net exporters of agricultural commodities. Such trends continue. In 1985-86, Egypt, with a population of 47 million, received more than 1.75 million tonnes of "wheat equivalent" (a way of combining wheat and non-wheat cereals in accounting). Ethiopia, with a population of more than 40 million, received not quite half a million tonnes. In fact, Egypt received more than half of the total cereal aid shipped to Africa, according to the international Food Aid Committee, whose figures represent 95 per cent of all international cereal aid shipments.

Given that 70 per cent of food aid is essentially given to governments to do with as they will, whether much or any of it reaches the hungry depends not so much on the policies of the donors as on the policies of the recipient governments. In most of the poorest and hungriest nations, the poorest and hungriest people are left right out of national political and economic systems. Governments are more interested in providing reasonably priced food for the small middle classes, the civil servants, the soliders and the police.

Those governments are also under considerable logistical pressure. In most African countries, the capital cities are either ports or at the end of the railways, often near the border. They are so situated because these nations hegan life as colonies serving Europe, and the capitals look outward rather than inward. It is much cheaper and easier to feed these cities with cheap imported grain than with harvests collected from widely dispersed farms and transported over bad roads from up country. And this is as true of countries with good agricultlural potential as it is of the drier Sahelian countries. President Abdou Diouf of Senegal was asked in 1985 why his country imported so much rice when it could grow all it needed. He replied frankly, "Because it is cheaper." What motivation can governments find to develop their agriculture, to pay farmers a reasonable price for the food they grow, when they can rely on cheap foreign shipments? This sort of food aid has produced "grain junkies," nations virtually addicted to their annual food fix from North America and Europe, and can find little reason to kick the habit.

By 1984, *before* the widespread drought bit, almost one in four Africans – 140 million of 531 million people – were dependent on imported grain. Not completely coincidentally, one in four Africans lives in cities. In that food aid allows governments to keep down the price of food in cities, and to keep down the prices paid farmers, then the poverty of African farmers has been institutionalized. Given that most Africans are farmers or in farming families, this means that African poverty has been institutionalized. There is little money in circulation among most of the population to buy the goods produced by fledgling urban industries. And poor farmers mean poor land, because they have neither the cash nor the incentives to engage in tree-planting, soil and water conservation work and small-scale irrigation necessary to sustain dryland agriculture. All of this puts the countryside – both environment and people – into a vicious downward spiral. None of this analysis is meant to blame all of Africa's problems on programme food aid, only to show that such food aid is more often a part of Africa's problems than of Africa's solutions.

There are also other problems associated with reliance on foreign grain. First, the food given as aid is often not the same food as that which local farmers can grow. In West Africa and the Sahel, city dwellers, and more and more people even in rural villages, are eating wheat and rice, while local farmers are growing millet, sorghum, cassava and various peas, beans and lentils. The urban poor may actually spend more of their money on food aid food than do the rich. One study by the International Food Policy Research Institute found that in the rainy season (before

"In 1985 Haitian peasants brandished machetes to stop US food aid from being landed by helicopter because of the disruptive effects that the food would have on local markets"

harvests) in the poor Sahelian nation of Burkina Faso, the wealthy in cities were spending 46 per cent of their cereals budget on rice; the middle-income groups 49 per cent; and the poor 54 per cent of their cereal budgets. Yet only five per cent of the nation's cultivated land is given over to rice production.

Recipient countries ae obviously putting themselves in a very insecure position by encouraging the consumption of foods that their farmers cannot grow, especially as recent history shows that they cannot rely on regular deliveries of set amounts of programme food aid. Food aid reached its highest levels, 167 million tonnes, in 1965-66. But its decline has not been a steady fall that reflects the increased food security of parts of Asia and Latin America. In 1973-74, higher world wheat prices meant that the wheat shipments of around ten million tonnes per year in the late 1960s dropped suddenly to four million tonnes, increasing hunger and starvation in the drought-stricken Sahel and Ethiopia. Should North American or European harvests falter, and a wealthy nation such as the Soviet Union experience a major shortage, then the normal programme-aid recipients will have to do without. Africa would suffer tremendously even in a year of "normal" rainfall and "normal" harvests.

Donors not only give away a lot of food that recipients do not particularly need, but they set rules on how the food can then be used. The World Bank notes that recipient governments could prevent the influx from dampening local production incentives if they resold it on the world market and bought back only what they

Triangular transactions

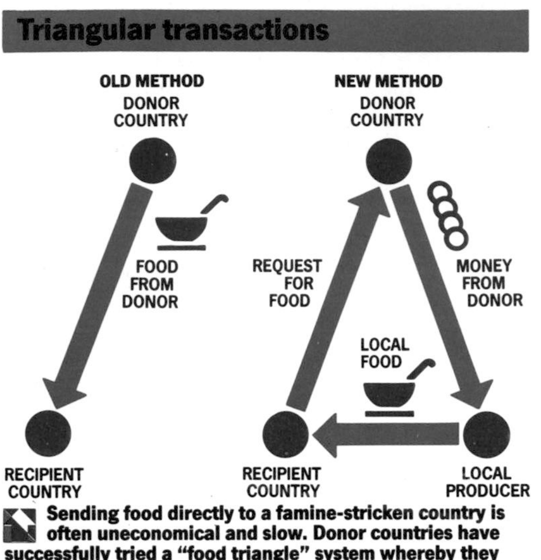

Sending food directly to a famine-stricken country is often uneconomical and slow. Donor countries have successfully tried a "food triangle" system whereby they pay for food to be shipped to a country in need from one of its neighbours.

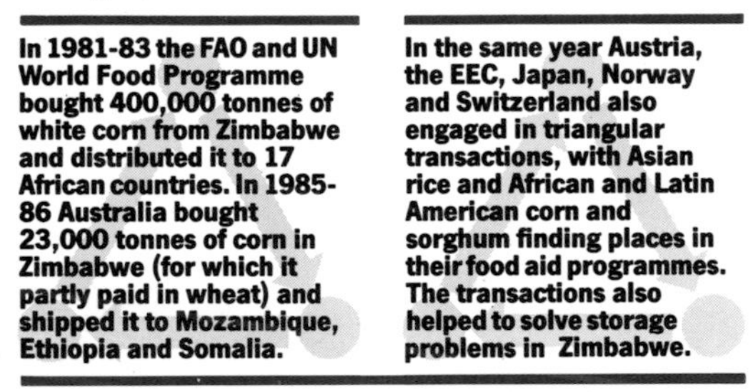

In 1981-83 the FAO and UN World Food Programme bought 400,000 tonnes of white corn from Zimbabwe and distributed it to 17 African countries. In 1985-86 Australia bought 23,000 tonnes of corn in Zimbabwe (for which it partly paid in wheat) and shipped it to Mozambique, Ethiopia and Somalia.

In the same year Austria, the EEC, Japan, Norway and Switzerland also engaged in triangular transactions, with Asian rice and African and Latin American corn and sorghum finding places in their food aid programmes. The transactions also helped to solve storage problems in Zimbabwe.

genuinely needed, or if they cut commercial imports by the amount of food aid. Donors set rules against recipient countries doing either, so that commercial markets for their grain exports are not hurt. (However, the Bank adds that these rules are rarely enforced.)

This sort of international food aid, then, provides recipient nations with no incentives for getting their own agricultural emergency storage and distribution systems right. But by providing a way of making expensive and embarrassing northern surpluses disappear, it also diminishes incentives to northern producers for rationalizing their own food production systems. There are "agricultlural crises" in both the rich and poor world, and the two are inextricably linked. The World Commission on Environment and Development, 22 policy-makers and scientists from 22 nations, spent three years researching these and similar development issues. Its final report, *Our Common Future* (also known as "The Brundtland Report", after the commission chairman, Prime Minister Gro Harlem Brundtland of Norway) firmly states:

"Production in industrialized countries has usually been highly subsidized and protected from international competition. These subsidies have encouraged the over-use of soil and chemicals, and the degradation of the contryside. Much of this effort has produced surpluses and their associated financial burdens. And some of this surplus [programme food aid] has been sent at concessional rates to the developing world, where it has undermined the farm policies of recipient nations."

Food triangles

Even in cases of famine relief, there are often better ways of getting food in than delivering it from the North. OXFAM, in funding some of its food-for-work schemes in Tigray, Northern Ethiopia, during the famine found it much more economical to send in one relief worker with cash to buy the grain locally in areas surprisingly near the famine-hit areas, and then transport it by donkey to the work areas. This saves OXFAM transports costs and also stimulated the local farming economy. Many other relief organisations engaged in similar exercises.

Northern governments, trying to move surpluses rather than spend money, have been slower to do this. Zimbabwe has, since independence, been doing precisely what northern development experts advised: paying peasant farmers reasonable prices for corn, moving marketing facilities into peasant areas, and providing advice, fertilizer and rural credit. The result has been the "corn miracle". In 1978-79, peasant farmers produced only 3.8 per cent of the nation's marketed corn harvest (the rest being produced mainly by white farmers on large farms); in 1984-85, the small-holders produced almost half, giving the nation a surplus of 800,000 tonnes in 1985, some of which it sent to Ethiopia as famine relief.

Zimbabwe was then and is still surrounded by seven food-poor countries, drought-stricken Botswana and famine-stricken Mozambique being in particularly deep need. Northern nations seriously interested in both helping the hungry in Mozambique and in supporting Zimbabwe's agricultural reforms would have engaged in what is called "triangular" transactions, buying the corn in Zimbabwe and transporting it to Mozambique. Indeed, a few nations have done this, but in small quantities. Australia in 1985-86 bought 23,000 tonnes of corn in Zimbabwe for shipment not only to Mozambique but to Ethiopia and Somalia as well, according to Food Aid Committee figures.

But the United States shipped more than 28,000 tonnes of corn to Mozambique during that period, and this was American yellow corn rather than the white variety that predominates in southern Africa. The avowed purpose of this was not only to move surplus American corn, but to encourage southern Africa to develop a taste for the yellow variety and to improve United States markets for this grain.

The big lessons, and ways out

Despite its problems, programme food aid is not inherently or necessarily evil, nor should rational governments – donors or recipients – cut off the supply next year. Its effects remain the subject of deep debate, as it is impossible to put accurate figures to its effects of farmers' incentives or the domestic prices in recipient nations. Only when the incoming food is additional to demand does it greatly affect prices or production. But some argue that even in this case the increase in supply and the fall in prices means that more hungry people get more food in times of shortage.

Robert Cassen, in his book *Does Aid Work?*, concludes optimistically that "food aid has gone through a complex history, and has had many areas of failure; but these are now largely well understood and largely avoidable." He does not conclude, however, that the failures are being avoided, and opines naively that programme food aid need not disrupt agricultural production incentives, provided that it goes to poor people who would not have bought anyhow or that market arrangements are made to maintain demand. But in few countries does either happen. (On project food aid, Cassen "can only hope that the lessons of failed and successful projects will be heeded," but then suggests that there is no comprehensive monitoring and assessment system for such aid.)

But the one truth about food aid that every northern taxpayer should keep firmly in mind is that the vast majority of it goes to people who can afford to buy food. The lesson that policy-makers should keep in mind is that hunger is rarely caused by a lack of food in a given area. It is almost always the result of people being too poor to buy food. In virtually every famine studied, food exports from the area have increased during the famine because local people could not buy the food being produced.

"There are many ways in which food aid could be reformed, but there will be little motivation for these reforms as long as northern surpluses are so large that governments are encouraged to dump them on the south"

Thus if northern nations were serious about alleviating poverty and hunger they would become serious about the development problems that lie at the root of the growing gap between rich and poor nations – issues such as the debt crises that forces nations to over-produce non-food commodities for export, and the "new protectionism" in the north that frustrates the efforts of developing nations to export more and to add value to their exports.

And, it would be in the interests of both north and south if northern nations restructured their systems of subsidies, incentives and agricultural protectionism – systems that have grown up *ad hoc* over the years and now often contradict real national interests. The European Community's Common Agricultural Policy has been threatening to bankrupt it by consuming about three-quarters of all its spending. Storing or destroying surpluses costs it about $225 million per week. It is odd that both northern and southern agricultural systems are out of balance because both support the interests of minorities – of the farmers in the north and of the city-dwellers in the south.

There are many ways in which food aid could be reformed, but there will be little motivation for these reforms as long as northern surpluses are so large that governments are encouraged to dump them on the south. It may sound peculiar, even dangerous, to encourage the north to produce less food in a hungry and famine-prone world, but keep in mind how little food aid goes to relieve either chronic or acute hunger. Lappé and her associates conclude that "surely the United States must reduce its own over-production before it can have a constructive aid policy. Currently US food aid helps to mask unsolved domestic farm problems, especially over-production." The same goes for Europe.

The Brundtland Commission also finds that "most industrialized nations . . . must alter present systems in order to cut surpluses, to reduce unfair competition with nations that may have real comparative advantages . . . ". But it also recommends that developing nations, besides turning the "terms of trade" to the advantage of the small farmer, build up national stocks in surplus years to provide reserves, as well as encouraging the development of food security at a household level. Recently, some "food security aid projects" have been initiated to encourage recipient governments to store grain under longer term agreements and use it in times of acute shortage or for price stabilisation. But the hard fact remains that Third World governments will find little reason to get involved in food security efforts or to take even more politically difficult steps – land reform, integrated rural development, etc. – until the grain floods from the north ease.

Most of the issues associated with international food aid point to one overwhelming conclusion – moving free or cheap food from country to country is a very inefficient way of seeing to it that the hungry eat.

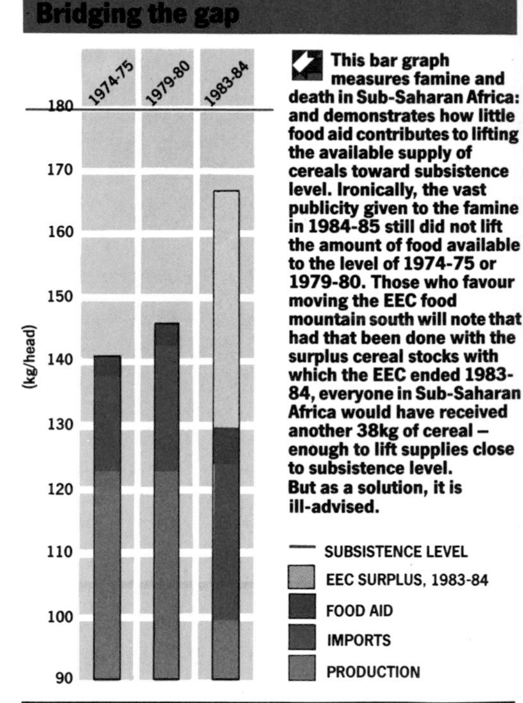

This bar graph measures famine and death in Sub-Saharan Africa: and demonstrates how little food aid contributes to lifting the available supply of cereals toward subsistence level. Ironically, the vast publicity given to the famine in 1984-85 still did not lift the amount of food available to the level of 1974-75 or 1979-80. Those who favour moving the EEC food mountain south will note that had that been done with the surplus cereal stocks with which the EEC ended 1983-84, everyone in Sub-Saharan Africa would have received another 38kg of cereal – enough to lift supplies close to subsistence level. But as a solution, it is ill-advised.

Peter Bunyard

"Where governments proceed with nuclear power they increasingly create a divided society, and where democracy is given its head the answer is to reject nuclear power. Its image is now tarnished, its brief moment of glory appears already to have passed"

At 1.23am on Saturday April 26 1986, a powerful explosion inside the number 4 reactor at the Chernobyl nuclear power station in the Ukraine blew aside a 1,000 tonne two-feet-thick steel lid and blasted through the surrounding concrete containment structure. Like an extraordinary firework display, bits of graphite, chunks of uranium fuel and pieces of control rods were strewn around the reactor building. Meanwhile, the graphite still inside the reactor burst into flame as air rushed in and, like coke in a steel furnace, began to burn vigorously.

Two men died in the first moments following the blast, one from falling masonry and the other from burns. Over the following months they were to be followed by some 30 others, most of them fireman who had battled heroically to prevent the fire from spreading to the reactor housed in the adjoining building. All died from radiation burns and sickness following exposure to gamma and beta radiation from the exposed reactor core. As well as receiving radiation externally, some had also breathed in radioactive particles. Those who died, as well as several hundred others who received large but not lethal doses of radiation, were to suffer horribly, their bodies ravaged by the effects of radiation.

Over the next few days the Soviet authorities marshalled their resources to evacuate some 135,000 people, all of whom lived within a 30-kilometre radius of the plant. Livestock had to be moved; and attempts were made to control the contamination that was covering soil and buildings alike. Meanwhile, the fire in the reactor had to be put out and the reactor itself smothered to prevent the escape of more radioactive material. Helicopters, flying around the clock, were used to dump some 5,000 tonnes of material onto the burning core: some 800 tonnes of dolomite to generate carbon dioxide to quench the fire; boron carbide to ensure that the chain reaction would remain shut down; some 2,400 tonnes of lead to blanket the exposed core and absorb the gamma

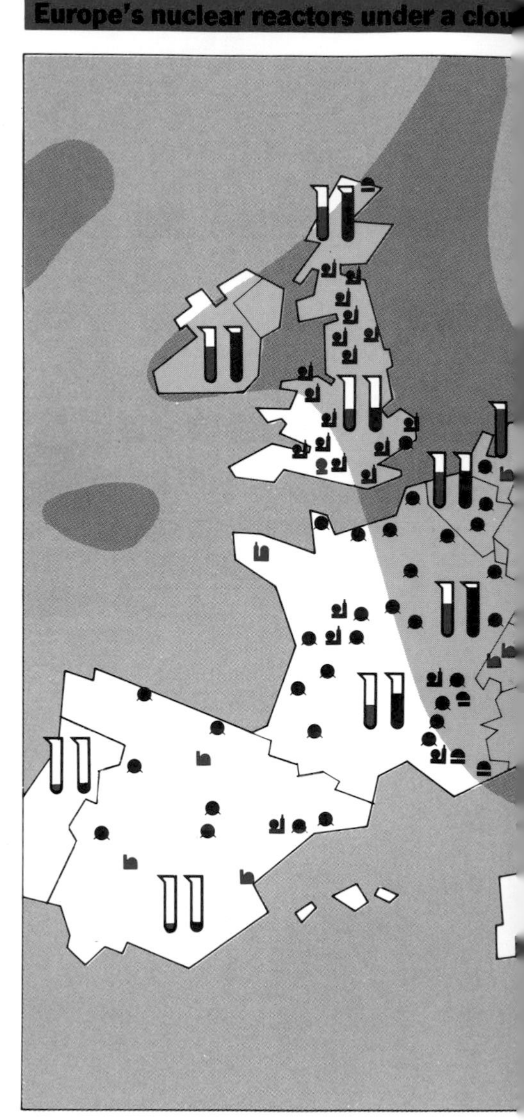

KEY TO REACTOR TYPES

GCR GAS-COOLED REACTOR
AGR ADVANCED GAS-COOLED REACTOR
HTGR HIGH TEMPERATURE GAS-COOLED REACTOR
MAGNOX MAGNOX-TYPE GAS-COOLED REACTOR

FBR/LWBR FAST/LIGHT WATER BREEDER REACTOR

GCHWR GAS-COOLED HEAVY WATER REACTOR
LWGR LIGHT WATER-COOLED GRAPHITE REACTOR

BWR BOILING WATER REACTOR

PWR PRESSURIZED WATER REACTOR

PHW PRESSURIZED HEAVY WATER CANDU
BLW BOILING LIGHT WATER CANDU
BHWR BOILING HEAVY WATER REACTOR
SGHWR STEAM GENERATING HEAVY WATER REACTOR
PHWR PRESSURE VESSEL HEAVY WATER REACTOR
LWCHWR LIGHT WATER-COOLED HEAVY WATER REACTOR

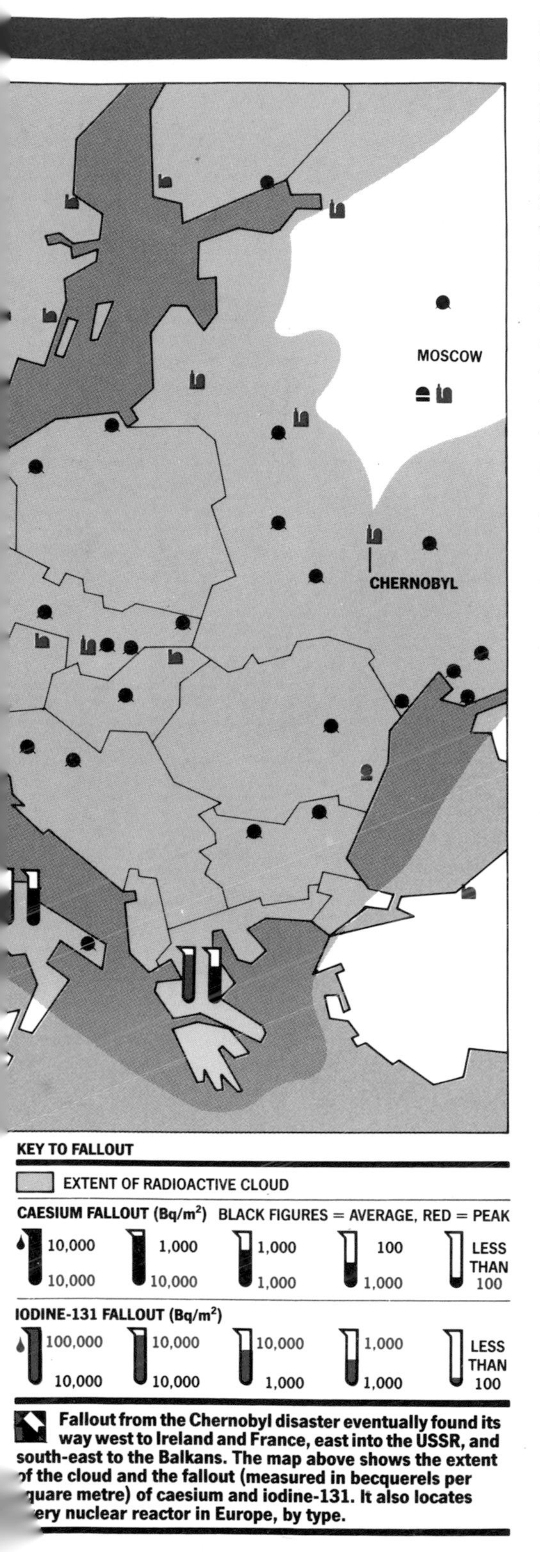

Fallout from the Chernobyl disaster eventually found its way west to Ireland and France, east into the USSR, and south-east to the Balkans. The map above shows the extent of the cloud and the fallout (measured in becquerels per square metre) of caesium and iodine-131. It also locates every nuclear reactor in Europe, by type.

radiation; and 1,800 tonnes of clay a sand to help seal off the fire. By May 6 th core temperature had fallen and the re-lease of radioactive materials dropped sharply. To help the cooling, liquid nitrogen was pumped through tubing underneath the reactor, and a heat exchanger was set up to keep the concrete containment beneath the reactor cool enough so that it would remain structurally firm. The Soviet authorities believed that between 30 and 50 million curies of radioactive substances escaped, amounting in all to a few per cent of the total inventory in the core. Nevertheless, the more volatile substances, such as iodine-131, caesium-134 and caesium-137, escaped in relatively large quantities, while the noble gases, such as krypton and xenon, escaped in their entirety. The estimate is that 20 per cent of the radio-iodine in the core escaped and about 12 per cent of the radio-caesium.

Considering the severity of the accident, the Soviet Union was initially fortunate insofar as the radioactive plume resulting from the explosion in the reactor core went up to heights of 1,500 metres and was carried away by the winds, first toward Eastern Europe and then to Scandinavia, where it was detected more than a day later. In fact, the Soviet authorities had sent aircraft up to seed an air mass on its way to the Chernobyl area so that rain would fall well away from the plume. The strategy may have succeeded, since the only deposition in the area around Chernobyl was from dry air.

Nevertheless, the fallout was sufficiently high over the first couple of weeks following the initial explosion to ensure that many thousands of people received substantial radiation doses from gamma and beta radiation. The authorities used a relatively large dose of 75 rems (750 millisieverts) as a criterion for evacuation. Toward the end of May, some villages and small towns in Byelorussia beyond the 30-kilometre zone around the stricken reactor had to be evacuated.

The Soviet authorities have now officially determined that the annual limit

Radioactivity – the basics of radiation and fallout

Types of radiation

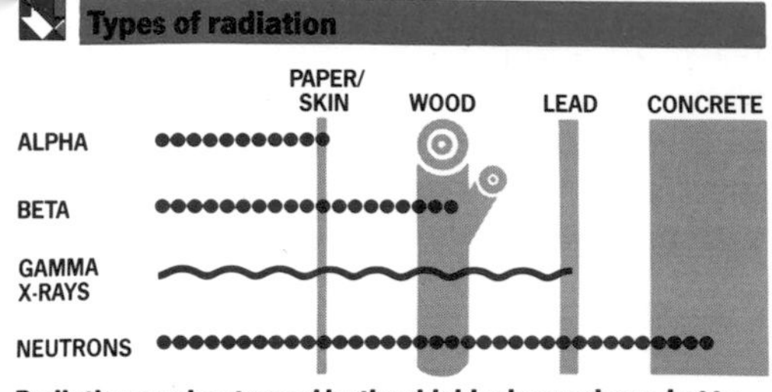

Radiation can be stopped by the shields shown above, but to protect living things the source must also be contained.

Measuring radiation

	OLD UNIT	NEW UNIT	CONVERSION
Amount of radiation given off by the source, be it milk, meat or nuclear fuel	CURIE (Ci)	BECQUEREL (Bq)	1Ci = 37,000 MILLION Bq
Amount of harm the radiation may cause depending upon the part of the body it hits	REM (rem)	SIEVERT (Sv)	100rem = 1Sv

The becquerel is a measurement of how radioactive something is, the sievert quantifies the threat that it poses.

Fallout – the different forms of radioactivity

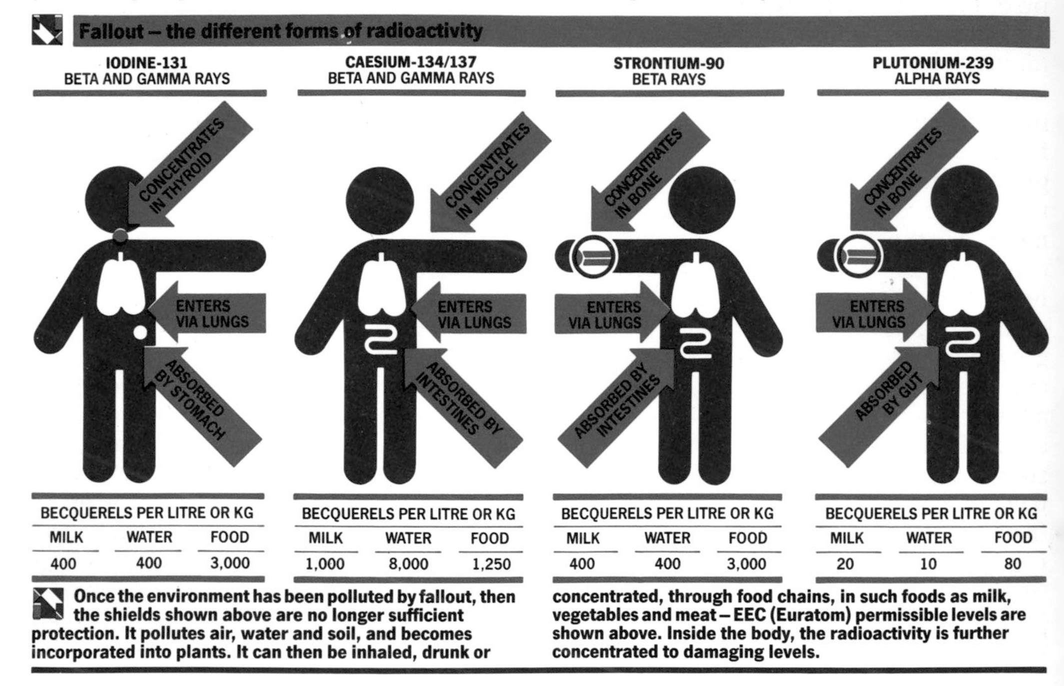

	IODINE-131			CAESIUM-134/137			STRONTIUM-90			PLUTONIUM-239		
BECQUERELS PER LITRE OR KG	MILK	WATER	FOOD	MILK	WATER	FOOD	MILK	WATER	FOOD	MILK	WATER	FOOD
	400	400	3,000	1,000	8,000	1,250	400	400	3,000	20	10	80

Once the environment has been polluted by fallout, then the shields shown above are no longer sufficient protection. It pollutes air, water and soil, and becomes incorporated into plants. It can then be inhaled, drunk or concentrated, through food chains, in such foods as milk, vegetables and meat – EEC (Euratom) permissible levels are shown above. Inside the body, the radioactivity is further concentrated to damaging levels.

for individual exposure to radiation is five rem (50 millisievert), which is ten times the limit set elsewhere in the world on the basis of recommendations of the ICRP (International Commission on Radiological Protection) and is equivalent to more than 25 times the natural background radiation dose. In the USA the maximum dose to which members of the public may be exposed to radiation from nuclear power is no more than 25 millirem per year (0.25 millisievert); hence 200 times less than the USSR post-Chernobyl level.

The radiation releases from Chernobyl continued over the first week and into the second. Heavy rains over parts of Scandinavia and Western Europe washed out considerable quantities of radioactive substances, some areas receiving several hundred times the fallout of those where it remained dry. Radio-iodine and caesium were by far the most important isotopes. Milk products and vegetables were quickly contaminated. In such places such as Gävle, some 100 kilometres north of Stockholm and several thousand kilometres away from Chernoybl, the fallout levels were so high – up to 200,000 becquerels of caesium per squar kilometre of soil – that livestock had to kept in for several months right into the summer. Hay and silage made from contaminated pasture also concentrated fal[illegible]out, leading to winter-feed problems [illegible] deed, in many parts of Europ[illegible] fallout had been high, as in B[illegible] West Germany, levels of radioa[illegible] tamination rose again sharply [illegible] winter because cattle were [illegible] taminated hay and silage.

In Britain, heavy rai[illegible] Wales, Cumbria and Sc[illegible] prisingly high deposi[illegible] caesium. Sheep in part[illegible] taminated and hundred[illegible] lambs that would otherwis[illegible] marketed for their meat, from som[illegible] farms, were found to have caesium levels greater that 1,000 becquerels a kilogramme, the level set by the British Government as the upper limit for sale and consumption. The Ministry of Agriculture, Food and Fisheries in the United Kingdom expected the levels to fall over the growing season, but had miscalculated, and the ban could be partially lifted only at the end of February 1987, ten months after the initial fallout. Meanwhile, suprisingly high levels of caesium were detected in vegetation in some parts of Cumbria, North Wales and Scotland during 1987 and the ban on lamb sales had to be kept in force there. By August 1987, 564 farms were still affected by the ban, including 39 new ones.

Undoubtedly those worst affecte[illegible] the fallout outside the USSR were [illegible] the Sami people – living prima[illegible] [illegible] products in the norther[illegible] [illegible] Sweden and Finland. [illegible]ation levels showed a [illegible]rds the end of the winter. [illegible]ing to a Swedish national [illegible] of February 12 1987, the [illegible] level in reindeer had risen [illegible] sevenfold from its September [illegible] of around 6,600 becquerels a [illegible]mme to more than 42,000 bec[illegible]s a kilogramme.

[illegible]e situation for the Lapps is particu[illegible]y acute since the lichens on which the [illegible]eindeer feed retain their contamination for several years. The reproductivity of the reindeer is also bound to be affected by such high levels of radioactive contamination. Matters are not made easier by controversy over acceptable levels of radioactive contamination. Until March 1 1987 the European Commission had suggested an upper limit for caesium contamination of 370 bacquerels a litre of milk and 600 bacquerels a kilogramme for other food products, the notion being that the average consumer would not exceed a maximum permissible level of five millisieverts a year given a normal appetite. Britain's level of 1,000 becquerels a kilogramme is clearly higher, but the Government had not anticipated that such high levels would be reached, let alone ex-

Fallout – second thoughts about safe levels

The USSR raised its limit for personal exposure to ten times the ICRP limit: from 5mSv to 50.

The EEC is proposing to raise its limit for caesium in food from 600Bq/kg to 1,250Bq/kg.

Official views of what were acceptable levels of radiation changed considerably in Europe. The UK set an upper limit for the sale and consumption of food of 1,000Bq/kg. Norway, as a concession to deer herders, set 6,000Bq/kg. In the USA the personal radiation limit remained at 0.25mSv, compared with the USSR's new limit of 50mSv.

Natural background radiation for one year = 1.9mSv

Lappland: "the radioactivity level in reindeer had risen more than sevenfold . . ."

ceeded, by the fallout from Chernobyl. In Norway, as a special concession to deer herders, maximum levels of 6,000 becquerels a kilogramme have been put in force and Sweden has considered similar action.

After considerable debate, with Britain and France calling for total radioactive contamination levels permitted in foodstuffs following the "next Chernobyl-type release" not to exceed 4,500 becquerels per kilogramme, the EEC decided on double the present ban levels. From November 1 1987 the EEC ban level will be 1,250 becquerel per kilogramme.

Controversy has raged and will continue to rage over the number of people in the Soviet Union and the rest of Europe, both East and West, who will die of cancer over the next 50 years as a direct consequence of radioactive contamination from Chernobyl. The United Kingdom National Radiological Protection Board, for instance, has estimated that approximately 1,000 people will die of cancer from Chernobyl fallout in the European Common Market countries over the next four decades. However, the precise figures depend on a number of imponderables, including the exact amount of radioactive fallout to settle out over the northern hemisphere, where it settles, how many people are affected and for how long, whether the longer-lived radioactive substances such as caesium get quickly locked away in the soil or instead continue to recycle through the food chain, and finally on the relationship between the radiation dose received and the induction of cancer. Given that scientists have different opinions on every one of these points, it is hardly surprising that the estimated cancer toll lies somewhere between a few thousand and well over a million.

At the same time the evidence from other studies of people exposed to radiation – as in nuclear installations – suggests that we are more sensitive to the effects of radiation than has generally been accepted by official bodies such the International Commission on Radiological Protection, from which many national radiological protection bodies, such as Britain's National Radiological Protection Board, take their cue. The problem, as always, will be to identify the Chernobyl-caused cancers among the many others that will occur. For instance, over the next 50 years it is estimated that in Britain there will be at least nine million cancer deaths, and not only will Chernobyl have to be taken into account, but also the United Kingdom's own nuclear power programme and the discharges from the large reprocessing works at Sellafield.

The cost of Chernobyl will run into thousands of millions of dollars. For the Soviet Union alone the clean-up of the site, the health care of the victims – including the several hundred that had to remain in hospital – the cost of the evacuation, the lost agricultural output and the replacement cost of the plant, will amount to more than $US5,000 million. Nor has that estimate accounted for the cancer cases to follow, or for the cost of dealing with inherited disorders that will affect generations to come.

Outside the Soviet Union the cost to farmers has also been considerable. Sheep farmers in the United Kingdom have claimed losses of $US15 million; in Sweden the costs of Chernobyl are put at least at $US145 million and the West German Government paid out some $US240 million for lost sales. The cost to the countries of Eastern Europe, as well as to Turkey, Greece and Italy, all of which were particularly hard hit, must also be taken into account. It is also quite clear that the Soviet Union will not even consider paying out compensation for contamination that crossed borders into other countries. Ironically, too, most countries will not pursue the question of compensation from the Soviet Union since many European governments remain strongly committed to nuclear power and do not want to draw continuing attention to what is considered to be an extremely embarrassing episode. For instance, despite opinion polls indicating a majority against the further development of nuclear power, in March 1987 the British Government granted permission for the Central Electricity Generating Board of England and Wales to order a 1,450MW pressurized water reactor (PWR) at Sizewell in Suffolk, to be followed by others elsewhere in the United Kingdom. The PWR, new to Britain, had been the object of a two-year-long public inquiry.

"The explosion at Chernobyl was the result of a runaway chain reaction of the kind that is used in an atomic bomb explosion"

Chernobyl was a profound shock to the nuclear industry. Although it was conceptually possible that a reactor could blow up as Chernobyl had done, nuclear engineers believed that they had designed into nuclear reactors adequate safety precautions and back-up systems that would preclude a nuclear runaway accident. Indeed, analyses of the probabilities of an accident such as occurred at Chernobyl are put in the order of one for every ten or even 100 million years of reactor operation: an almost negligible quantity. And while the public may have got used to hearing of core meltdowns as a result of the partial core meltdown at Three Mile Island in Pennsylvania, it had been assured that nuclear power stations would not explode, and certainly not like an atomic bomb. Yet the explosion at Chernobyl was the result of a runaway chain reaction of the kind that is used in an atomic bomb explosion. In addition, the fallout from Chernobyl was many times greater, possibly by a factor of a thousand, than that which would follow a Hiroshima-size atomic bomb explosion. In fact, the reactor core at Chernobyl disrupted after a surge in the chain reaction that took the power to several hundred times

Nuclear reactors – the types currently in use

The principle behind nuclear reactors is as old as the steam engine: fuel burns, produces heat that turns water into steam, and the steam turns turbine blades to drive electrical generators. But unlike wood, coal or oil, the fuel is inherently unstable and dangerous. Designers have tackled the problem of separating the fuel core from the turbines in various ways. The RMBK is exceptional in using the water coolant that surrounds the core to drive the turbines. Other designs isolate the heat exchangers from the coolant.

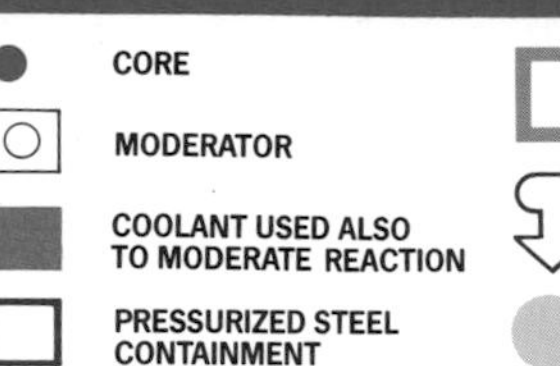

Advanced gas-cooled reactor

Fuel	**Enriched uranium dioxide**
Coolant	**Carbon dioxide at 40 bars**
Moderator	**Graphite**

Weaknesses

- **Immediate shutdown must follow failure of circulatory system or explosion may breach pressure vessel. Result would be explosion greater than Chernobyl**
- **Vibration**
- **Steam generators subject to corrosion. Could release water and steam under pressure into pressure vessel, producing explosions of various kinds (including surge of reactivity similar to Chernobyl)**

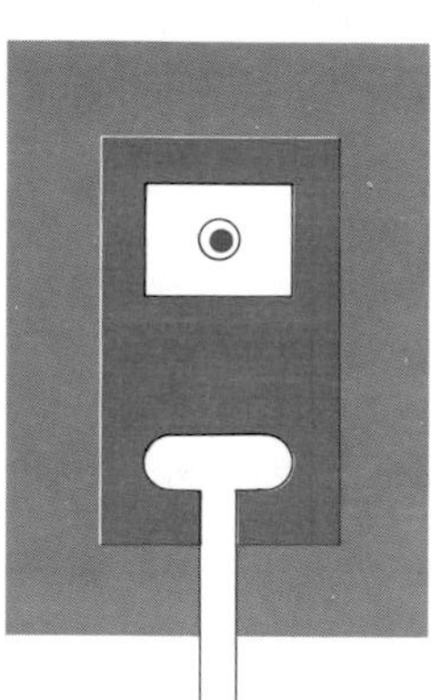

Pressurized water reactor

Fuel	**Enriched uranium dioxide**
Coolant	**Specially treated water**
Moderator	**As coolant**

Weaknesses

- **Coolant/moderator is prevented from flashing into steam by pressure of 150 bars. Pressure vessel may burst because of faulty manufacture, embrittlement through neutron bombardment or thermal shock**
- **Even if reactor is shut down, coolant pumps must be kept working**
- **A steam explosion, caused by the coolant hitting molten fuel, was involved in the RMBK version of a PWR at Chernobyl**

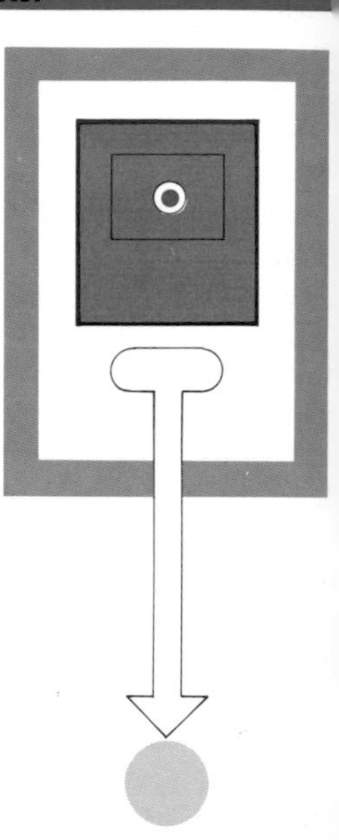

Fast breeder reactor

Fuel	**Plutonium and uranium oxides**
Coolant	**Liquid sodium**
Moderator	**None**

Weaknesses

- **Positive void coefficient (see RMBK)**
- **Fuel melting or bowing can lead to sharp rises in reactivity**
- **Nuclear explosions theoretically possible, and would be disastrous because of plutonium in core**
- **Sodium will interact with water and generate hydrogen, or catch fire in air**

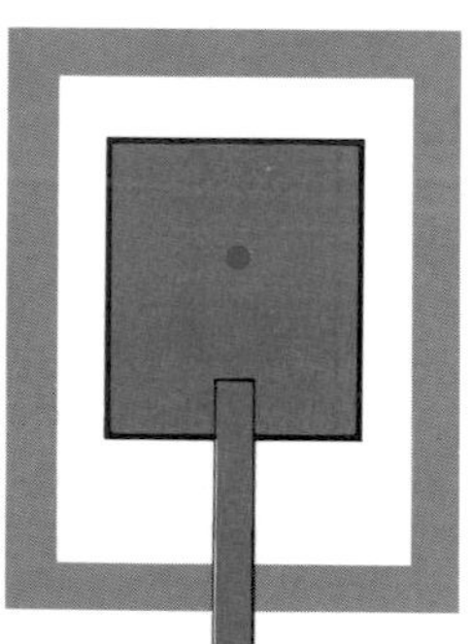

Soviet RMBK

Fuel	**Enriched uranium dioxide**
Coolant	**Water**
Moderator	**Graphite**

Weaknesses

- **Positive void coefficient (Reducing the coolant density leads to an increase in reactor power, which is normally balanced by a rise in fuel temperature. At levels below 20 per cent of full output, an imbalance occurs and control systems that respond rapidly are needed. At Chernobyl an operator dropped below 20 per cent with various automatic safety systems switched off)**
- **Direct connection between core and generating turbines**

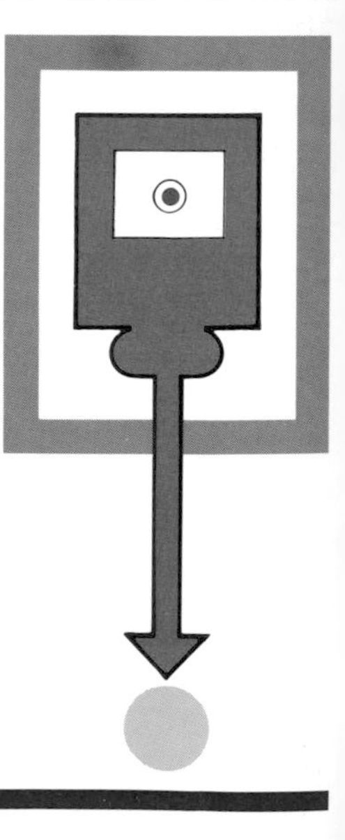

normal operating power.

What happened at Chernobyl? In a remarkably candid account to an international meeting in Vienna in August 1986, held at the headquarters of the International Atomic Energy Agency, the Soviet delegation stated that the prime cause of the accident was a number of violations of operational practice by technicians at the plant. Ironically, they had set out to test a system for improving plant safety, wishing to discover whether the spinning momentum in a running-down turbine would be sufficient to provide power for a short space of time to the pumps that circulated coolant water through the reactor core. All nuclear plants take power from the electricity grid to operate their safety systems, a major part of which is maintaining coolant flow. Should that power from the grid be lost, then back-up generators, usually diesel-powered, start up, but only after a short time lag. The Chernobyl experiment was designed to ensure that operators could fill the gap in time between the loss of power from the grid and the start-up of the diesels.

"The prime cause of the accident was a number of violations of operational practice by technicians at the plant"

Operating any reactor is a balancing act, with just enough neutrons being made available from the fissioning of uranium and plutonium in the fuel to keep the chain reaction going at the desired level. Should the population of neutrons begin to fall, the power goes down, as do the chances of the reactor snuffing out. If there is a sudden rise in the population of neutrons, the chances are that the operators will have a runaway nuclear reaction on their hands. During operation the reactor is basically controlled through inserting and withdrawing control rods made up of substances that absorb neutrons. Reactors are designed so that the full insertion of the control rods between the packages of fuel will stop the chain reaction in its tracks.

Standard operating procedures demand that the power should not fall below 20 per cent of full power. Unfortunately, in the experiment that led to the disaster, the operating technician in charge let the power fall too far, down to some seven per cent of full power, at which point he had to juggle with the reactor controls to prevent the reactor shutting down altogether. In the meantime he had disconnected various automatic safety devices, including the shutdown scram mechanism, since their intervention would have spoiled the experiment.

To raise the power from its low level the technician operated the control that lifted the control rods out the core and reduced the flow of water through the pressure tubes. The reactor was extremely sluggish at this stage and hardly responded. But as the operator continued manipulating the controls, the chain reaction slowly began to build up, so raising the power and generating more heat. The water began to flash into steam, that process in itself tending to make more neutrons available for the chain reaction. Within seconds the operator had a runaway situation on his hands with the power building up to more than a thousand times what it had been moments before. He tried to scram the reactor by dropping in the control rods, but by then they had been raised out of the core and needed many seconds to become fully inserted – far too slow for the reaction that had been unleashed. The explosion blew the fuel apart and effectively brought the chain reaction to a violent end.

That fearsome increase in power at Chernobyl was the result of what nuclear physicists call a "positive void coefficient", with "void" meaning that the coolant vaporizes and takes up more space for a given mass, and "positive coefficient" meaning that reactivity increases as voiding continues. In the immediate aftermath of Chernobyl, nuclear experts, notably Lord Walter Marshall, chairman of the United Kingdom Central Electricity Generating Board, claimed that such an accident was highly improbable in the West because reactor designs such as the RBMK would never receive a licence. He dubbed the RBMK reactor a "chimera" and "hybrid", explaining that Britain's nuclear power stations did not suffer from the drawback of having a positive power coefficient.

Other experts quickly assumed that the Chernobyl reactor that exploded had no containment structure built around it to contain any radioactive release in the event of an accident. They referred in particular to the differences between Chernobyl and Three Mile Island, where most radioactive substances, and especially caesium, were confined to the space within the containment building, even though they had escaped from the reactor.

Closer study of the RBMK proved such experts wrong. Chernobyl had a containment of the type used at many nuclear plants in the West, for instance to contain boiling water reactors and even some pressurized water reactors. Moreover, many of the gas-cooled graphite moderated reactors used in Britain and France – the Magnox reactors – had no containment whatsoever. At the Vienna meeting in August 1986, the Soviet delegation questioned whether any containment structure in place in the world would have been able to withstand the explosive power that tossed aside a 1000-tonne steel lid. Furthermore, the stronger the containment the worse the explosive event should it burst. At Chernobyl, as is true of many Western nuclear stations, reactors are sited next to each other, and there is always the danger that should one explode it will set off the others. The French electricity board, EdF, for instance, operates six pressurized water reactors at Gravelines facing the English Channel. Most of France's PWRs are in groups of four. Nor was Walter Marshall being candid when he said that reactors licensed in the West did not have positive void coefficients. Both the experimental fast reactor at Dounreay in Scotland, and the large 1,450MW Super-Phénix fast reactor at Creys-Malville near Lyons in France, have positive void coefficients, whereby the reactivity tends to increase should the liquid sodium coolant start boiling, as may occur should there be a blockage in part of the core.

Aberrant reactivity effects can occur in many different reactor systems under abnormal operating conditions. Even Britain's gas-cooled reactors have reactivity coefficients that could cause fuel meltdown and violent reactivity surges. As in RBMK reactors, such power coefficients are normally kept under control, and it would take abnormal circumstances for them to become effective.

Normally, too, reactors will shut themselves down automatically if abnormal conditions inside them are detected, such as a rise in temperature of the fuel or coolant. But what if the flow of coolant were suddenly lost and the reactor failed to scram? The nuclear physicist Dr. Richard Webb, carrying out an independent analysis of Britain's gas-cooled reactors, has discovered on the basis of reactor physics that such an accident in an advanced gas reactor (AGR) could lead within seconds to an explosion every bit as big as that at Chernobyl. The explosion would follow a sequence of events involving melting of the steel cladding that encases the fuel, then melting of the fuel and its frothing as fission gases such as Xenon bubble out, vaporization of the fuel and its escape from the core. Meanwhile enormous explosive pressures will build up from the accelerating fission process, beyond the capacity of the concrete pressure vessel to contain them.

In March 1987 a tonne of water escaped into the reactor core after a boiler tube burst in the steam generator of the AGR at Hartlepool. The water put out of action half the gas circulators, therefore jeopardizing coolant flow. In fact, the reactor shut itself down as it was designed to do. Yet the breakdown of the four circulators was already one step toward disaster, and Chernobyl has demonstrated graphically how rapidly a situation can deteriorate.

Positive power coefficients are not necessarily implicated in reactor accidents, pressurized water reactors can suffer serious accidents even after emergency shutdown, as happened at Three Mile Island in 1979.

PWRS were originally designed as compact, powerful reactors for submarines. They had the advantage from a design point of view that the water used as coolant for transferring the heat from the core to the steam generators also doubled up as moderator. To prevent the water in the reactor boiling at the relatively high temperatures generated in the reactor core, it must be under considerable pressure – normally of the order of 160 atmos-

Three Mile Island: seven years later, the debris has still not been cleared up

Chernobyl: Soviet workers decontaminate the wrecked plant

pheres. The reactor pressure vessel and the remainder of the circuit taking water to and from the steam generators back to the reactor must be of sufficient tensile strength to take the strain.

One of the great concerns with PWRS is that the pressure vessel should suddenly fail and burst, discharging steam at high pressure and metal fragments, some of which may breach the concrete containment that enshrouds the reactor. A considerable quantity of radioactive debris could then escape into the environment, the amount depending on the size of the breach. For many years a debate has raged over the likelihood of pressure vessel rupture, given that the metal is welded together and defects and cracks can be covered over. The nuclear industry has claimed that the chances of a pressure vessel break are one in a million years for each reactor. So far, pressure vessels have held, though serious doubts are now being raised over the quality of certain reactor vessels, such as that at Stade near Hamburg in West Germany, which was of inferior metal. Analysis indicat the metal has become brittle – a reactor were suddenly to cool, emergency shut down, the fall in ture might precipitate its sudden

Another major worry over th PWRS is related to accidents in from the core is lost, whe pipe fractures or through Even when shut down dur cy, a PWR generates consider confined space, enough to and the steel of the pre molten fuel could al through the thick con lying the reactor and surface and groundw built dykes as well beneath the number to try to prevent ra tion of essential wat of Kiev, some

PWRS are theref cy core cooling

water into the core should coolant be lost. Yet the problem (which surfaced during the accident at Three Mile Island in March 1979) is to provide instrumentation that can accurately indicate what is happening in the core during an emergency. Pump in too much water and the system becomes overpressurized and liable to burst apart at the seams. Pump in too little and the core is uncovered and quickly melts. At Three Mile Island the operators were trained to worry about putting in too much water and making the reactor go "solid". They throttled back the emergency core cooling system, not realizing that a faulty valve in the water circuit had stuck open, and thereby initiated a core meltdown. At the temperatures reached in the core, the water still remaining in the reactor began to react with the zircaloy cladding used to contain the uranium oxide fuel and hydrogen was generated, which later exploded, but luckily with insufficient force to burst the containment.

"Now, more than seven years later, debris inside the Three Mile Island reactor has not yet been cleared up . . . The cost will exceed $US1,000 million"

Had the explosion been sufficient to rupture the containment, the United States would have had a major accident on its hands. More hydrogen did build up in the days following the beginning of the accident at Three Mile Island but much of it disappeared, probably through reacting with zircaloy to form a brittle metal hydride compound which then disintegrated within the reactor core.

Now, more than seven years later, the debris inside the Three Mile Island reactor has not yet been cleared up, despite a round-the-clock clean-up process involving thousands of workers and the development of special robotic machines to work in highly contaminated areas. The cost will exceed $US1,000 million and money has been sought from Japan as well as from other United States power company utilities.

By 1986 there were more than 190 PWRS in operation out of a world total of commercial reactors of nearly 380. At the same time there were some 110 PWRS under construction.

Before the Three Mile Island accident, scientists had set out to predict the probability and consequences of a major nuclear accident. In 1957, the United States Atomic Energy Commission, then in charge of both promoting and regulating nuclear power, came up with its WASH 740 report on the effects of a major accident to a nuclear reactor. The conclusion was that a massive radiation release from a 200MW reactor sited 30 miles from a large city would lead to 3,400 deaths and 43,000 injured, together with up to $US7,000 million worth of damage.

Seven years later, in 1964, the AEC set out to update the WASH report to take account of the large reactors then being built and better analytical techniques. The expectation was that the result would be more reassuring. Though the results were never made officially public, they were released after a request had been made through the Freedom of Information Act. They indicated that an accident to a 1,000MW reactor sited inside a city might lead to as many as 55,000 immediate deaths and 70,000 injuries, the disaster encompassing an area the size of Pennsylvania.

The United States Nuclear Regulatory Commission then instituted another study, known as WASH 1400, which was published in 1975 and was specifically limited in its scope to the 100 light water reactors likely to be operating between 1976 and 1980. The conclusions of WASH 1400 were intended to be reassuring, since the team of scientists, led by Professor Rasmussen of the Massachusetts Institute of Technology, found the risk of a major accident in which large quantities of radioactivity were released to be one chance in every thousand million years of reactor operation: a highly improbable event. The Rasmussen team nevertheless accepted that a major release of radioactivity from a 1,000MW reactor might lead to 45,000 deaths, quarter of a million injuries and property damages from $US14,000 million upwards.

To measure probabilities the scientists used the technique known as "fault tree" analysis in which the consequences of each malfunctioning and its likelihood are through to a logical conclusion. A major criticism of such methodology is that it fails to take account of a "multiple mode failure" in which several distinct functioning parts simultaneously break down. In 1975, for instance, a fire in a cable channel started by workmen searching for air leaks with a candle destroyed the control systems of two twin reactors at the Browns Ferry plant in Alabama. Luckily both reactors responded to manual control and were shut down, yet the accident could have led to a common mode failure.

The Union of Concerned Scientists in the United States was strongly critical of the conclusions of WASH 1400, believing not only that the consequences of a major accident could be 100 to 1,000 times greater, but that fault tree analysis seriously underestimated risk. The UCS pointed out, as an example, that a serious malfunction which occurred at the Dresden 2 reactor in 1977 involving problems with the control rods had a probability of occurring, according to WASH 1400, of such a tiny number as to be inconceivable: yet the accident occurred. Moreover, in terms of magnitude and risk, the Three Mile Island core meltdown came close to the one in a thousand million probability. It, too, involved common mode failures.

The United Kingdom nuclear establishment has been equally phlegmatic about the risks of radiation release from reactor accidents. For instance, in one analysis, R. R. Farmer of the United Kingdom Atomic Energy Authority suggested that while releases of up to 1,000 curies of radioactivity might be expected at most every hundred years, one involving releases of a million curies would be expected at worst every million years of reactor operation. Yet many millions of curies were released from Chernobyl some 20 years after Farmer's evaluation of risk.

"After the accident the percentage against nuclear power went up to the seventies and even the eighties, with the exception of France . . . "

How has Chernobyl affected the fortunes of nuclear power in the world? Before the accident in the Ukraine the public in many coutries tended to be more against nuclear power than for it, according to opinion polls. After the accident the percentage against went up to the seventies and even the eighties, with the exception of France where it became more or less equally divided. In the United States, for instance, 67 per cent were opposed to the building of additional nuclear power plants prior to Chernobyl and 78 per cent after; in West Germany the change was from 46 per cent to 83; and in the United Kingdom from 65 per cent to 83 per cent. Yugolsavia also showed a switch from 40 per cent to 74 per cent. In Finland, meanwhile, some 4,000 women activists stated that they would not bear children until the government turned away from its nuclear power policy. The psychological effect of Chernobyl was particularly strong in the Soviet Union and in countries of the Eastern bloc, which for the first time saw the beginnings of an opposition to the further development of nuclear power.

In the Soviet Union the immediate consequence of Chernobyl was the shutting down of the country's remaining 27 RBMKS. But by August 1986, and the

Vienna meeting of the Soviet delegation with delegates from some 113 member countries of the International Atomic Energy Agency, the Soviet Union was explicit in its plans for the future development of nuclear power. "By 1990," said A. Petrosyants, chairman of the Soviet State Committee on the Utilisation of Atomic Energy, "we expect the country's nuclear power stations to produce 360,000 million kilowatt-hours (360 terawatt-hours) as compared with 170,000 million in 1985." Nor was he thinking only of thermal reactors such as the Soviet Union's pressurized water reactors and the RBMKs. "By the year 2000," he continued, "fast reactors will have joined the system and will gradually supplant the thermal reactors. . . ."

"Even before the accident at Three Mile Island, there had been growing concern in the United States over the safety of its reactors . . . Chernobyl has reinforced the American public's antipathy to nuclear power"

Even the RBMKs, of which seven more were under construction at the time of Chernobyl, were to be brought back on line, but with certain modifications, such as additional control rods and increased enrichment of the uranium-235 in the fuel, the purpose being to increase the reactivity at low power and to overcome the reactor's inherent sluggishness. By 1987 all RBMKs, save for the Chernobyl number 4 reactor and its twin number 3 reactor were back on the grid.

Though the United States has more nuclear power stations operating, as well as under construction, than any other single nation, it is a country where the fortunes of the nuclear industry have been at a low ebb for at least ten years. Since 1977 not one new nuclear power station has been ordered and more than one hundred that had been ordered since 1973 have been cancelled. Those that have survived being axed have taken ten years or more to complete and their costs have escalated accordingly. Undoubtedly the most spectacular rise in cost was for five nuclear plants ordered by the Washington Power Supply System in the United States that were originally estimated to cost $US4,000 million in the early 1970s. Within a decade the costs had risen to $US26,000 million, bankrupting many local authorities in Washington State that had invested in the project and leading to the mothballing of three of the new reactors when they were only half-built.

Even before the accident at Three Mile Island, there had been growing concern in the United States over the safety of its reactors and the Nuclear Regulatory Commission had begun demanding that new safety measures be implemented. Many reactors had to be re-designed and back-up safety systems retrofitted during the course of construction. As was shown in many studies, but particularly by Charles Komanoff and his associates, nuclear power was becoming considerably more costly as a source of electricity compared with coal, even when coal-fired stations had to comply with ever-more stringent regulations concerning sulphur and nitrogen oxide emissions. Utilities also began to discover the virtues of energy conservation measures and in some areas exhorted their customers to save electricity rather than consume more – a fundamental switch in thinking from the growth-for-growth's-sake mentality of the 1960s.

Chernobyl has reinforced the American public's antipathy to nuclear power. The public can hardly have been encouraged by estimates within the NRC that the chances of a major nuclear accident in the United States is fifty-fifty over the next 20 years. Applied to the 380 commercial reactors now operating in the world, that accident rate amounts to a probability of a bad accident every eight years, a prospect that appears to fit in with the nuclear record over the past 30 years.

And whereas the population density in the Chernobyl area was low, the same does not necessarily hold for areas around nuclear power stations in other parts of Europe and the United States. For instance, some five million people live within 50 kilometres of the two nuclear power plants at Biblis on the River Rhine near Mannheim in West Germany, while 1,500,000 are within the same sort of range of the Indian Point plant near New York City. An accident such as occurred at Chernobyl could lead to financial losses of more than $US300,000 million, according to estimates carried out by the Sandia National Laboratories. And should such an accident occur in the United States or in Western Europe, would the public tolerate running the remaining nuclear power plants?

Outside the Soviet Union and its satellite countries, France has had the least debate over its nuclear power programme. Instituted in 1973, just after the Yom Kippur war, the French nuclear power programme consisted of a production line manufacture of Westinghouse type pressurized water reactors (PWRS), the aim being to embark on construction of five to six new reactors each year. That programme has come to fruition, as planned, and at present some 70 per cent of France's electricity is generated within nuclear power stations, there being 40 PWRS in operation, with others still being built. The programme planners foresaw France becoming increasingly electrified, and they anticipated an annual growth in electricity consumption of seven per cent per year – a doubling every ten years. That growth has not been forthcoming, industry in particular having resisted further electrification, and the result is that France is moving rapidly to the time when it has five or more nuclear power stations surplus to its needs. The programme has now been slowed down with between one and two new nuclear plants being ordered each year.

Ostensibly, the French nuclear programme was established to give France

Checking for xenon near Bugey nuclear station

energy independence and certainly to free the country from its dependenc on OPEC oil. However, because petroleum is needed primarily for transport and for petrochemicals, France's nuclear power programme has made a relatively small dent on oil imports, and no more than has been achieved by neighbouring countries such as West Germany and the Netherlands. In return for nuclear power, France has acquired an international debt of $US35,000 million, putting it on a par with Mexico and Brazil. EdF, the electricity board, has stated that it will never be able to repay the capital borrowed.

Nor has France confined itself to PWRS. Its fast reactor, Super-Phénix, is the largest of its kind in operation, and apart from its cost, which is more than double that of an equivalent PWR, critics have questioned the wisdom of operating a machine which is fuelled with five tonnes of plutonium and has 5,000 tonnes of liquid sodium running through the core as coolant. Since the early spring of 1987 sodium has leaked at a rate of some 500 litres per day from the fuel storage container at Super-Phénix. Having failed to locate the leak, the Edf operators have had to shut down the reactor. A number of cities, including Geneva, are all within easy range of Super-Phénix. Moreover, unlike other reactor types, the mechanisms do exist in fast reactors for atomic-bomb-type explosions. A major disaster at a fast reactor could leave an enormous swathe of countryside permanently contaminated with fallout rich in plutonium.

From France's point of view, a major advantage of a fast reactor is that it not only consumes plutonium but breeds a grade of plutonium that is excellent for nuclear weapons. France's Force de Frappe nuclear deterrent is a major political consideration.

"A siege situation now exists at Wakersdorf, with protesters being met with thousands of riot police armed with water cannon"

France, like Britain, has embarked on commercial reprocessing of reactor spent fuel. It is expanding the throughput of its plant at Cap de la Hague in Normandy, just as Britain has been expanding its own at Sellafield in Cumbria. But in contrast to Britain, France has achieved a much cleaner plant in terms of discharges to environment; after years of negligence British Nuclear Fuels, the operators of Sellafield, have been ordered by the United Kingdom Nuclear Installations Inspectorate to improve safety and reduce discharges if they are to continue to receive a licence to operate. Certainly the appalling practices over the years at Sellafield, in addition to the discovery of a significant increase in childhood cancer close to the plant, have led to public opposition to nuclear power.

Despite the growth of opposition to nuclear power in West Germany, the Bavarian State Government has insisted on proceeding with the construction of a reprocessing plant at Wackersdorf, close to the border with Austria. Austrians are particularly incensed since they have voted in a referendum not to have nuclear power, not even the nearly-complete plant at Zwentendorf. A siege situation now exists at Wackersdorf, with protesters being met with thousands of riot police armed with water cannon. Austrians have also been forbidden by the Bavarian State Government to cross the border, and anti-nuclear meetings in West Germany have been disrupted by the authorities, the claim being that they are subversive activities.

An equally acrimonious situation has also developed between France and its neighbours, particularly Luxembourg, over the completion of Cattenom, a site of four 1,300MW reactors. Luxembourg, like Austria, has no nuclear power, and argues that an accident at Cattenom could make its entire country uninhabitable. Denmark, too, is concerned over the Swedish nuclear plant at Barseback, just across the water from Copenhagen.

In the Third World, meanwhile, the fortunes of nuclear power have not changed significantly because of Chernobyl. The hopes for nuclear power in the early 1970s, and the spate of ordering that followed, have now given way to the economic realities of building costly power plants that would have to be imported almost in their entireties. Mexico, for instance, ordered some 20 reactors, but has now restricted itself to two; and both Brazil and Argentina have cut back on their nuclear plans, though both countries have several plants under construction. In the Phillipines, one of the first acts of the new President Corazon Aquino was to cancel the almost complete Westinghouse PWR that had been ordered by Marcos. Bribery and corruption had played a big part in the original ordering, and the plant's safety has been undermined by shoddy, irresponsible construction.

The Daya Bay plant, to be built by France for the Chinese just across the border from Hong Kong, has led to considerable concern in the British colony, and a million signatures against the project have been collected. Despite Chernobyl, the Chinese government appears determined to proceed. However, in 1986 it announced that it would be cancelling eight of other ten plants planned. Japan, with 33 plants operating in 1986, is in common with France one of the few countries that has remained determined to proceed with nuclear power. It has, however, reduced its plans to two orders per year, the reason given being a slowdown in the economy. Meanwhile it has 11 plants under construction and will proceed with their operation.

The mighty hopes for nuclear power that came into being with the 1955 Atoms for Peace plan have virtually collapsed. The International Atomic Energy Agency extravagantly forecast in the 1970s that the world would have nearly 4,500 large nuclear plants in operation by the year 2000, enough to provide for double the world's 1986 electricity capacity. Such demand for electricity has failed to materialize, and its projections have since been cut by nearly a thousand-fold and even then they are unlikely to be realized. The Worldwatch Institute, based in Washington DC, projects that the world will probably have around 380,000MW by the end of the century, and therefore barely 350 large nuclear plants.

Even so, the industry is brazening out its setbacks, helped to some extent by politicians such as Britain's Energy Secretary Peter Walker, who a month after Chernobyl proclaimed that "If we care about the standard of living of generations to come, we must meet the challenge of the nuclear age and not retreat into the irresponsible course of leaving our children and grandchildren a world in deep and probably irreversible decline."

Such statements are undoubtedly out of step with public opinion and can hardly make sense to those who have lost their homes or even their livelihoods as a result of one nuclear accident. Despite assurances by governments and nuclear experts, the public had always feared that a reactor would one day explode, and having felt the consequences of one such accident is hardly likely to want to risk another. Moreover, it has become aware that nuclear power leaves a legacy of wastes which have to be disposed of safely. It has firmly rejected the notion of dumping wastes into the seas and everywhere is resisting the disposal of such wastes on land. Where governments proceed with nuclear power they increasingly create a divided society, and where democracy is given its head the answer is to reject nuclear power. Its image is now tarnished, its brief moment of glory appears already to have passed.

James Lovelock

"The Gaia hypothesis sees the evolution of the species of living organisms so closely coupled with the evolution of their physical and chemical environment that together they constitute a single and indivisible evolutionary process"

The seeds for today's ruthless exloitation of the natural world were undoubtedly present at the beginning of the neolithic revolution, some 12,000 years ago, when mankind first embarked on permanent agriculture and created large, fortress-like settlements to keep out nomadic invaders. Whereas all the members of the tribal forbears, the hunter gatherers, had participated fully in the venture of survival, the new settled agrarian way of life enabled a hierarchy to develop of labourers in the field and of overseers who could become rich through the toil of others.

New attitudes gradually prevailed; the world was there to be dominated and mankind saw itself increasingly separated from the rest of nature, which was there for husbanding and exploitation. With the move out into space, the ultimate manifestation of that worldview is already emerging; we have left the Earth and instead of looking back to see how beautiful she is, and how we are a part of her, we see Earth merely as a place from which to conquer the universe. Inherent in our culture is a modern evolutionary theory, a distortion of Darwin's great vision: the notion that living organisms are in deadly competition one with another, scrabbling for limited resources, the ultimate goal being to leave behind successful progeny. With the rational process of thought that developed from Descartes, the world and the universe of which it was part came to be seen as a kind of clockwork machine obeying fundamental laws of physics. Living organisms, though subject ultimately to the same laws, were deemed essentially separate from the environment in which they lived, reacting to it and, through the laws of Darwinian natural selection, adapting as best as possible to it, but affecting it superficially and ephemerally. Life was therefore seen as a kind of skin with a tenuous grip on the surface of the Earth, to whose varagies it had to adapt if it was to continue to survive; truly a struggle for existence.

A living redwood – but 99 per cent dead

Yet conventional thinking on the relationship between organisms and their environment, the discipline known as ecology, has left some fundamental questions unanswered. What, for instance, keeps oxygen at its remarkably constant level of around 21 per cent in the atmosphere, or prevents the seas becoming saltier each year as they receive a new burden of minerals from the constant erosion from the land? Is there some chance mechanism at play that enables organisms to adapt in time through the basic mechanisms of evolution to fundamental changes in the environment? Or, on the other hand, could the planet somehow be alive in the sense that this veneer of life on its surface, like the physiological mechanisms in an organism, somehow regulates the flow of nutrients in the system, restoring balances and controlling temperature? And before such a notion is dismissed as absurd, insofar as the Earth is almost all rock, most of it molten and incandescent, it is worth reflecting on a wonderful living organism – the mighty redwood tree, alive, yes, but 99 per cent of it dead wood, with just a thin layer of living tissue on the surface.

Paradoxically, it was the search for life on Mars that gave clues as to the answers to such earthly conundrums. In the 1960s NASA was preparing its unmanned space missions to Mars, including the orbiting Mariner spacecraft, which viewed the surface of Mars as it circled around, and later the two Viking probes, which soft-landed on the planet and were able to sample both surface atmosphere and the nature of the ground. In 1961 I was asked to join the team of scientists who had been brought together by NASA, to devise ways of investigating whether life existed on Mars. An obvious way, particularly once the Viking robots had soft-landed, would be to scoop up "soil" and test it for organic matter that might have a living origin. But such an approach would be a hit and miss affair; indeed, what would be one's conclusion on testing for life on Earth if a Martian spacecraft soft-landed in the midst of the Sahara sands or in the Antarctic wasteland? The philosopher Dian Hitchcock, employed by NASA to assess the logical consistency of life-detecting experiments, was also sceptical about such attempts to probe for life, and together we decided that the best, if not the only way, to detect life elsewhere was to analyze the composition of the gases in the planet's atmosphere. We concluded that life on a planet would be obliged to use the atmosphere and oceans as conveyors of raw materials and depositories for the products of its metabolism. The process of life would therefore change the chemical composition of the atmosphere so as to render it recognizably different from the atmosphere of a lifeless planet. Infra-red telescopy from Earth showed that Mars' atmosphere was dominated by carbon dioxide and that the composition of gases on Mars was close to the equilibrium one would expect from the laws of physics and chemistry. Such evidence strongly suggested to us that Mars was lifeless.

By comparison the Earth has an extraordinary atmosphere, far removed from chemical equilibrium and apparently kept in that state from one year to the next. Thus whereas Mars has 95 per cent of its atmosphere composed of carbon dioxide and Venus 96.5 per cent, Earth has 0.03 per cent. Similar sharp differences are found when comparing the planetary compositions of nitrogen and oxygen. Thus Mars has 2.7 per cent of its atmosphere composed of nitrogen and Venus 3.5 per cent. Earth on the other hand has 79 per cent. And while Mars has 0.13 per cent oxygen and Venus just traces, Earth has 21 per cent. Equally surprising the Earth's atmosphere has some methane, while the other two planets have none. In essence, the Earth's atmosphere contains oxidizing and reducing gases at the same time. For instance methane reacts in sunlight with oxygen to produce carbon dioxide and water vapour, and unless methane were constantly introduced into the atmosphere it would soon vanish. For the balance to be maintained some 500 million tonnes of methane must get into the atmosphere each year and 2,000 million tonnes of oxygen to make up that lost in oxidizing the methane. Both those gases are the metabolic products of living organisms, either of methanogenic bacteria or of photosynthesizers.

"If Earth were a dead planet and there was no life to inject specific gases into the atmosphere, then . . . the atmosphere would have a similar composition to that of Mars and Venus"

If Earth were a dead planet and there was no life to inject specific gases into the atmosphere, then knowing the likely outcome of chemical interactions between the atmosphere, rocks and water, the atmosphere would have a similar composition to that of Mars and Venus, with carbon dioxide predominating, with neither oxygen nor methane, and with a small percentage of nitrogen.

Just as significant, the surface temperature of Earth without life would lie between that of Venus and Mars. Moreover, it would be uncomfortably hot for life as we know it. Thus, whereas Venus has an average surface temperature of 459°C, and Mars -53°C, Earth without life would have a surface temperature of 290°C +/- 50°. With life, contemporary Earth has an average surface temperature of 13°C (55.4°F).

This spectacular difference between a burning hot, boiling planet and one with comfortable temperatures for life led me to the conclusion that life itself has evolved the capacity to regulate the exchange of heat between the Earth's surface and outer space. But first one has to understand the reason why such a discrepancy exists between the two temperatures, that of a living and dead Earth. Two fundamental facts have to be taken into account. The first is that the sun, similar to other stars, is becoming hotter, and is now 25 per cent more luminous and energy-giving than it was 3,800 million years ago when life probably came into being. The second is that the quantity of heat retained by a planet bathed in sunlight depends not just on the distance from the sun, but also on the reflectivity – called the albedo – of the Earth's surface and on the nature of the atmospheric gases. A dark surface, such as the oceans or bare igneous rock, will absorb heat as will the dark green vegetation associated with tropical forests; whereas clouds, snow or the pallid colour of a desert like the Negev will tend to reflect heat back into space. Meanwhile, gases such as carbon dioxide, or indeed methane and water vapour, have what are known as "greenhouse" properties. Thus they will let through light but will retain the shorter waved, less energetic heat radiation, just as does a greenhouse.

For life to have begun on Earth not only would the chemical precursors have had to be available in suitable concentrations – amino acids, nucleosides, polysaccharides – but the temperature would have had to be right for the "prebiotic" soup of chemicals to remain together in the kind of relationships that we now associate with the living organism. Since the sun was considerably cooler – some 25 per cent – a gaseous greenhouse blanket would have been required to trap sufficient to keep the mean surface temperature somewhere between 0 and 50°C.

Carbon dioxide is an effective greenhouse gas. When life began, the quantities of carbon dioxide in the atmosphere may have been as much as 1,000 times greater compared with today. Though surface temperatures would also depend on the distribution of water as oceans, snow, ice, clouds and water vapour, estimates suggest that the mean surface temperature of Earth would have been around 23°C; therefore not significantly different from now.

But with the sun heating up perceptibly since that Archean era, a blanket as thick as that then required to provide warmth would have become a dangerous encumbrance as far as life was concerned. One way around the problem of maintaining an equitable temperature would have been to shed the blanket, not in one fell swoop, but gradually, and that is precisely what appears to have happened.

Life as a whole reduced the size of the blanket by taking it from the atmosphere and dumping it in the form of limestone into the oceans, where we now see relics of that continuing process in the massive chalk cliffs that overlook the English Channel at Dover. The process by which life managed to dump carbon dioxide was part of an extraordinary revolution in the then-existing biochemical repertoire of living organisms. Central to it was photosynthesis, involving the capture of light energy to break the strong chemical bonds that bound oxygen to hydrogen and to carbon. Prior to photosynthesis, the first cells that had come into existence on Earth probably derived sufficient energy from the abundant organic chemicals lying around, but there were undoubted

The White Cliffs of Dover – one of Gaia's carbon dioxide dumps

limits to the quantities of living biomass that could be supported on such nutrients. Photosynthesis linked life to an abundant inexhaustible source of energy in the sun.

If the consumption of carbon dioxide in photosynthesis had been simply a one-way process, then in a matter of a few million years the blanket essential then for keeping the planet warm would have gone. Earth would have been plunged into an inhospitable, frozen state, bringing much of life to a short-lived end. However, just as bacteria such as the cyanobacteria had discovered how to tap and exploit the sun's energy, so other bacteria evolved ways of scavenging and decomposing the excretions and corpses of the photosynthesizers. In a world without free oxygen those decomposers produced methane and carbon dioxide, both of them greenhouse gases, thereby helping to maintain some balance between nutrient and energy requirements and the retention of a greenhouse blanket. The successors of those methane-producing bacteria exist today in the methanogenic bacteria which survive in oxygen-free environments such as in swamps, septic tanks and in the guts of animals, particularly herbivorous ones such as cows.

Though oxygen was released once photosynthesizing organisms began using carbon dioxide as a source of carbon for building up organic structures and providing energy, in the Archean era it was immediately mopped up by sulphides, iron, ammonium, hydrogen and other reducing agents that were initially present in substantial quantities. Indeed, it was probably crucial for life at the beginning that free oxygen was lacking, otherwise few organic chemicals would have survived long enough for incorporation and use by the first tentative organisms. Finally, some 2,300 million years ago, sufficient quantities of oxygen built up to the point when oxygen could escape freely into the atmosphere. That had a two-fold consequence: first, it hastened the evolution of respiration whereby organisms could use oxygen to metabolize carbohydrates completely down to water and carbon dioxide,

Gaia – life adapting to, and adapting, its environment

How it happened... But...

1 When the Earth was formed it was unable to sustain life. Vast quantities of carbon dioxide (CO_2) and water (H_2O) as steam were released by volcanic activity.

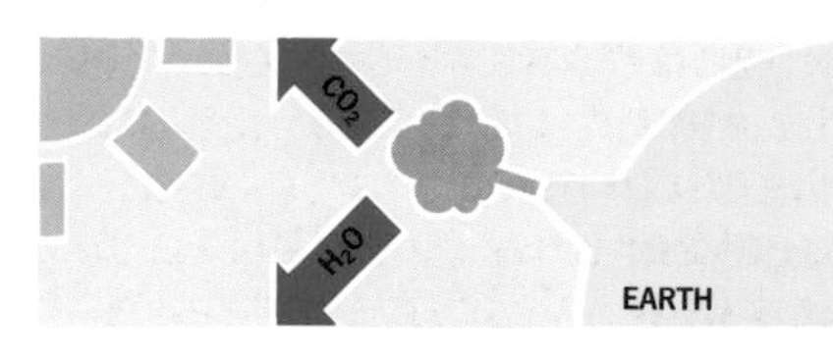

2 As the Earth cooled the steam condensed, forming the oceans. The increasing amount of CO_2, a greenhouse gas, trapped the sun's rays (which were 25% weaker than today), warming the atmosphere.

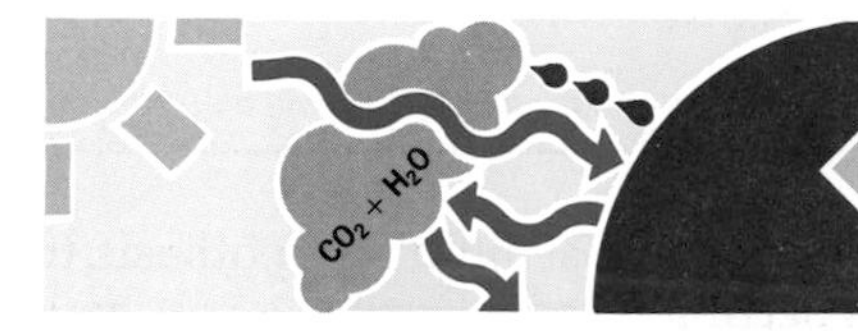

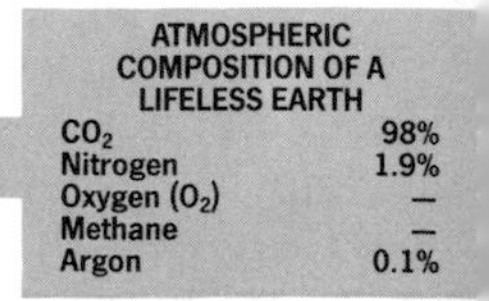

ATMOSPHERIC COMPOSITION OF A LIFELESS EARTH

CO_2	98%
Nitrogen	1.9%
Oxygen (O_2)	–
Methane	–
Argon	0.1%

3 The conditions were now right for life. The basic building blocks were present, along with a suitable temperature. Early life began mobilizing nitrogen, sulphur and carbon.

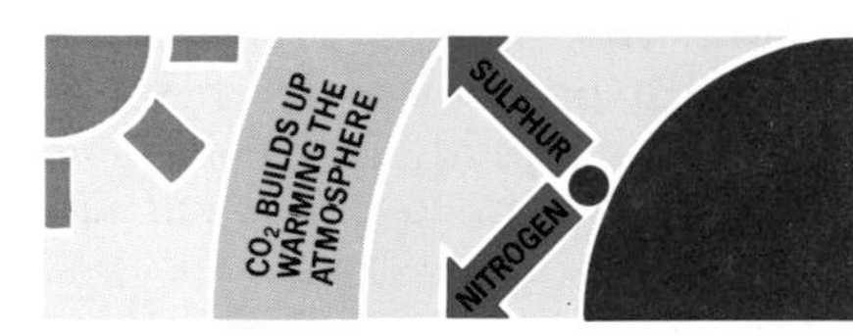

...as the sun warmed up, the Earth would have overheated had the amount of CO_2 not been reduced.

4 The amount of CO_2 was reduced by photosynthesis (which consumes CO_2 and releases O_2) and by the deposition of limestone by organisms. All the O_2 was initially absorbed by O_2 "sinks".

...if the CO_2 level had continued to fall, the greenhouse effect would have been lost.

5 Bacteria decomposing the products of photosynthesis released CO_2 and methane, both greenhouse gases, and thus maintained a stable temperature for life.

...if the steady loss of hydrogen had not been controlled, life would have been under threat.

6 Photosynthesis had now produced enough O_2 for it to build up in the atmosphere. This free O_2 combined with the hydrogen to form water. It also permitted the evolution of respiring animals.

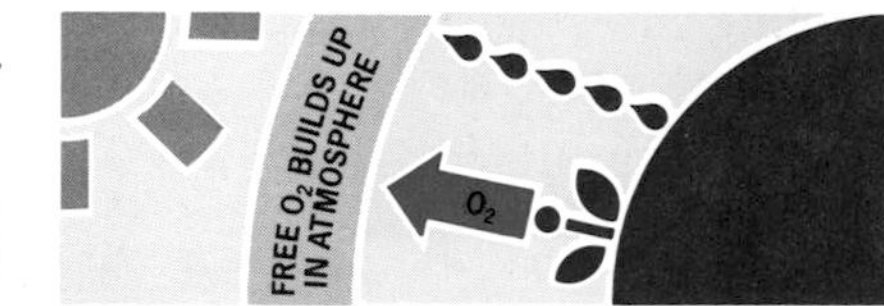

...if CO_2 had been used up too fast, then the greenhouse effect would have been lost.

7 Respiration restored the CO_2 balance by consuming O_2 and releasing CO_2, thus maintaining the greenhouse blanket. Life had established a dynamic balance adapting to, and adapting, its environment.

ATMOSPHERIC COMPOSITION OF EARTH TODAY

CO_2	0.03%
Nitrogen	79%
Oxygen	21%
Methane	–
Argon	1%

8 That dynamic balance is being disrupted by the products of our industrial way of life, such as CO_2, methane, sulphur dioxide, nitrogen oxides and chloroflourocarbons (CFCs). How will Gaia respond?

thus enabling a far more efficient use of resources; and secondly, by its readiness to form water (H_2O) it prevented the further escape of vast quantities of hydrogen into outer space, a process which if it had continued as during the anoxic period would have left the planet as arid as both Venus and Mars were to become.

Undoubtedly organisms have shown themselves to be extraordinarily opportunistic, evolving with great rapidity mechanisms that would enable them to exploit new potential habitats in the environment. But, as we have begun to glimpse, life in its entirety had a profound impact on the environment, transforming it completely. Thus the evolution and the development of the planet into the multihued sphere that we have seen projected onto our television screens has been a consequence of life, not apart from it.

In 1972, after my work for NASA and my preliminary considerations about the atmosphere of a living as distinct to dead planet, I postulated the Gaia hypothesis, the name having been suggested to me by the novelist William Golding in acknowledgement of the ancient Greek name for the Goddess of the Earth. In principle the hypothesis stated that the Earth has remained a comfortable place for living organisms, for the entire 3,500 million years since life began, despite a considerable increase in the heat output from the sun. Moreover, the atmosphere, despite being composed of an unstable mixture of reactive gases, oxygen and methane for instance, retains remarkable constancy at levels that appear to suit the high metabolic needs of today's organisms. The claim is that we live on "the best of all possible worlds" with living organisms actively and right from the beginning keeping the planet fit for life. Furthermore, the Gaia hypothesis sees the evolution of the species of living organisms so closely coupled with the evolution of their physical and chemical environment that together they constitute a single and indivisible evolutionary process.

"The Gaia hypothesis sees the evolution of the species of living organisms so closely coupled with the evolution of their physical and chemical environment that together they constitute a single and indivisible evolutionary process"

In fact, the Gaia hypothesis took on a more complete form as a consequence of the supprt it received from the eminent biologist Lynn Margulis. Her major work has been to show that symbiosis and cooperation among organisms has always been integral to successful existence, as well as a spur to evolution. Indeed, many of the cellular components essential to cell function in multicelled eukaryote organisms such as mammals and trees – for instance, the mitachondria for carrying out the oxidation pathway for carbohydrate metabolism, or the plastids for photosynthesis – appear once to have been free-living bacteria that, following a period of close symbiosis, became wholly incorporated into another organism.

The implication that life is bountiful and rich not simply because it has luckily found a suitable planet but because it has used the special resources of the Earth to create a planet capable of harbouring living forms, has evoked sharp criticism from many other scientists. Their immediate reaction to the Gaia hypothesis is that planetary self-regulation would require foresight and planning on the part of living organisms. For instance, the Canadian molecular biologist Ford Doolittle fails to see how the global altruism needed to regulate the Earth's surface temperature or the salinity of the seas could ever have evolved through the cold, dispassionate mechanism of natural selection.

However, far from being a half-baked theory, the Gaia hypothesis has proved to be eminently testable. I have established a series of models to show that without any foresight or planning living organisms can regulate the Earth's surface temperature simply through altering surface reflectivity (albedo) and greenhouse gas levels. For instance, in Daisyworld two species of daisy, one with black flowers, the other with white, grow on an imaginary planet. However, as on Earth, the daisies cease to grow should temperatures fall below 5°C or rise above 40°C; meanwhile, as with Earth, the sun gradually becomes hotter. Once the temperature reaches 5°C the daisy seeds germinate, but the black variety quickly benefits from its ability to absorb radiation and at the end of the first season it has left more seeds behind than has the white. The population of black daisies then increases rapidly as does the planet's surface temperature. In time, as the sun continues to heat up, the white daisies come into their own, increasingly gaining the competitive edge over the black. However, the precise proportions of either black or white daisies at any time are perhaps less interesting than the effect of the two daisy populations on surface temperature. Between the time that the seeds first germinate until that when the sun gets too hot, the mean surface temperature remains remarkably steady and close to 20°C. Moreover, the greater the number of colour varieties of daisy, the better the regulation of temperature. Indeed, the model shows that diversity in terms of numbers of species is at its greatest when temperature regulation is most efficient and falls away when the system is stressed, as it is at the extremes of the early chilly planet and the later overheated one.

The models can be increased in complexity, putting in organisms that graze the plants and others that are predators on the grazers, yet as long as feedback is kept between the mass of living organisms in the system and an important environmental feature, such as surface temperature, stability is maintained. The system can also be shown to recover from massive perturbations, similar in effect to the crashing into the Earth of asteroids, such

Life's thermostat

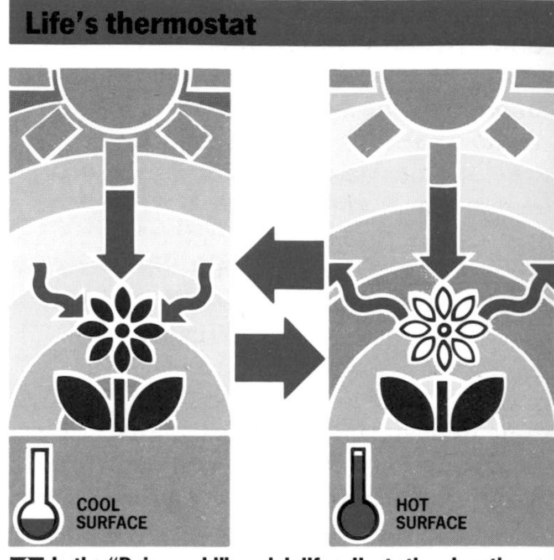

In the "Daisyworld" model, life adjusts the planet's temperature to suit itself. A basic assumption is that on Daisyworld, as on Earth, the sun is getting progressively hotter. Black daisies grow best when the surface temperature is low, because their petals absorb radiation. As the sun's heat increases, the white daisies begin to flourish. Their petals reflect radiation, thus helping to lower surface temperature. Between the time the seeds first germinate and when the sun gets too hot, the plants maintain a steady surface temperature by adjusting the planet's albedo.

are known to have occurred during the Earth's history.

With my colleagues Andrew Watson and Michael Whitfield of the Marine Biological Laboratory at Plymouth, the mathematical basis of these models has been firmly established.

My first versions were amateur numerical models on a home computer.

We have continued this collaboration and have shown that in the real world albedo effects on surface temperature are only part of the story and control through the greenhouse gases is of equal if not greater importance. That, too, can be modelled once the basic mechanisms involved have been unravelled. Thus, the control of carbon dioxide in the atmosphere is brought about through life effecting the rate at which rocks are weathered. Carbon dioxide in the form of the mild acid, carbonic acid, interacts with calcium silicate in rocks, forming calcium bicarbonate and silicic acid, both of which are soluble and move through groundwaters to streams and rivers and eventually to the sea. Burrowing organisms, such as soil bacteria, worms, termites, and trees and other plants through their extensive root systems both bring and release carbon dioxide in close proximity to soil and rocks, thereby enhancing the rate of weathering. Meanwhile, in the sea marine organisms such as corals and algae take up calcium bicarbonate and use it to make their shells. When they die the shells fall to the bottom and gradually build up as sediment into limestone mountains. The movement of the Earth's crust in tectonic plate movement may then force the limestone deposit out of the sea so that it forms land. Though very much conjecture at this stage, it is possible that limestone formation enhanced if not brought about tectonic plate movement by exerting downward pressure on the edge of continental shelves, causing one to subduct under the other. Moreover limestone acts as flux material, lubricating the movement of crustal material. It is surely of relevance that while Earth, Venus and Mars have volcanism, only Earth appears to have a well-established tectonic plate movement.

The burial of carbon dioxide as limestone, while affecting the Earth's heat balance, would not *per se* have any effect on oxygen levels. That came about as a result of photosynthesis and the burial of carbonaceous matter as in fossil fuels. At present about 100 million tonnes of carbon are buried each year, equivalent to the release of 133 million tonnes of free oxygen gas to the air. Not that the content of oxygen in the atmosphere is increasing, since there are plenty of sinks for mopping up any excess, volcanic materials for instance, or reduced materials in soils.

In fact, as much as 97.5 per cent of the products of photosynthesis are consumed by oxygen-breathing consumers, leaving just 2.5 per cent for anaerobes such as methanogenic bacteria. It would appear that the amount of carbon buried each year has been steady throughout life's history. Therefore one must assume that in the Archean, anoxic organisms such as the methanogens consumed most of the products of photosynthesis, but that as free oxygen built up in the environment the place of such organisms was taken over by respiring consumers.

In the last 700 million years oxygen has been at its present relatively high level. Through photosynthesis there is clearly a direct relationship between carbon dioxide and oxygen levels in the atmosphere. The amount of carbon dioxide now left in the atmosphere, if not replaced, would provide green plants with only a few years' supply of photosynthetic precursor. Therefore without the consumers to burn off the surplus oxygen and generate carbon dioxide, the world of green photosynthesizing plants would be short-lived. Nevertheless, plants have embarked on a plethora of different strategies to prevent themselves being totally consumed, including the production of spines, toxic alkaloids or even of extraordinary relationships with protector ants. Thus, once again, a balance is maintained, and one which appears to be self-adjusting; indeed we have not needed to give a thought as to whether tomorrow's air would still be breathable, rather we have accepted it as a fact of life.

Nonetheless, there seem to be good reasons why oxygen is now present at no more nor less than 21 per cent. When free-swiming eukaryotes appeared in the early Proterozoic some 2,000 million years ago levels of oxygen in the air of as little as 0.1 per cent would have been sufficient to meet their metabolic requirements. Large, even massive organisms such as the dinosaurs that appeared in the Phanerozoic could have existed only in a much richer oxygen environment. Too little oxygen, certainly less than 16 per cent, and the metabolic requirements of large amphibians or reptiles would not have been possible, let alone of flying organisms and warm-blooded animals and birds. Too much oxygen, perhaps no more than 25 per cent, and fires would constantly be raging, even in highly humid tropical conditions. Oxygen is also a highly toxic substance and organisms that deal with it must equip themselves with all manner of detoxifying agents, such as the enzyme catalase for eliminating hydrogen peroxide, one of its metabolic products in the cell. Again a satisfactory compromise has been reached and maintained.

Inherent to the Gaia hypothesis is the notion that the flow of essential nutrients through the system is promoted and regulated through the activity of living organisms. Gases are thus circulated through the atmosphere and various nutrient substances such as sulphur or nitrogen are oxidized and washed down with the rains to become again part of the terrestrial and marine ecology. The entire Gaia system has many features in common with the physiology of warm-blooded animals, the atmosphere acting as a global lungs, and the water system, including rivers and the oceans, acting like the blood in a circulatory system. All the living components of the system are like cells and tissues, each fulfilling unconsciously some role in the regulation process. With animal physiology in mind I have called this process geophysiology, echoing the notion, stated by the geologist James Hutton in the late 1790s that the Earth was a superorganism and that its proper study would be physiology.

"The entire Gaia system has many features in common with the physiology of warm-blooded animals, the atmosphere acting as a global lungs, and the water system, including rivers and the oceans, acting like the blood in a circulatory system"

The nitrogen cycle is a classic example of nutrient interchange between terrestrial and aquatic organisms and the atmosphere. Nitrogen is an essential component of amino acids, proteins, and nucleosides, and the capture of nitrogen from the air is carried out by bacteria such as *Rhizobium* which live symbiotically in association with leguminous plants. According to estimates by Sôderland and Svensson in 1976, terrestrial nitrogen-fixers capture some 139 million tonnes each year, and aquatic organisms up to 120 million tonnes. The fixed nitrogen is in the form of highly soluble nitrate, which is taken up by plants and converted into useful organic compounds such as proteins. When the plant dies, bacteria and fungi decompose proteins and other nitrogen-containing protoplasmic compounds into amino acids and then to ammonia, which either escapes into the atmosphere, where it undergoes oxidation to the gaseous oxides of nitrogen, or is converted again by bacteria into first nitrites and then to nitrates. If those nitrates are not taken up by plants then they may be reduced to nitrogen by denitrifying bacteria. Without the intervention of human beings, the cycle was in balance, nitrogen-fixation being counterbalanced by denitrification.

The use of artificial fertilizers adds at least 55 million tonnes to the terrestrial nitrogen balance and combustion as in power stations and motor vehicles another 24 million tonnes. Those additions combined with other man-made perturbations of the sulphur cycle are beginning to cause fundamental changes – most of them deleterious – to a broad range of ecosystems. Clearly the Gaian balances that have been established over aeons of evolution are delicately maintained.

Sulphur has proved particularly interesting from a Gaian point of view. Conventional wisdom had it that more sulphur was washed down to the oceans than was returned to the land. The same, indeed, was believed for another essential element, iodine. The sulphur was supposed to return, as much as did, in the form of hydrogen sulphide. But hydrogen sulphide is rapidly oxidized and where was the smell of rotten eggs? In the 1950s Professor Frederick Challenger had discovered the substance dimethylsulphide to be a product of marine algae, and that gave me the clue as to the missing link for the flow of sulphur back to the land.

Accordingly I went with the research vessel R.V. Shackleton on its voyage from England to Antarctica and I was able to detect dimethylsulphide wherever I looked.

But why should marine algae waste energy on recycling sulphur back to the land for the benefit of trees and giraffes? In fact, marine algae carry substances in their cells to help them maintain osmotic pressure and prevent drying out. Dimethylsulphonio propionate is one such involatile neutral solute, and it is relatively easy for algae to make since sulphur is plentiful in the sea. When the algae die the solute breaks down to dimethylsulphide, which is highly volatile and escapes into the atmosphere. Onshore breezes carry the gas inland. Meanwhile, as the gas travels, it is transformed through oxidation processes, mainly involving a highly active oxygen derivative known as the hydroxyl radical, into sulphur oxides and is washed down to Earth in the rains.

"Without question, man's industrial activities are causing a ripple of perturbations that are pushing elaborate, often barely discerned balances out of kilter"

The process is quintessentially Gaian. On land the sulphur promotes the growth of land organisms, not only as a vital element but also through its acidity bringing an increase in weathering and mineralisation. In return, terrestrial organisms speed up the flow of nutrients into the sea, thus favouring those communities of algae which include members that make dimethylsulphonio propionate to prevent desiccation. The methyl iodide also produced by marine algae could well be part of a similar Gaian process.

The demethylsulphide of marine origin has now been shown to have another vital function. Sulphur compounds, particularly after oxidation, can act as condensation nuclei so that water vapour aggregates around them to form clouds, which will later bring rain. With my scientist colleagues, Robert Charlson, and Andreae and Stephen Warren, we have proposed that the sulphur compounds emanating from marine algae play a critical function in cloud formation over extensive areas of the planet and therefore may contribute powerfully to the hydrological cycle, helping to bring rain to landmasses. The cloud formation also has a powerful Gaian effect by serving to keep the planet cool, by reflecting sunlight back into space, just as did the white daisies in Daisyworld.

In general, biologists believe that the salinity (saltiness) of the solution inside the tissues of land-based organisms, including ourselves, reflects evolutionary origins from a marine environment when the sea was that salty. Again, such thinking reflects an old-fashioned attitude concerning the adaptation of living organisms to a kind of detached, even hostile, environment, rather than questioning whether organisms have once more sought the "best of all possible worlds" for their tissues. Nevertheless, the sea is now some four times more salty compared with the solution in our tissues; moreover, the sea's saltiness is close to the limits at which most normal exposed cells can survive. But why is the sea not many times more salty than it is – so much so that, except for a few bizarre salt-loving organisms such as those found in the Dead Sea, nothing living could survive in it? Indeed, estimates of salt run-off from the land suggest that the present degree of saltiness would be arrived at after only 80 million years of weathering and erosion, so that today the sea would be much saltier, too salt for life. What has happened is that from Achean times to the present enormous basins of ocean have been shut off by laying down mineral deposits, such as the stromatolite reefs of ancient bacterial communities, the limestone mountains of marine algae (such as still operate today in the Atlantic at the edges of the continental shelf) or the coral reefs of tropical waters. Exposed to the sun, those shut-off basins evaporate, leaving behind enormous salt deposits. But life often goes further; the bacteria of the mat communities actually coat the salt that has crystallized out of solution with a varnish to prevent its dissolving back again with the rains and tides. Geologists are now finding that all major salt deposits on land, the relics of ancient evaporite basins, are all contained within a fossilized limestone barrier.

The ramifications of all such processes are extraordinary. The laying down of calcium carbonate has helped solve the potential over-heating problem caused by the Sun's increasing luminosity while simultaneously acting as a way to offset excessive salinity in the oceans. Furthermore, the limestone deposits may be the trigger to tectonic plate movement and the shifting of continents. Not least, the same organisms that are depositing limestone, for instance the algal coccolithophorids, may be primarily resonsible for closing the sulphur cycle and for the production of clouds over the oceans, with all the consquences for climate cooling and for stimulating the growth of terrestrial life.

But how is man affecting Gaia? Without question, his industrial activities are causing a ripple of perturbations that are pushing elaborate, often barely discerned balances out of kilter. The most-discussed change is to carbon dioxide levels which, through fossil fuel burning and the destruction of forests, have risen by 30 per cent since the early 1800s and are likely to double within a century. The effect of that doubling of carbon dioxide on the global energy balance is some 80 times greater than the heat generated by the burning of fossil fuels and forest destruction that has given rise to the carbon dioxide in the first place. Scientists estimate that such a rise may lead to an average 2 to 3°C increase in surface air temperatures, with the effects becoming more exaggerated towards the Poles. The sea level will undoubtedly rise, threatening low-lying areas, but will it be

Greenhouse effect

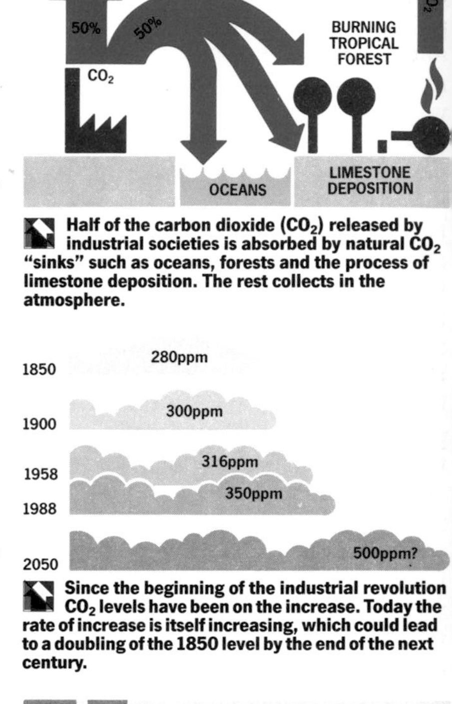

Half of the carbon dioxide (CO_2) released by industrial societies is absorbed by natural CO_2 "sinks" such as oceans, forests and the process of limestone deposition. The rest collects in the atmosphere.

Since the beginning of the industrial revolution CO_2 levels have been on the increase. Today the rate of increase is itself increasing, which could lead to a doubling of the 1850 level by the end of the next century.

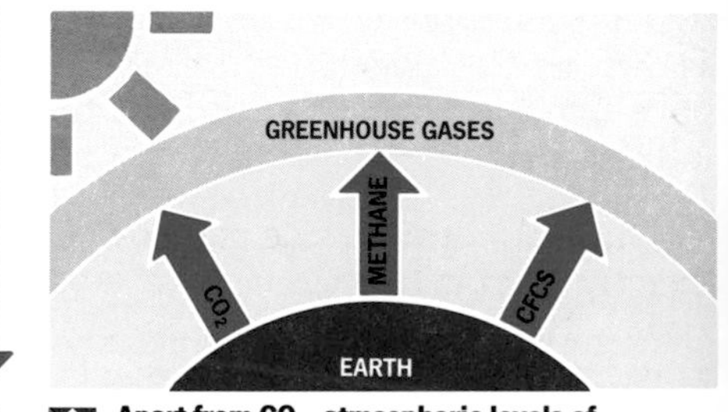

3 **Apart from CO_2, atmospheric levels of methane and chloroflourocarbons (CFCs) are also on the increase. All three are greenhouse gases, warming the atmosphere by trapping the heat radiated back into it from the Earth's surface.**

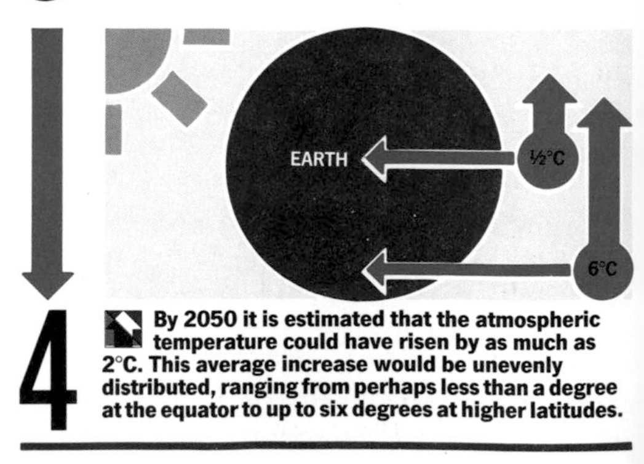

By 2050 it is estimated that the atmospheric temperature could have risen by as much as 2°C. This average increase would be unevenly distributed, ranging from perhaps less than a degree at the equator to up to six degrees at higher latitudes.

followed by the melting of the ice-caps? What will happen to cloud cover, will it increase and so offset some of the heating effects? Will some parts of the globe, such as the semi-arid areas become drier still and other areas wetter? The climate will undoubtedly change, but perhaps in Gaian ways that will tend to reduce the impact. But the problem is not just carbon dioxide. Our activities are also increasing the build-up of other greenhouse gases. Methane, for instance, is now building up at the rate of one per cent per year as a result of forest destruction and its replacement by cattle ranches, as in Central and South America, or rice paddy. The CFCs, the chlorofluorocarbon gases now famous because of the controversy over stratospheric ozone depletion, are actually extremely potent greenhouse gases, some 10,000 times more than carbon dioxide. In fact, living organisms, such as certain fungi, produce gases very similar to CFCs insofar as they also potentially interact with ozone and bring about its destruction. In terms of the quantities involved to date, the manufactured CFCs are unlikely to have had significant effect on stratospheric ozone, but there should be concern that levels do not build up to the point where significant greenhouse effects occur.

There is considerable controversy over exactly what is happening to stratospheric ozone. In my view the problem at present is not ozone depletion but ozone build-up in the lower atmosphere brought about through fossil fuel combustion. Nitric oxide, one of the gases generated during combustion, particularly in motor vehicles, acts as a catalyst in the production in the atmosphere of both ozone and hydroxyl. Hydrocarbons, such as methane and unburnt fuel, are part of this catalytic process, and the overall consequence is that the atmosphere over industrial areas is becoming increasingly oxidizing and therefore acidic as sulphur and nitrogen compounds are converted into sulphuric and nitric acids. To exacerbate the acid problem still further the flow of excess nutrients into the oceans, as a result of urbanization and intensive agriculture, is bringing about a massive increase in marine algae and as a consequence an increase in the production of dimethylsulphide.

The acidification of Scandinavian, Scottish and North American rivers, lakes and soils is therefore the result of accelerating a number of different, but interrelated processes, involving both oxidation pathways and nutrient cycles. The dying of forests in the northern hemisphere is undoubtedly associated with such changes and must surely alert us to the dangers of continuing our present industrial life style with its emphasis on high-energy-consuming activities. Ironically, with carbon dioxide and other greenhouse gases rising in concentration we need our forests intact as perhaps never before. Yet, apart from having brought sickness and death to our forests, we are now engaged in an all-out attack on the magnificent tropical forests encircling the Equator. That such forests exist at all on such depleted, even toxic soils, is evidence of the subtle feedback mechanisms and symbiotic relationships that have evolved through life's innate creativity. Our attempt to replace such forests with cattle ranching and monoculture plantations has proved in the main a disaster to the locality and as we now begin to detect far further afield. In higher latitudes we depend on tropical forests to keep us warm through the circulation of water vapour and to help maintain an atmosphere that supports life as we know it. Perhaps as much as 90 per cent of all species on Earth are to be found in tropical areas, yet within the space of decades we are tearing all apart. The rate of extinction of species today, as a result of our interference, may be many hundred times greater than in the recent geological past. To destroy such a large chunk of the living ecosystem when we do not properly understand how it all works is like pulling apart the control system of a modern aircraft while in mid-flight.

In many respects the three Cs, cars, cattle and chainsaws (or if you like,

Caterpillar bulldozers) have become prime weapons against Gaia. However, as the ancient Greeks and other early civilizations realised full well, our own well-being depends first and foremost on how we treat the Earth. Gaia would reward mankind with her bounty when treated well, but equally she would revenge abuse. As Chief Seattle told the US Government in 1854 when his people were betrayed once again: "If men spit upon the ground, they spit upon themselves. This we know, that the Earth does not belong to man; man belongs to the Earth . . . Man did not weave the web of life, he is merely a strand in it. Whatever he does to the web, he does to himself."

"To destroy such a large chunk of the living ecosystem when we do not properly understand how it all works is like pulling apart the control system of a modern aircraft while in mid-flight"

Acid Rain and Forest Decline

Don Hinrichsen

"Acid rain spares nothing. What has taken humankind decades to build and nature millennia to evolve is being impoverished and destroyed in a matter of a few years – a mere blink in geologic time"

After nearly two decades of intensive scientific research regarding the acidification of aquatic and terrestrial ecosystems, coupled to equally intensive squabbling over the results and implications, it is clear that the acidification of the environment remains one of the indusrialized world's most intractable problems.

Despite the collective efforts of a host of scientific bodies and well-meaning governments, the scourge of acid rain continues to plague the industrialized regions of the northern hemisphere. And it is now spreading to the rapidly industrializing areas of the south as well.

Acid rain spares nothing. What has taken humankind decades to build and nature millenia to evolve, is being impoverished and destroyed in a matter of a few years – a mere blink in geologic time.

In Poland, for example, acid deposition (which includes both wet and dry acidic substances) erodes iron railway tracks in the highly polluted Upper Silesian industrial area, limiting the speed of trains to 40 km per hour. In nearby Krakow, the ornate facades of historic buildings slowly disintegrate from this domestic form of "chemical warfare". Acid rain, along with other airborne pollutants, is dissolving Greece's classic past, eating into the marble of such priceless monuments as the Parthenon in Athens. It is also responsible, in part, for the sad state of the Cologne Cathedral, which is literally falling apart in a hail of masonry. Even the exquisite Taj Mahal in India is under assault from airborne acids.

Nature fares little better. In Scandinavia – one of the world's most "acidified" regions – 20,000 of Sweden's 90,000 lakes are acidified to one degree or another, and 4,000 of these are said to be totally devoid of fish life.

The situation is even worse in Norway where 80 per cent of the lakes and streams in the southern half of the country are either technically "dead" or on the critical list. Norwegian authorities say that fish have been wiped out in more than 13,000 square kilometres of lakes.

Researchers in the Federal Republic of Germany have identified acid rain as one of the prime suspects in the *Waldsterben* (tree death) syndrome currently affecting 38,000 square kilometres, or 52 per cent of the country's forests. Acid deposition is thought to be one of the principal culprits in the decline and death of Swiss forests as well, particularly those in the central alpine region, where 43 per cent of the conifers are dead or severely damaged. Acid rain is under investigation in southern Sweden where about ten percent of the conifers are beginning to show *Waldsterben*-line symptoms of decline.

On the other side of the Atlantic the picture is not much different. Acid rain, in combination with other airborne pollutants, has stricken the Canadian Parliament building in Ottawa with a creeping blight that is turning the facade black. Meanwhile, officials report that more than 300 lakes in the province of Ontario are estimated to have pH values below 5, with an additional 48,000 lakes (roughly three per cent of the total number) designated as "acid sensitive". Similarly, in Nova Scotia, nine rivers have average pH values below 4.7 and are no longer capable of supporting salmon or trout reproduction.

United States scientists claim that thousands of lakes on the eastern seaboard, especially in the Adirondack Mountains, are so acidic as to be virtual "fish graveyards". The U.S. National Surface Water Survey discovered that at least ten per cent of the lakes in the Adirondack region have pH values below 5.

Airborne acids have also been implicated in the deterioration of eastern North America's higher elevation coniferous forests, stretching up the spine of the Appalachian Mountains from Georgia to New England. And Canada's sugar maples are dying out over wide areas; due in large measure to the increasing acidification of soils in eastern Canada.

While the causes of acid rain are more or less understood, its effects are still hotly contested. Certainly acid deposition is responsible for the death of fish in thousands of lakes and streams across

The problem spreads

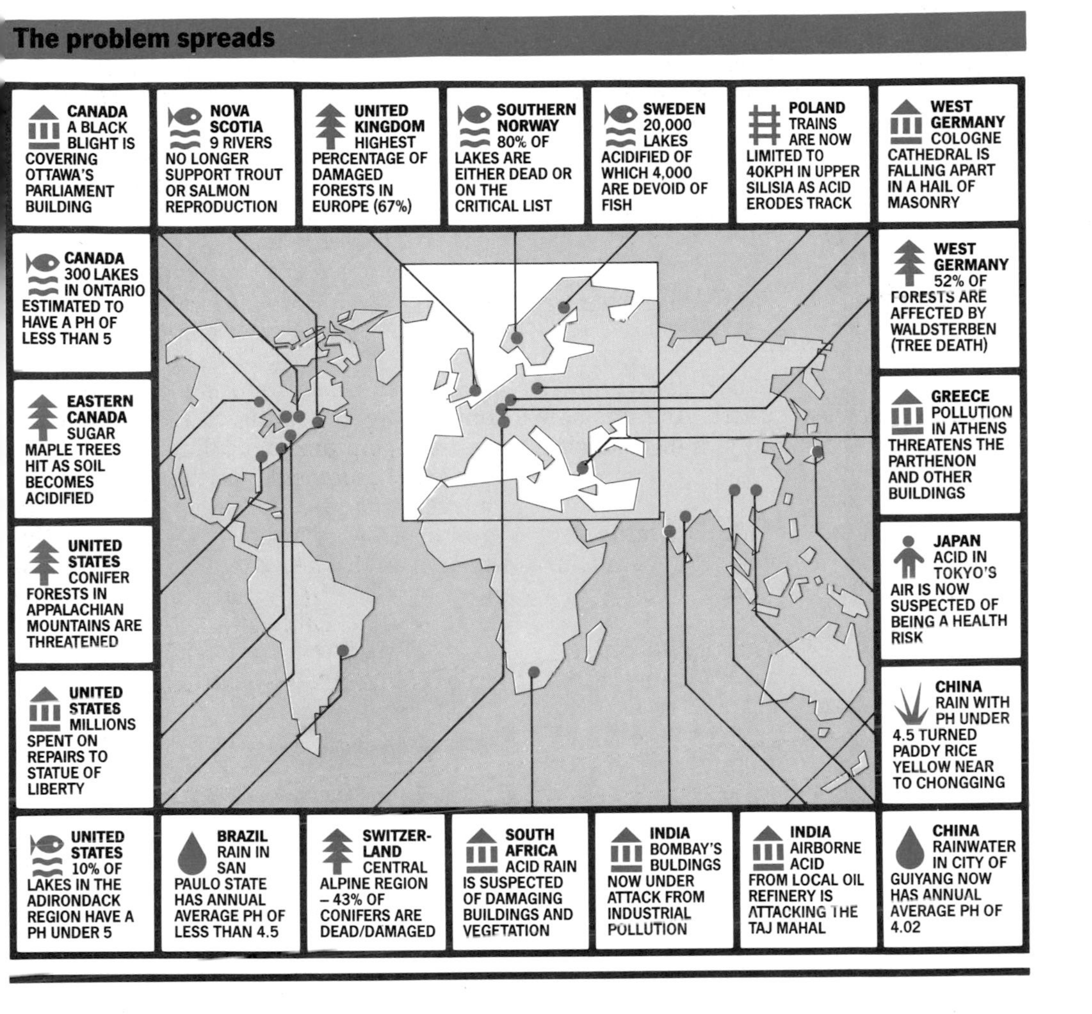

substantial areas of northern Europe and North America. However, its effect on forests and croplands is a grey area that has spawned scientific controversy. And its economic consequences are even more difficult to pin down.

It is generally agreed that acid deposition is caused mainly by sulphur and nitrogen emissions from the burning of fossil fuels, such as coal and oil in power plants, and from various industrial processes such as metal smelting. Nitrogen oxide emissions also originate about equally from motor vehicle exhuasts.

Ironicially, one of the reasons acid rain is so widespread is due to the policy "mind-set", prevalent in the 1960s and 1970s, which proclaimed that "the solution to pollution is dilution." Consequently, over the past three decades power plant and industrial smokestacks were built much higher, so emissions wouldn't pollute the immediate environment. Instead, sulphur dioxide and oxides of nitrogen are carried by prevailing winds over considerably greater distances. There are documented cases of sulphur dioxide compounds covering 1,000-2,000 kilometres in a period of three to five days. The nickel and copper smelting plant at Sudbury, Ontario, makes a good example. It has the dubious distinction of being the largest single source of sulphur dioxide pollution in the world. One 400-

metre high stack belches out more than 650,000 tonnes of sulphur dioxide a year, acidifying lakes and forests hundreds of kilometers away. By comparison, Sweden's total yearly sulphur dioxide emissions from all sources amount to around 300,000 tonnes.

When these high-flying pollutants combine with water vapour, sunlight, and oxygen in the atmosphere, they create a diluted "soup" of sulphuric and nitric acids. In some heavily industrialized regions, hydrogen chloride gases are mixed up in this atmospheric soup kitchen to produce hydrochloric acid, which can also be an ingredient in acid rain.

It has been estimated that, in the northeastern US, 65 per cent of the acid rain is due to sulphuric acid, 30 per cent to nitric acid and five per cent to hydrochloric acid.

"There are two phases involved in the formation of acids – dry and wet," writes John McCormick in *Acid Earth*. "In both, sulphur dioxide and nitrogen oxides are converted to sulphate and nitrate. Dry or 'gas phase' conversion predominates in the vicinity of emission sources, whereas wet or 'acqueous phase' conversion involving reactions within water droplets is more predominate at a greater distance. Dry phase sulphur dioxide, nitrogen oxides, sulphates and nitrates return to earth by direct deposition on surfaces in the form of gases and particles, especially in the vicinity of the sources (i.e., within 300km). This is known as dry deposition. The rest of the oxides, converted to acids by wet or 'aqueous phase' reactions, eventually return to earth as acid rain, hail, snow, sleet, or fog. This is known as wet deposition."

After this witches' brew settles to earth – washed out of the atmosphere by rain, encapsulated in snow crystals or in the form of dry particles – it increases the acidity of freshwater lakes and streams (and in some cases terrestrial ecosystems) by *decreasing* the pH values. The pH scale is used to express the extent of acidity or alkalinity of a solution, and is based on a solution's concentration of hydrogen ions.

Dispersing the problem

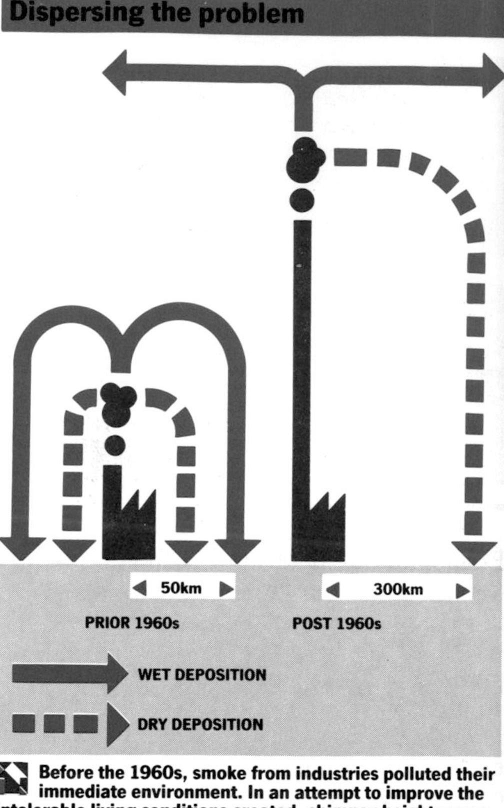

Before the 1960s, smoke from industries polluted their immediate environment. In an attempt to improve the intolerable living conditions created, chimney heights were increased, the view being that the pollutants would thus be dispersed harmlessly in the atmosphere. The result of that miscalculation was international acid rain.

Some scientists define acid rain as any precipitation with a pH value below 5.6, others push it down to pH 5.

Acid rain is not a new environmental phenomenon brought to us by "progress". It is, in fact, as old as the Earth itself and can be triggered by volcanic eruptions and forest fires, among other natural processes. Nevertheless, allowing for a certain amount of acidity from natural causes, this does not account for today's apparent creeping acidification of large chunks of the northern hemisphere.

Nature's own "doses" of sulphur and nitrogen oixdes are dwarfed by man's industrial-based pollution. Every year, somewhere between 110 and 115 million tonnes of sulphur dioxide are spilled across Europe and North America, while the countries in the Organization for

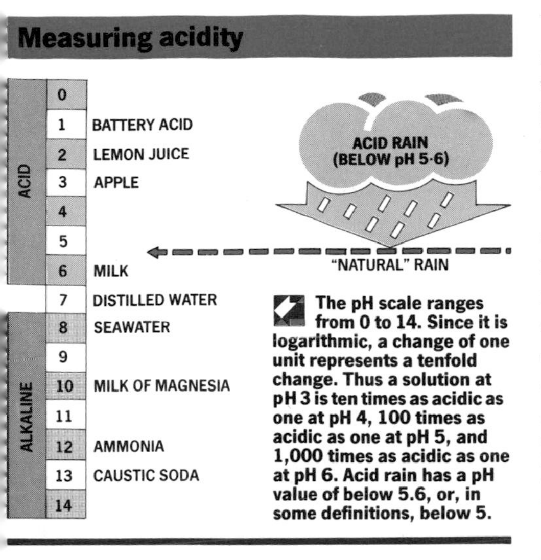

The pH scale ranges from 0 to 14. Since it is logarithmic, a change of one unit represents a tenfold change. Thus a solution at pH 3 is ten times as acidic as one at pH 4, 100 times as acidic as one at pH 5, and 1,000 times as acidic as one at pH 6. Acid rain has a pH value of below 5.6, or, in some definitions, below 5.

Economic Co-operation and Development (OECD) as a whole generate about 37 million tonnes of nitrogen oxides.

Scientists agree that on a global basis probably 50 per cent of the atmospheric sulphur is from natural sources, but in industrialized regions, like Europe and eastern North America, more than 90 per cent of the deposited sulphur is from man-made emissions.

Data gathered by the European Monitoring and Evaluation Programme (EMEP), which now consists of 82 monitoring stations in 23 countries, show that the average pH levels in central Europe are 4.2 or below. And according to the OECD, polluted areas in Scandinavia, Japan, central Europe, and eastern North America have annual pH values ranging from 3.5 to 5.5. Furthermore, the sulphate content of rainfall in these same regions varies from one to 12 milligrammes per litre, while nitrate concentrations average 0.5-6 milligrammes per litre.

Nitrate deposition in Europe has doubled since the 1950s. And this rise in deposition corresponds to increasing emissions of nitrogen oxides. Currently, annual nitrogen deposition in central Europe is 30-40 kilogrammes per hectare, levels at which forest ecosystems can expect to suffer from nitrogen saturation. Bengt Nihlgard, a plant ecologist at the University of Lund in Sweden, has surmised that "nitrogen-saturated forests would begin to appear after 20 to 25 years if the nitrogen deposition rate was in the range of 30 kilogrammes per hectare per year."

Ironically, it was an observant English chemist named Robert Angus Smith who first discovered a relationship between the increasingly sooty skies of industrial Manchester and the acidity he found in precipitation. The year was 1852! One of the earliest modern accounts of the effects of acid rain on fish dates from 1926 and was reported by the Inspector for Freshwater Fisheries in Norway. He noted that the rather sudden death of alarming numbers of newly hatched salmon fry was linked to water acidity. However, these early warnings went largely unheeded.

In the late 1950s acid rain was detected in Belgium, the Netherlands and Luxembourg. A decade later it was showing up in West Germany, France, Great Britain and southern Scandinavia, while at the same time silently spreading over the entire eastern half of the USA and Canada.

Today, the effects of acid rain (and dry deposition) are impossible to ignore. Last year Norway experienced rainfall that was so acidic it might just as well have been lemon juice. And the southern part of the country has suffered through acidic snowstorms that deposited a sickly black film instead of the usual white powder. In North America, precipitation as acid as vinegar has fallen on Kane, Pennsylvania, and "rainfall" a notch away from battery acid once poured on Wheeling, West Virginia.

If such staggering amounts of sulphur dioxide and nitrogen oxides are being pumped into the atmosphere every year, then why aren't more lakes and rivers in North America and Europe acidified? And how does acid rain actually kill fish and damage soils?

In the first place, every lake and river is different, and soil types vary considerably from region to region. Just as nature induces acid rains, it also provides natural "buffers" for some regions. Alkaline soils,

Top emitters of NO_2

The countries listed are the top ten emitters of nitrogen dioxide (NO_2), a major component of acid rain. But figures are not available for all countries and a notable omission is the USSR, which is suspected of being among the worst offenders. In the same way that sulphur dioxide is exported, NO_2 finds its way across frontiers. Much of it comes from motor vehicle exhausts, but other sources include power stations and industry. The OECD countries generate about 57 million tonnes of nitrogen oxides each year and, as nitric acid, they form a third of acid rain. In cities, nitrogen oxides pose a direct threat to human health, penetrating the lining of the lungs which affects breathing. The EEC wants emissions cut by 40 per cent by 1995.

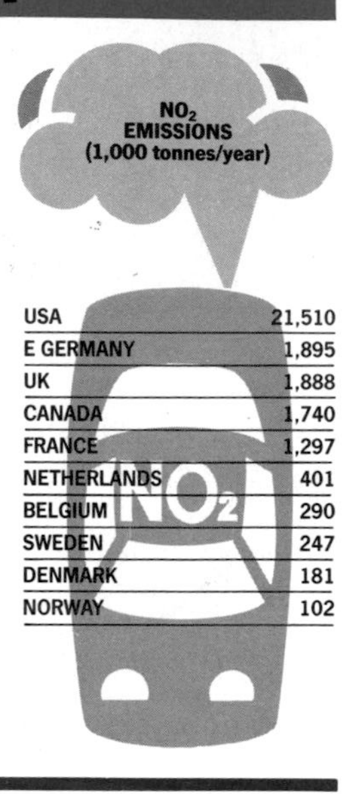

like those covering most of the US midwest, can tolerate higher amounts of acid fallout. So, too, can lakes that are cradled over beds of limestone and sandstone (for example, those found in southern England, parts of France, and in the Allgeheny Mountains, USA).

On the other hand, areas where lakes and soils sit atop thin glacial tills or thick slabs of granitic rock – as in most of Scandinavia, Scotland, and central Europe – the buffering capacity is greatly reduced. Nature's defenses soon break down. And it is these sensitive areas that are hardest hit by acid deposition.

When researchers finally found the "smoking gun" – linking biological effects with acid deposition – they discovered that acid rain is not the real killer. Something else pulls the trigger. After seven years of detective work, Norway's largest natural science research project "Acid Precipitation: Effects on Forests and Fish" (known as the SNSF project), concluded that no fish have probably died as a *direct* result of acid rain. Instead, sensitive fish like salmon, trout, roach, minnows, and artic char – which begin to die out with just slight decreases in pH – succumb to the lethal water chemistry that acid rain fosters. Investigators have determined that low pH levels are associated with elevated concentrations of heavy metals such as mercury, aluminum, manganese, lead, zinc, and even cadmium. However, it is aluminum – the most common metal found in soils – leached into lakes and streams that delivers the final *coup de grace.* Aluminum toxicity depends on water pH and appears to have a maximum around pH 5, where it is most lethal to fish because it precipitates in their gills as aluminum hydroxide, reducing the oxygen content in the blood and causing internal salt imbalances. As to be expected, the problem is most acute where the organic content in the water is low, a state characteristic of many acid-sensitive lakes. Experiments carried out in the United States demonstrated that brook trout could tolerate water with a pH of 4.9 for some days, but as soon as minute quantities of aluminum were introduced, 50 per cent of the exposed fish died within two days.

It was further noted that the overall toxicity of heavy metals increases in acidified waters. Hence, at a pH of 5 or below, nearly all of the game fish and other top predators will be extinct. And this, in turn, gives rise to a drastic reduction in species composition and diversity. Such changes in a lake's ecology impoverishes the entire ecosystem. In the end the only things that seem to survive are some species of water beetles and old, resistant eels. Most everything else may be wiped out.

In addition to the long-term, chronic effects of acid deposition on freshwater ecosystems is the sudden and deadly effects of "pulses" of acidity. These can result from precipitation with a very low pH, or from the rapid melting of snow during spring thaws. "In either case," states *World Resources, 1986*, "the volume and rapidity of the pulse can overwhelm the catchment's ability to neutralize or absorb the acid input so that the depressed pH values generate elevated levels of toxic metals. When the acid pulse is from

The bitter-sweet smell of success: a drophead Cadillac and a war-surplus gas mask . . .

spring snowmelt, it often occurs at the most vulnerable time in fish life cycles. Snow normally contains more nitrate than do summer rains, so the spring shock can exceed a system's ability to absorb nitrogen. In this way, nitrogen deposition can be more important in some fish kills than is suggested by its relative contribution to total acidity."

Until recently, the effects of acid deposition on soils and forests was a grey area of research. Within the last three years, however, scientists investigating *Waldsterben* – which has now claimed more than 70,000 square kilometres of forests in 15 European countries – have made some important discoveries. Not only does acid rain mobilize heavy metals in poorly buffered soils, increasing their toxicity, but it also leaches calcium, magnesium, and potassium from the soil, depriving trees and other vegetation of these essential nutrients for growth.

Scientists investigating *Waldsterben* have discerned three ways in which acid rain affects forests: through the foliar leaching of nutrients; from the leaching of nutrients out of the root zone; and through the mobilization of phytotoxic concentrations of aluminum in the soil. These "agents of destruction" can work alone or in combination.

A number of studies in North America have shown that acid rain in the range of pH 2.3-5.0 leached potassium, calcium, and magnesium from the leaves and needles of sugar maples, yellow birch, and white spruce. When considering the fact that many forest ecosystems in North America and Europe regularly receive up to 30 times more acidity than is deposited in "pristine" environments, it is not surprising that nutrient loss occurs.

Hand-in-hand with leaching from the canopy also comes nutrient leaching from the root zones. Research on 194 forest sites in the Federal Republic of Germany confirmed that trees attempt to compensate for the loss of nutrients in their leaves or needles by taking up more nutrients from the soil. If sufficient stocks of soil nutrients are not available – due perhaps to the effects of acid deposition – then the trees are rendered more susceptible to climatic stresses like frost and winter damage.

Dr. Bernhard Ulrich at the University of Göttingen, West Germany contends that acid deposition accelerates normal

In 1970, he bought himself a holiday retreat in the beautifully wooded Harz Mountains of Germany . . .

soil acidification processes. Dr. Ulrich's work on the Solling Plateau in West Germany has conclusively demonstrated that acid deposition does leach essential nutrients like calcium from the soil, while at the same time mobilizing toxic quantities of aluminum, which then damage the tree's fine feeder roots. Obviously, thin soils with more restricted root zones are more vulnerable to the effects of acid rain than richer, deeper soils containing higher concentrations of buffering agents such as potassium and calcium.

Just as freshwater lakes and streams experience acid "pulses" during spring snow melts, so do soils. During such episodes huge quantities of aluminum and other heavy metals are liberated at the same time, soaking into soils and leaching into water courses. Dr. Ulrich found that aluminum concentrations of as little as one to two milligrammes per litre in soil solutions was enough to damage root systems. At Solling he found six milligrammes of aluminum per litre under beech forests and 15 milligrammes of aluminum per litre under spruce forests – well above his threshold levels for injury. Dr. Ulrich's work raises some warning flags concerning soil acidification and aluminum toxicity, which appear to be serious problems in parts of Europe, particularly those areas characterized by mineral soils with poor buffering capacities.

It is also known that soil acidity has a profound effect on the mobilization and utlization of lead, as well as other heavy metals. Controlled studies carried out in the USA and Europe show that soil microorganisms exhibit marked metal toxicity as the soil pH is lowered. When heavy

. . . and by the time he had finished restoring it in 1985, acid rain had destroyed the beautiful woods

metals interfere with soil microbial processes, beneficial fungi and bacteria are inhibited in their efforts to break down organic matter into the nutrients trees need, thus altering normal biogeochemical cycling patterns.

Mounting evidence from Eastern Europe suggests that alterations in soil chemistry are taking place on a wide scale. In Czechoslovakia, large portions of the Erzgebirge Mountains northwest of Prague now resemble a wasteland. Ecosystem studies carried out by Czech geochemist Tomas Paces found that acidification of this mountainous region has altered the soil's ability to even support a forest. He found that losses of magnesium and calcium from the forest soils averaged, respectively, 6.8 and 7.5 times greater than from an undamaged forest area used as a control. The runoff of aluminum, which normally remains bound to soil minerals, was found to be 32 times greater than from the undamaged "control" forest.

Meanwhile, a European-wide forest damage survey compiled in 1987 by the UN's Cooperative Programme for the Monitoring and Evaluation of Long-Range Transmission of Air Pollutants in Europe, highlighted some startling results. Involving 15 European countries, the report flagged the United Kingdom as having the highest percentage of damaged forests in Europe! In all, some 67 per cent of Britain's conifers exhibit slight to severe damage, while 28.9 per cent have "moderate to severe" damage. This puts the UK well ahead of countries like West Germany (52.7 per cent), Switzerland (52 per cent), Czechoslovakia (49.2 per cent), the Netherlands (59.7 per cent), and Yugoslavia at 38.8 per cent – countries thought to suffer far more forest damage from air pollution and acid rain than the United Kingdom. The announcement came as an embarrassment for the government and the Forestry Commission, which were insisting that *Waldsterben*-like forest decline symptoms had not yet begun to show up in Britain.

"It is not a very comforting notion that, depending on where you live, your next rainfall might have the pH of lemon juice, vinegar, or something stronger"

Another alarming trend is the increasing acidification of groundwater. Acid precipitation, percolating through the soil, can leach heavy metals into groundwater reservoirs. It is already a serious problem in Belgium, Finland, the Netherlands, Norway, Sweden, and the Federal Republic of Germany. In Holland, for example, the worst affected areas have acidified groundwater down to a depth of ten metres. Because of the over-use and misuse of nitrogenous fertilizers, much Dutch soil is saturated with nitrogen and there are lakes with extremely high ammonium levels.

In the province of Bohuslän on Sweden's west coast, 49 per cent of the water wells tested had pH values below 5.5, and a number of people complained about coroded water pipes. Scientists from the Swedish Water and Air Pollution Research Institute observed that the tap water in many houses of the area contained high concentrations of copper, zinc and, in some cases, cadmium. This finding only confirmed what the residents had already experienced – the corrosion of aluminum cooking pots, diarrhoea in young children, foul-tasting water, and in several cases, blond hair that turned green from washing in water with high amounts of copper. Chicken farmers also reported that hens exposed to acid groundwater laid thin-shelled eggs. A large scale survey carried out during 1986 revealed that the acidification of groundwater in Sweden is much more serious than previously thought: 15 to 20 per cent of all private wells dug or drilled in granitic rock contain acidified groundwater. At pH values below six, zinc, copper, and lead begin to corrode. Nearly 70,000 of these 100,000 wells supply people with water all year round (the rest are used by summer homes or temporary dwellings).

Many environmental scientists fear that this is only the beginning – much more serious long-term effects, as yet undetected, may be at work. And it is these subtle effects, especially on soils, crops, and forests, that require substantial research efforts. Indeed, the World Commission on Environment and Development, chaired by Gro Harlem Brundtland, the Prime Minister of Norway, concluded in its final report *Our Common Future*, "Europe may be experiencing an immense change to irreversible acidification, the remedial costs of which could be beyond economic reach."

Damage to metals, building exteriors, and painted surfaces alone cost the 24 member countries of the OECD some $20,000 million a year. If the costs of dead and dying forests, acidified lakes and streams, and crop losses were factored in to the equation, the price might very well prove astronomical.

"Damage to metals, building exteriors, and painted surfaces alone cost the 24 member countries of the OECD some $20,000 million a year"

In the meantime, a number of emergency measures have been launched to save acidified lakes, rivers, and forests. Spreading lime on affected lakes and forests appears to be one method that is getting results,

A witches' brew with lethal effects

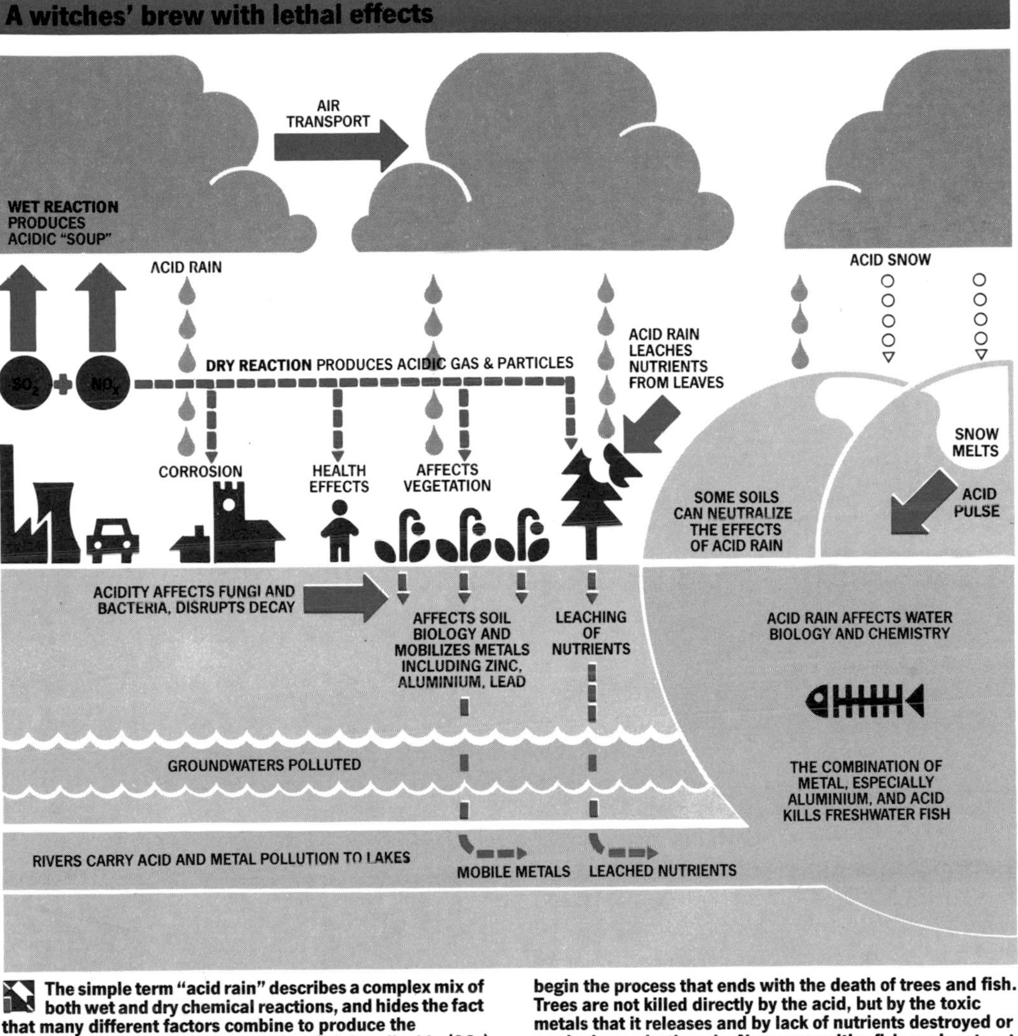

The simple term "acid rain" describes a complex mix of both wet and dry chemical reactions, and hides the fact that many different factors combine to produce the devastating effects that we see today. Sulphur dioxide (SO_2) and nitrous oxides (NO_x), released by our modern industries, begin the process that ends with the death of trees and fish. Trees are not killed directly by the acid, but by the toxic metals that it releases and by lack of nutrients destroyed or washed away by the rain. Nor are sensitive fish, such as trout and salmon, killed by acid alone, but by an acid-metal mix.

however temporary. By 1984 some 3,000 Swedish lakes had been limed at a total cost of $25 million. Liming programmes have also been carried out in Norway, the Federal Republic of Germany, Austria, Czechoslovakia, Poland, and the Soviet Union. Unfortunately, liming damaged ecosystems is at best a stopgap measure. In the long term, only the imposition of stringent limitations on the amounts of sulfur dioxide and nitrogen oxides discharged into the atmosphere will have the desired effect– that of drastically reducing damage from acid pollution.

Similarly, West German scientists have discovered that fertilizing damaged conifers with appropriate amounts of calcium, magnesium, potassium, zinc, and manganese, reversed the decline syndrome. Sick trees dosed with these chemicals recovered in a matter of weeks. However, this method of treatment works only for those trees suffering from acute nutrient deficiencies. It certainly does not work in all those forests that have been afflicted with *Waldsterben*.

Net exporters of SO_2

Sulphur dioxide (SO_2) is a main component of acid rain. The countries listed export it – that is, the thousand of tonnes shown cross their frontiers and are deposited elsewhere each year. With two notable exceptions – the UK and Poland – they have all joined the 30 per cent club, and have committed themselves to reducing SO_2 emissions by 30 per cent by 1993. Export figures are not available for Canada and USA – those given are for total emissions. The USA is not a member of the 30 per cent club: it claims the benefits are scientifically uncertain.

SO_2 EXPORTS (1,000 tonnes/year)	
USSR ✓	7,922
UK	3,750
E GERMANY ✓	2,888
POLAND	2,576
W GERMANY ✓	2,338
CZECHOSLAVAKIA ✓	2,100
FRANCE ✓	2,042
ITALY ✓	1,804
HUNGARY ✓	1,178
BELGIUM ✓	638

TOTAL EMISSIONS FROM	
USA	24,600
CANADA ✓	4,610

✓ COUNTRIES TICKED ARE MEMBERS OF THE 30% CLUB

Since Scandinavia was the first region to document ecosystem damage from acid deposition, it is not surprising that Norway and Sweden spearheaded international efforts to bring the "percursors" of acid rain – sulphur and nitrogen oxides – under the heel of international regulation. Since Sweden estimates that 80 per cent of all the sulphur dioxide falling on the country every year comes from abroad (and Norway pushes that figure up to 90 per cent), it is obvious that national control policies in both cases would have merely cosmetic effects. Only concerted international action will reduce the emission and deposition of acidifying substances in Europe and elsewhere. Sweden and Norway were instrumental in persuading the member countries of the UN's Economic Commission for Europe (ECE) to sit down at the conference table. The end result of a tiring process was the signing of the Convention on Long-Range Transboundary Air Pollution (LRTAP) in Geneva in November 1979. Still, the convention did not come into force until March 1983, when the requisite number of signatories ratified it. Dismayed with the slow progress shown by most parties to the convention in carrying out its terms, Norway and Sweden were also instrumental in forming the "30 per cent Club", launched in Ottawa, Canada in 1984. Initially consisting of ten countries – those more polluted than polluting – the "club" called for the reduction of sulphur dioxide levels by at least 30 per cent by 1993 (using 1980 emissions as the baseline). By April 1985, 19 countries had agreed to lower their emissions of sulphur dioxide by 30 per cent, along with unspecified reductions in other pollutants, mainly nitrogen oxides.

A United States mining community in the 1930s?

When the members of LRTAP met in

ɫo, a smokeless fuel plant in South Wales, 1985

Helsinki in July 1985 to further discuss the implementation of the convention, another important milestone was reached. The convention itself was given some real teeth. A protocol mandating sulphur dioxide reductions of 30 per cent by 1993 – under the same terms as the "30 per cent Club" – was opened for signature. Not surprisingly, it was immediately endorsed by the 21 members (representing 19 countries) of the "30 per cent Club", and was finally ratified by sufficient countries and came into force in September, 1987.

However, two of the largest emitters of both sulphur and nitrogen oxides – the United States and the United Kingdom – refused to sign the protocol on the grounds that scientific uncertainty about the benefits made further reductions problematic.

The European Community (EC) is also moving to reduce pollution loads in member countries. In 1983, the Commission of the European Community proposed a Council Directive calling for significant cuts in three categories of emissions by

1995 – 60 per cent reduction for sulphur dioxide, 40 per cent for nitrogen oxides, and 40 per cent for dust. The main targets were large fossil-fuel-fired power plants.

At the same time, the commission also recommended that all members of the EC introduce unleaded gasoline by 1989, and that catalytic converters be fitted to all new cars by 1995.

In any case, it is not going to be an easy road to travel. An OECD survey did show that Belgium, Denmark, West Germany, France, and the United Kingdom are net exporters of pollution, while countries like Norway, Sweden, Finland, and Switzerland are net importers. Nevertheless, knowing where the pollution comes from is one thing, quite another is cajoling the exporters into reducing it, especially since control technologies are expensive.

No doubt acid rain will continue to be one of this decade's most troublesome international environmental problems – one that is bound to worsen before it gets better. How, for example, will compensation schemes be worked out? Who will pay what to whom and how much?

Meanwhile, as scientists debate decimal points and administrators are paralyzed by "inconclusive" data, a rain of acid continues to spread across the Northern Hemisphere and perhaps elsewhere. It is not a very comforting notion that, depending on where you live, your next rainfall might have the pH of lemon juice, vinegar, or something stronger.

Armin Maywald
Barbara Zeschmar-Lahl
and Uwe Lahl

"Rigorous conservation methods are necessary to protect both our groundwater reserves and our surface waters . . . To achieve this considerable changes are required in the operation of our industrial system"

"In the Third World it is scarcity that is the key . . ."

Of the Earth's total supply of water, only about three per cent is fresh, and the bulk of that is locked up in polar ice caps or as groundwater so deep that it is inaccessible. Even so, there are something like 4,300,000 cubic kilometres (a million cubic miles) of usable fresh water, and none of it is ever lost: there is exactly the same amount of water on Earth today as there was when it was born.

But water can be misplaced. What evaporates in Singapore doesn't necesarily fall there as rain. And it can be polluted, possibly beyond our abilities to purify it.

Whether the world has enough water fit to drink depends on how much is available and where it is; how much we use; and its quality. And the drinking water problems of the industrial nations and developing countries are quite different. In industrialized countries the threat to the quality of water is of prime importance; in the Third World it is scarcity that is the key. In 1975 the World Health Organization estimated that some 60 per cent of the people in developing countries did not have enough water. Between nine and 22 million children less than five years old die each year because of lack of

"Between nine and 22 million children less than five years old die each year because of lack of water, inadequate sanitary facilities, and waterborne diseases"

water, inadequate sanitary facilities, and waterborne disease. Ironically, reservoirs, irrigation and drainage systems, as well as inadequate separation of drinking water and waste disposal systems, ease the way for disease.

Fulfilling demand for water in industrial nations is not difficult – when demand increases, rivers are damned, pipelines laid, or new groundwater areas tapped by drilling. Even so, demands are high (see illustration below) and increasing. The authors of the American report on the environment, Global 2000, expect annual increases of four to five per cent by the end of this century. Water consumption in urban households will quadruple between 1967 and 2000; demands made by industry and mining will increase fivefold. An example of just how much water can be used by industry is given by the Federal Republic of Germany. There, electricity production and mining account together for up to 80 per cent of the water requirement (see illustration on p.86). Seventy per cent of it, in turn, is needed to cool down heat-generating stations. Though this water is for the most part re-directed into the rivers, this leads to an increase in the temperature of the Rhine, Weser, Elbe and even coastal waters. The chemical, food, paper, pulp and metal-working industries account for up to two-thirds of the remaining 30 per cent.

Though fulfilling demand is not difficult, keeping water pure is. The waste water re-directed into the rivers contains pollutants. Those generated by the paper industry (non-degradable organic and chlorine-based substances), the metal-working industry (heavy metals such as chromium, nickel, zinc, cadmium, lead) and the petro-chemical industry (minerals oils, phenols) are particularly damaging to rivers, lakes and canals (see illustration on p.84).

Modernized processing methods, the use of closed water cycles, and legislation, together with research and development programmes, have helped to reduce pollution caused by waste water. But in spite of stricter laws and the expenditure of more money, water reserves are still considerably at risk from pollutants. Non-degradable industrial chemicals such as organohalogen compounds (e.g., chloroform, trichloroethylene and tetrachloroethylene); polycyclic aromatic hydrocarbons and pesticides (e.g., DDT and HCH); and polychlorinated biphenyls (PCBS) have not only been detected in sewage but also in surface water, seawater and rainwater, and even in filtrates at the water's edge and in ground and drinking water.

The pollution burden of a river tends to be made up of three main factors. The first is the population density of its catchment area, the second the level of economic activity, and the third the volume of water carried by the river. Thus, though the Rhine carries a much lower volume of water than the Mississippi, the theoretical burden of waste it has to carry is much higher, in spite of the fact that the Mississippi's catchment area is relatively highly populated and the agricultural and industrial output of the area is high.

Future water consumption

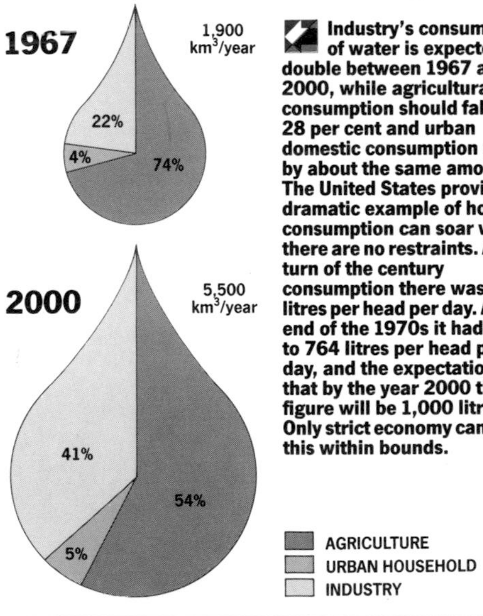

1967

1,900 km³/year

22%
4%
74%

2000

5,500 km³/year

41%
5%
54%

Industry's consumption of water is expected to double between 1967 and 2000, while agricultural consumption should fall by 28 per cent and urban domestic consumption rise by about the same amount. The United States provides a dramatic example of how consumption can soar when there are no restraints. At the turn of the century consumption there was 45 litres per head per day. At the end of the 1970s it had risen to 764 litres per head per day, and the expectation is that by the year 2000 the figure will be 1,000 litres. Only strict economy can keep this within bounds.

AGRICULTURE
URBAN HOUSEHOLD
INDUSTRY

The dirtiest river in Europe

About one-fifth of the world's chemical production takes place on the banks of the

"Without intensive purification of the Rhine's effluent, the river could degenerate into little more than a sewer and drinking water could be recovered only at considerable risk"

Rhine. Nonetheless, 20 million Europeans still obtain their drinking water supplies from the river. There is practically no difference between the conflict facing the West German authorities over the use of the Rhine as a sewer on the one hand and as a drinking water reservoir on the other, and the current situation in many developing countries. In the latter the inadequate separation of drinking water and waste water has led to considerable hygiene problems. Without intensive purification of the Rhine's effluent, the river could degenerate into little more than a sewer and drinking water could be recovered only at considerable risk.

In spite of numerous improvements in the treatment of waste water, and the employment of the most up-to-date reprocessing techniques by waterworks, more and more pollutants are managing to seep into drinking water. This is true of readily volatile chlorinated hydrocarbons, in particular the solvents trichloroethylene and trichloroethane. Cornelius van der Veen, head of the Dutch water works in the Rhine catchment area, is quoted as saying with some resignation "Even well-thought-out purification and reprocessing systems mean that just about every substance present in untreated water is also to be found in drinking water".

In an attempt to produce drinking water from the polluted Rhine, waterworks use chemical reagents such as chlorine, chlorine dioxide and ozone, which pose new health problems. Investigations in the USA found increased cancer rates among people supplied with chemically polluted and chlorinated water, hardly surprising as numerous organic chlorine compounds formed during chlorination are suspected of being carcinogenic. It has to be said, however, that the USA permits far higher concentrations of these compounds in drinking water than the EEC.

Chemical re-processing of drinking water must be viewed with due scepticism. The population's drinking water supply should preferably be drawn from unpolluted groundwater. In spite of various anti-pollution measures adopted by industry, vast amounts of chemicals are still being dumped into wells and waterways from which drinking water is derived. Already about ten per cent of drinking water in West Germany is contaminated with nitrate fertilizer exceeding the permitted limit of 50 milligrammes/litre.

Pesticides also penetrate deep into ground water – the worst culprit being atrazine. This herbicide, principally used in maize production, has been detected in

One consequence of the Sandoz fire

Drinking waste water

More than half the population of OECD countries drink water that has passed through waste water treatment plants – and in the 12 countries listed, the figure is more than 70 per cent.

There are three types of waste water treatment. Primary treatment is mechanical, designed to remove solids and sediment; secondary is biological, which removes organic matter and bacteria; and tertiary is biological and chemical, the chemical treatment usually involving the use of chlorine, chlorine dioxide and ozone. Sweden is the only country of those listed that has switched entirely to tertiary treatment; France the only country that does not use it – or did not when the survey was made in 1983.

SWEDEN	100%
DENMARK	90%
W GERMANY	84%
UK	83%
SWITZERLAND	81%
NETHERLANDS	72%
USA	70%
FINLAND	69%
FRANCE	64%
CANADA	56%
NORWAY	51%
JAPAN	33%

various groundwater samples in the West German states of Baden-Wuerttemberg, Bavaria, Hessen, Lower Saxony, North Rhine-Westphalia, Rhineland-Palatinate and Schleswig-Holstein. The highest concentration found in groundwater was 2.5 microgrammes per litre. However, atrazine is only the tip of the iceberg. Up until now 38 different pesticides have been detected in groundwater, drinking water or filtrates at the water's edge including aldrin, an established carcinogen, and the fungicide hexachlorobenzene (HCB). In 1987, 8.620 milligrammes of the nematicide 1.3 dichloropropene were detected per litre of ground water in Schleswig-Hostein. As early as 1982 one milligramme of the herbicide mecoprop was discovered in a spring in Baden-Wurttemberg. Similarly in 1987 17 microgrammes (0.017 milligrammes) of the notorious chemical 2,4,5-T, known as "Agent Orange", were found. The waterworks in question were not forced to close down because West Germany's legislation on pesticide limits (i.e., 0.1 microgrammes per litre for single substances and 0.5 microgrammes per litre for the total amount of pesticide) does not come into force for almost two years, at the beginning of November 1989.

Italy had another way of dealing with the matter. In recent years the amount of atrazine detected in drinking water in the rice-growing areas of the Po plateau in northern Italy (0.1 microgrammes per litre) has from time to time exceeded the legal limit by 20 times. As a result, the authorities closed the springs in question and the population received its water supply from water carts. However, this situation was not allowed to continue for long. Without further ado the government established a new legal limit of 1.7 microgrammes of atrazine. Thus drinking water was legally "drinkable" again. The list of pesticides so far detected in ground and drinking water can be increased arbitrarily. Methods of analysis are at present in their infancy and there are still no methods of detection for many of the agrochemicals that are suspected of posing a threat to water quality.

"One thing is certain: the importance of an impeccable water supply is recognized only when it is almost too late"

It seems the quantity and acceptable quality of water has become a worldwide problem. In the meantime, one thing is certain: the importance of an impeccable water supply is recognized only when it is almost too late. This is certainly always true in developing countries and is becoming more evident in Western nations.

Exploitation in the United States

At the beginning of this century, the Tennessee Valley region of the United States was systematically drained. Whole forests were cleared and groundwater reserves destroyed. With the introduction of intensive agricultural methods, the vulnerable soils were rapidly eroded, leading to reduced yields and, indeed, to the abandonment of agriculture in certain

Industrial pollutants in domestic water

Selected major industries and manufacturing processes that are potential or actual sources of water pollution

	CHLORINATED ORGANIC COMPOUNDS	MINERALS & OILS	PHENOLS	NITROGEN	PHOSPHORUS	MERCURY	LEAD	CADMIUM
FERTILIZER INDUSTRY	X	X		X	X	X	X	X
PESTICIDE INDUSTRY	X				X	X		
PETROLEUM REFINING		X	X	X				
PETROCHEMICAL INDUSTRY	X	X	X	X		X	X	
CHEMICAL INDUSTRY	X	X	X	X	X	X	X	X
PULP/PAPER INDUSTRY	X	X	X	X	X	X		
METAL/METAL PLATING INDUSTRY	X	X	X	X		X	X	X
TEXTILE INDUSTRY	X	X		X	X			
IRON/STEEL INDUSTRY		X	X	X		X	X	X

Chlorinated organic substances pose a chronic health risk. Most of them are similar to fat, and can accumulate in the food chain. Many of them can cause cell mutations and cancer, liver damage, weakness of the immunological system, or loss of fertility. In large amounts, minerals can cause diarrhoea, hypertension, and susceptibility to heart attacks. Phenols can cause liver damage, and some are suspected of being carcinogenic. As nitrite, nitrogen leads to oxygen deficits and death. Though phosphorus salts are essential minerals, in large doses they are suspected of producing hyperactivity in children; mercury can damage the nervous system, the kidneys, and the foetus; lead can cause anemia and disturb the central nervous system; and cadmium can cause functional disturbances in the kidneys. Cadmium chloride has produced carcinogenic effects in animals.

areas. The valley would have become a desert had it not been for the work of the Tennessee Valley Authority, which led to its ecological rehabilitation.

America, however, did not seem to learn much from the Tennessee incident. Water continues to be used and abused on a large scale. At the turn of the century about 200,000 million litres (52,800 million US gallons) of water were consumed daily in the USA. This had increased to 1,911,000 million litres (504,800 US gallons) at the end of the 1970s. Individual consumption rose from 45 to 764 litres (12 to 202 US gallons) per day. Without strict economy measures, every American citizen will consume more than 1,000 litres (264 US gallons) of water per day by the year 2000. Many cities such as New York or Denver have never even heard of water meters – everyone is allowed to use as much water as he or she wants. (In most European countries household and small industry usage has fallen in the last 20 years.)

The water resources of other areas in the USA are also exploited without restraint. Many eastern rivers are so polluted by industrial chemicals that they can no longer possibly serve as sources of drinking water, while the water of other less polluted rivers is treated at immense cost without any guarantee that it will provide an uncontaminated water supply.

In many places in the USA groundwater is already contaminated. In 1982 poisonous industrial chemicals were detected in the groundwater of 35 states. The most common contaminants were found to be the solvent trichloroethylene (TCE) and the pesticide dibromochloropropane (DBCP). Wells in 25 states had to be closed due to serious contamination.

To now, more than 700 chemicals have been detected in US drinking water, 129 of which are considered to be particularly dangerous by the United States Environmental Protection Agency (EPA). Nonetheless, drinking water is regularly tested for only 14 of these contaminants. Another EPA study conducted in 1983 showed that in 52 out of 186 towns and municipalities with more than 10,000 inhabitants, traces of toxic chemicals were detected in the drinking water. One of the reasons for this is that supplies are often from sources near former industrial sites, or sites where hazardous wastes have been dumped. This problem first came to light with the Love Canal scandal. That canal, near the Niagara Falls, was used in the 1940s and 1950s as a waste dump. Some 22,000 tonnes of chemical wastes were dumped in it and then it was filled in and houses built on it.

In 1978 it was found that the people living there were succumbing to mysterious diseases and many children were born with various physical defects. It did not take long to establish the cause of those defects and the most contaminated area was evacuated. The incident prompted the EPA to carry out a survey of other sites used for the disposal of chemical wastes. This revealed that the Love Canal was but the tip of an iceberg. By 1984 17,500 such sites had been identified, 546 being regarded as particularly hazardous to the health of the local inhabitants. The scandal led to the setting up by the Federal Government of a "superfund", originally of $1,500 million, to carry out investigations and clean up the worst sites. In 1986 it was decided to clean 703 sites. Another 248 sites were listed as so hazardous they will have to be cleaned too. The original "superfund" of $1,500 million was raised to $8,500 million. In the meantime, chemicals are seeping from many of these sites into the nation's groundwater reserves, which in many areas are now seriously contaminated.

"In many places in the USA groundwater is already contaminated. In 1982 poisonous industrial chemicals were detected in the groundwater of 35 states"

Great Lakes – problems with pollution

It is not just America's groundwater reserves that are being poisoned, but also its surface waters. The Great Lakes, for instance, covering an area of 242,000 square kilometres (9,340 square miles), form the largest fresh water reservoir in the world. That ecological system, however, has been irreversibly degraded by human settlements, the introduction of modern agriculture and industrialization – all on a massive scale. Forty million people live in the area adjoining the Great Lakes, 80 per cent of them in large conurbations.

Right up to the mid-1960s untreated

"Right up to the mid-1960s untreated sewage was pumped directly into Lake Erie and Lake Ontario. These lakes were also seriously contaminated with agrochemicals"

sewage was pumped directly into Lake Erie and Lake Ontario. These lakes were also seriously contaminated with agrochemicals, which included nutrients – in particular phosphates – which have led to serious eutrophication, which in turn has contributed to the decimation of many fish species. Heavy metals and industrial chemicals have added to the problem.

In 1983 the International Joint Commission (IJC) published a list of 8,000 alien substances identified in lake water.

In the 1960s priority was given to measure to halt eutrophication. The construction of water treatment plants was extended and they were equipped with phosphate precipation plants.

In 1982 the annual amount of phosphates introduced into Lake Superior and Lake Michigan had fallen below the permitted level. The amount introduced into the other lakes is also approaching the annual permitted level. This, however, is not sufficient. Phosphate fertilizers are still being washed into the Great Lakes at an unacceptable rate. However, there are a few hopeful signs. The use of DDT, aldrin, dieldrin and heptachlor was banned at the beginning of the 1970s. The use of mirex, chlordane and PCBs was also banned by the end of the decade. However, since none of these substances is easily degradable, and as they tend to accumulate as they move up the food chain, fish taken from the lakes are still badly contaminated, and their regular consumption would be a serious health hazard. Thus in 1980 PCB concentrations of 5.3 parts per million (ppm) were measured in sea trout. The goal of achieving a PCB level of no more than 0.1ppm as set by the water quality agreement will almost certainly not be achieved this century.

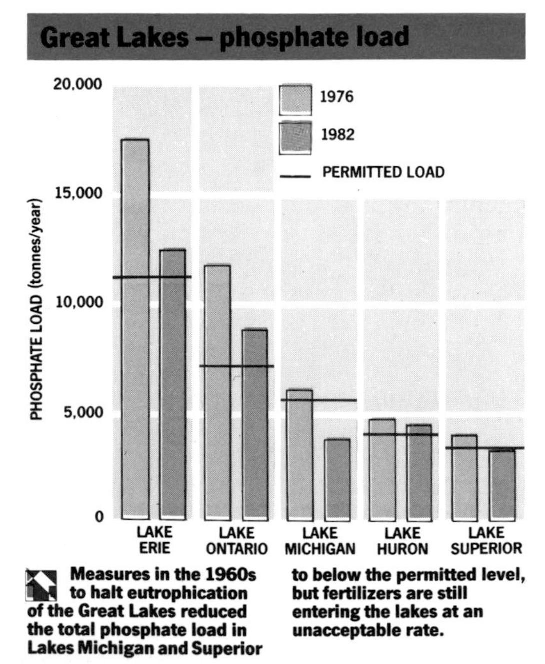

Measures in the 1960s to halt eutrophication of the Great Lakes reduced the total phosphate load in Lakes Michigan and Superior to below the permitted level, but fertilizers are still entering the lakes at an unacceptable rate.

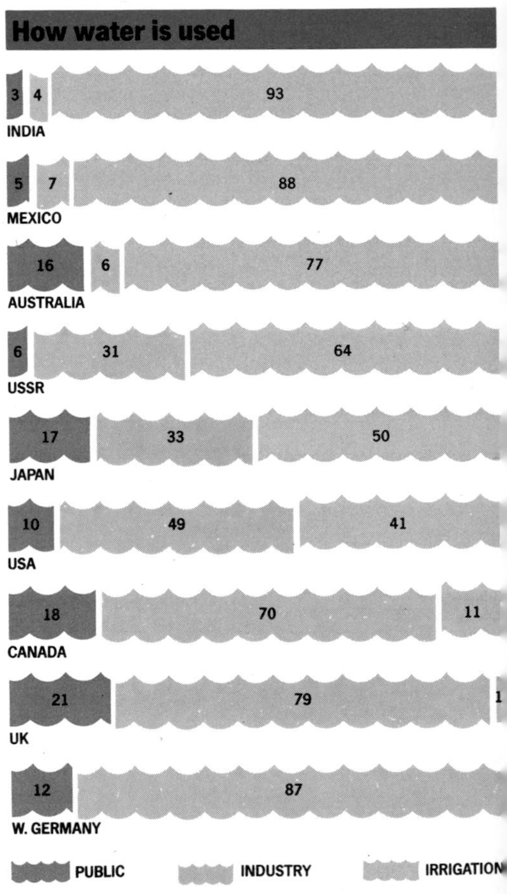

The amounts of water devoted by individual countries to public use, industry and irrigation (shown above as percentages of available supply) reflect both the extent of their industrial development and their rainfall. Developing countries such as India and Mexico devote almost all their water to irrigation; developed countries such as Canada, the UK and West Germany use most for industry and rely heavily on rain to irrigate crops.

Holland opens flood gates to let the polluted Rhine flow directly into the North Sea

Emergency situation in Holland

The Netherlands is among the countries of Europe with the lowest reserves of fresh water and is therefore experiencing great difficulties in making sufficient drinking water available. Good quality drinking water is rare and is obtained chiefly from the dunes in the Gooi region (a sandy area to the east of Amsterdam), in Veluwe (in the province of Gelderland), and from the quaternary and tertiary sandy deposits in the east and south of the country. Up until 1955, both the Hague and Amsterdam extracted groundwater from a lentil-shaped fresh water deposit which was floating on the underlying salty groundwater. However, since the rate of extraction was far in excess of recharge, those cities were forced to switch to alternative sources from 1955, that is, from the Rhine and the Maas.

Both these rivers are, to a considerable extent, polluted with all sorts of synthetic organic industrial chemicals. It is not only the routine releases (which are only permitted in certain areas) all along the Rhine in Switzerland, France and Germany that are responsible for this, but also large-scale releases caused by the inevitable accidents that occur at only too regular intervals.

It is usually the Dutch scientists who, when monitoring the quality of water supplies, detect the sudden increases in the pollution levels. They then raise the alarm and suspend drinking water recovery programmes. In this way high levels of at least the more dangerous pollutants can be prevented from contaminating the Dutch drinking water supplies.

The routine examination of untreated sewage water and the suspension of drinking water recovery when there is a high concentration of contaminants present cannot, however, prevent all poisons from the waters of the Rhine and of the Maas from reaching the consumer. Rhine water includes carcinogens, and mutagens such as benzene, trochloroethylene and bischlor isobutylether. Every litre of drinking water recovered from the Rhine contains about ten milligrammes of synthetic organic substances and it is estimated that some 90 per cent of them are harmful.

Conclusions

We seem to have limitless faith in technology as a means of solving any problem. This faith is unfortunately unjustified. It is more than questionable whether technical expertize can ever remove all the undesirable pollutants that have seeped into our groundwater reserves. It is equally unlikely that water purification techniques will enable us to derive from highly contaminated sources, on a sufficient scale, and at an economic price, the pure drinking water that we require. Nor can one conceive of any technological means of overcoming the water shortages that are likely to occur as a result of the destruction of the world's tropical rainforests. It follows that wherever possible drinking water supplies should be derived from clean and unpolluted water. This means that rigorous conservation measures are necessary to protect both our groundwater reserves and our surface waters. In particular, it means that agricultural and industrial chemicals must not be allowed to enter our water supplies. To achieve this considerable changes are required in the operation of our industrial system. Dangerous chemicals must be used only in closed circuit systems so that they cannot enter into contact with the biosphere. The production and use of persistent chemicals that cannot be used in this way should be prohibited and modern intensive agriculture, depending as it does on nitrates, phosphates and synthetic organic pesticides whose use must inevitably contaminate groundwater reserves and surface waters, must be phased out.

"We seem to have limitless faith in technology as a means of solving any problem. This faith is unfortunately unjustified"

Our water reserves, even where they are abundant, must be used with the utmost economy and treated with the utmost respect. If they are not, the availability of clean and unpolluted water in the next decades will decrease, with the most serious consequences for our health and the health of living things in general.

THE YEAR IN PERSPECTIVE

A roundup of some of the events that affected the health of the planet, for better or worse, in the 12 months from late 1986 to late 1987.

AGRICULTURE

PLUSES

- USA permits first-ever release of genetically-engineered microbe, designed to prevent the formation of ice on plants (p.154).
- First illegal trial of genetically-engineered material takes place in USA (p.154).
- Report shows 12 to 24 million hectares of farmland in Europe are at risk from erosion due to unsound agricultural practices (p.142).
- Hedgerows in Britain continue to disappear at an alarming rate (p.162).
- Incidents involving farm pollution in Britain increase, rising by 25 per cent in 1985.

MINUSES

- EEC proposes severe restrictions on the use of artificial nitrogen fertilizers (p. 101).
- EEC bans the use in cattle of Zeranol and trenbolene acetate, both synthetic anabolic growth hormones (p.159).
- Select Committee of the House of Commons recommends sweeping changes in the way that pesticides are controlled in Britain (p.197).
- Heinz bans the use of the so-called "dirty dozen" pesticides.

Erratum: the blocks of text above are transposed

CONSERVATION

PLUSES

- US conservation group, Conservation International, buys $600,000 Bolivian debt for $100,000 and agrees to write it off in exchange for 1,600,000 hectares of rainforest being set aside as a national park (p.225).
- Ivory Coast announces that it will ban timber exports in order to protect its 400,000 remaining hectares of tropical forests.
- Panama passes a decree making it illegal to fell any tree older than five years.
- Sarawak tribes start blockade to prevent their lands from being logged.
- The International Whaling Commission (IWC) passes a resolution condemning "scientific whaling" (p.231).
- USSR announces that it will stop commercial whaling to allow stocks to recover.

MINUSES

- EEC governments found guilty of failing to implement the Convention on International Trade in Endangered Species.
- US Secretary of the Interior recommends permitting oil drilling on 60,000 hectares of the 7,600,000 hectare Arctic National Wildlife refuge in Alaska (p.95).

HEALTH & FOOD

PLUSES

- Britain reviews the safety of some suspect food additives (p.92).

MINUSES

- BHA, an antitoxidant used in food, is linked to cancer.
- EEC pushes for legislation to permit the importation and sale of irradiated food in member countries (p.150).

NUCLEAR

PLUSES

- British Government abandons its search for a land-based nuclear waste site (p.188).
- France's Super Phénix fast breeder reactor is shut down following a sodium leak.

MINUSES

- British Government gives the go-ahead for a new pressurized water reactor to be built at Sizewell on the Suffolk coast (p.214).
- US Nuclear Regulatory Commission estimates that there is a 10 to 50 per cent

chance of a core meltdown at one of the 100 nuclear plants now operating in the USA (p.187).
● US Department of Energy report suggests that Chernobyl may kill 39,000 people, 21,000 of them in Europe (p.118).
● Report links a rise in the death rate among old people living downwind of the Three Mile Island reactor in the USA to the 1973 accident at the plant (p.221).
● Cattle deaths in Bavaria are linked to fall-out from Chernobyl (p.34).
● French military research station deliberately releases 7,000 curies of radioactive tritium into the environment as an experiment (p.202).
● Report estimates that 200,000 homes in the USA have excessive levels of radon.

POLLUTION

PLUSES

● International protocol agrees to reduce the production of chlorofluorocarbons by 50 per cent by the turn of the century to reduce the threat to the ozone layer (p.120).
● California votes in favour of Proposition 65 (p.202).
● British Government bans the use of tributyl tin (p.225).
● Iowa places tax on pesticides.
● EEC introduces new controls on car exhausts (p.114).
● West Germany announces that it will ban low octane leaded petrol (p.114).
● A House or Lords Select Committee urges the British Government to ban the pesticides aldrin, dieldin and endrin (p.96).

MINUSES

● Rhine disaster. A 100–200km stretch of the river is rendered lifeless after a chemical spill from a warehouse near Basel, Switzerland (p.206).
● US Environmental Protection Agency (EPA) fails to impose an outright ban on sales of the carcinogenic pesticides chlordane and heptachlor (p.119).
● US National Academy of Sciences warns that pesticides in food are a major cause of cancer (p.197).
● Imports of hazardous waste to Britain rise to 40,000 tonnes – eight times the figure for 1984 (p.161).
● Greenpeace warns that the North Sea could be irreversibly damaged within five years (p.185).
● US report blames lead pollution for reducing the learning abilities of city children (p.173).
● World Health Organization warns of the increasing impact of pollution on human health.
● The number of river pollution incidents in Britain is up 2.5 per cent on 1985/86 (p.109).
● Report discloses evidence of "widespread damage to wildlife from acid rain in Britain" (p.66).

THIRD WORLD

PLUSES

● The World Bank announces a new "green image". The bank's president, Barber Conable, tells the press: " I believe we can make ecology and economics mutually reinforcing disciplines." (p.233).

MINUSES

● World population passes 5,000 million mark, with 90 per cent of the growth taking place in the Third World (p.199).
● Indian Government gives the go-ahead for the Narmada project (p.184).
● Indian Government gives the go-ahead for the giant Tehri Dam in the Himalayas (p.219).
● The number of hungry people grows by eight million – rising to 730 million (p.146).

A-Z

Acceptable daily intake(ADI)
The number of milligrammes of a chemical that can safely be consumed every day by an average human, for each kilogramme (2.2lb) of body weight.

There are three major institutions setting ADIs, and the numbers they set often, but not invariably, coincide: the Joint Expert Committee on Food Additives of the United Nations Food and Agriculture Organization (FAO) and the World Health Organization (WHO); the EEC's Scientific Committee for Food; and the US Food and Drug Administration (FDA).

The label of ADI is inappropriate for any chemical that causes cancer by a mutagenic (see *Mutagen*) route – and perhaps by other routes too – because with those chemicals we cannot assume that there is any threshold level below which they can safely be used. (See also: *Additives* Ⓛ Ⓟ, *Cancer* Ⓛ, *Carcinogen* Ⓛ, *Junk food* Ⓛ, *Teratogen* Ⓛ.)

Acid rain
Term used to describe fallout of industrial pollutants, sometimes literally as acidified natural rainfall and sometimes as dry deposition. Most of the pollutants are produced by burning fossil fuels, though a significant number come from vehicle exhausts. Acid rain damages forests, plants and crops; acidifies lakes, rivers and groundwater; and corrodes building materials. In Europe it has been falling for more than 100 years; and its effects are cumulative. (See also: essay on *Acid Rain and Forest Decline*, *Car emissions* Ⓟ, *Catalytic converter* Ⓟ, *Coal* Ⓔ Ⓟ, *Deforestation* Ⓒ ③, *Scrubber* Ⓟ, *Tree diseases* Ⓒ.)

Additives
Chemicals that are deliberately added to food products in the course of industrial processing. They are distinct from the chemicals of which foods are composed, the chemicals introduced deliberately in farming, or the contaminants that may inadvertently enter the food supply.

Food technologists identify at least 23 different sorts of additives, but they can conveniently be grouped into five major types.

First, there are the preservatives (*qv*), that inhibit or prevent the growth of micro-organisms which would otherwise cause the food to rot and become poisonous.

Secondly, there are antioxidants (*qv*) that prevent oils and fats from turning rancid, and thereby extend the shelf lives of products containing those ingredients.

While those two groups are important, they together account for less than one per cent of the total use of chemical food additives. The largest single group of chemicals is the cosmetics. These are additives that modify the taste, appearance and texture of food products. About 88 per cent of the money spent on all additives is spent on cosmetics. Another eight per cent is accounted for by processing aids, which are as useful in the processing as in the products. Nutrients such as vitamins, calcium and minerals account for the remaining fraction.

In 1986, the UK food industry spent about £225 million ($380 million) on more than 200,000 tonnes of additives – equivalent, per person, to about 8g (0.3oz) of chemicals in a kilogramme (2.2lb) of food – while the industry in the USA spent $2,250 million buying 1,250,000 tonnes of additives, which corresponds to 9g (0.3oz) per person in a similar quantity of food. There are about 3,850 additives in use in the UK, 3,500 of them being designed to enhance taste. Along with enzymes and modified starches, such additives are not regulated in the UK.

The British government says that it approves additives only if they are needed and safe. In recent years, the application of both these criteria has been challenged. While the usefulness of preservatives is generally agreed, the need for cosmetics, especially food dyes, is far more controversial. No one doubts that they can be useful for industry, but many argue that consumers have no need for them. The standards by which safety has been judged have also been challenged. There is increasing evidence that some

Additives – the risks in dyeing food

Artificial dyes	UK	Denmark	Australia	Belgium	France	West Germany	Switzerland	Sweden	Austria	USA	Norway	Used in	Hazard
Ponceau 4R (red) No: E124	+	+	+	+	+	+	+	+	+	X	X	Ice cream, salami, beefburgers, packet soup, strawberry whip	2 5
Amaranth (red) No: E123	+	+	+	+	+	+	+	+	X	X	X	Tomato ketchup, ice cream, swiss roll, black cherry yogurt	1234
Allura Red No: E129	•	X	—	X	X	X	X	X	X	+	X	Only available in American foods	2 5
Carmoisine (red) No: E122	+	+	+	+	+	+	+	X	X	X	X	Yogurt, blackcurrant jam, canned fruit, pickled beetroot	3 5
Erythrosine (red) No: E127	+	+	—	+	+	+	+	+	+	+	X	Sausages, frozen pizza, fish paste, glacé cherries, nougat	12 5
Red 2G No: 128	+	X	X	X	X	X	X	X	X	X	X	Chocolate whip, sausages, processed fish, scotch eggs	123
Tartrazine (yellow) No: E102	+	+	+	+	+	+	+	+	X	+	X	Canned peas, frozen fish in sauce, chocolate drink, lime jelly	2
Yellow 2G No: 107	•	X	+	X	X	X	X	X	X	X	X	Sweets, chocolate, canned vegetables	2 5
Sunset Yellow No: E110	+	+	+	+	+	+	+	+	+	+	X	Smoked fish, ice cream, fish paste, tangerine jelly, sweets	2 5
Quinoline Yellow No: E104	+	+	X	+	+	+	+	+	+	X	X	Potato crisps, bitter lemon drink, lemon cake filling	2
Green S No: E142	+	+	+	+	+	+	+	X	+	X	X	Plum jam, fruit cocktail, sweets, lime jelly, lime cordial	2 5
Indigo Carmine No: E132	+	+	+	+	+	+	+	+	+	+	X	Chocolate biscuits, cake decoration	2 5
Brilliant Blue FCF No: 133	+	+	+	X	X	X	X	X	X	+	X	Ice cream, jelly, canned vegetables, blackcurrant drink	12
Patent Blue No: E131	+	+	X	+	+	+	+	+	+	X	X	Ice cream, sweets, processed peas	2 5
Brown FK No: 154	+	X	X	X	X	X	X	X	X	X	X	Sausages, smoked fish, potato crisps, packet rice	123
Chocolate Brown HT No: 155	+	X	+	X	X	X	X	X	X	X	X	Chocolate, ice cream	2
Black PN No: E151	+	+	+	+	+	+	+	+	+	X	X	Black cherry jam, blackcurrant jelly, chocolate, cheesecake	2

KEY: X BANNED; + PERMITTED; — NO LEGISLATION; • IN 1987 THE UK FOOD ADVISORY COMMITTEE RECOMMENDED THAT YELLOW 2G SHOULD BE BANNED AND ALLURA RED PERMITTED

1 Carcinogen Causes or suspected of causing cancer

2 Allergen Causes allergic or intolerant reactions

3 Mutagen Causes or suspected of causing genetic damage

4 Teratogen Causes or suspected of causing damage to the foetus

5 Animal Carcinogen Causes or suspected of causing cancer in animals

additives are provoking acute intolerance in a small but indeterminate proportion of the population. For example, some food dyes (*qv*), preservatives, artificial sweeteners (*qv*) and monosodium glutamate (*qv*) have been implicated as a cause of hyperactivity and asthma.

The use of additives is defended by industry and government on the grounds that they can be used to reduce the incidence of bacterial food poisoning, lengthen shelf lives, and make food products more attractive and cheaper. Consumer groups have criticized their use not just because they may pose a hazard to health, but also because they can be used to embellish cheap or inferior ingredients and enable them to be presented as premium quality goods, and because their use can contribute to the over-consumption of fats, sugars and processed starches. (See also: *Cancer*Ⓛ, *Coal-tar dyes*Ⓛ, *Emulsifiers*Ⓛ, *Junk food*Ⓛ.)

Advanced gas-cooled reactor
The UK is the only country in the world that operates advanced gas-cooled reactors (AGRS), having five stations with twin reactors in operation and two other stations under construction. The AGR uses carbon dioxide gas as coolant and has a moderator (*qv*) made up of a loose assembly of graphite bricks. The fuel in an AGR consists of uranium oxide ceramic pellets, contained within one-metre-long (39 in), small bore steel pins, 36 of which are bundled together into a graphite sleeve to make a fuel stringer. AGRS are designed to be refuelled while operating on load.

All the commercial 660MW AGRS use a prestressed concrete pressure vessel which, in addition to the core, contains the steam generators and the ducts to and from the reactor.

From an economic point of view the building of the first series of AGRS was an unmitigated disaster, brought about by intrinsic design problems not foreseen when scaling up from the prototype. All the AGR stations underwent long delays in construction. Dungeness B, which opened in 1985, took just less than 20 years to come into service, instead of the five years projected.

When all the costs are taken into account, including excess construction costs and interest charges, then the AGR is proving to be between 40 and 70 per cent more expensive to operate than a coal-fired plant of similar vintage.

A major problem in the AGR is the severe rattling and vibration in the core, brought about by the flow of coolant gas. The danger is that either the fuel stringers or the control rods will collapse, leading to localized overheating or to difficulties in shutting down the reactor. Tests on the two new AGRs being built at Torness in Scotland and Heysham in Lancashire indicate that damage to the control rods can occur after operating the reactors at full load for no more than a few hours.

The AGR's steam generators are subject to corrosion, both from the hot carbon dioxide gas and from water getting into the steam "superheat" section of the reactor. Should a number of boiler tubes fail simultaneously, then considerable quantities of water and steam will flow under pressure into the pressure vessel.

If that happened, a potentially dangerous situation would be created. The pressure vessel might become over-pressurized and breach, thus letting the coolant escape. Or steam explosions might occur, through the reaction of hot fuel and water. Hydrogen might also be generated, as a result of the interaction of steam and graphite, again leading to the risk of an explosion. Finally, water and steam might bring about a sudden surge in reactivity similar in many respects to that which caused the Chernobyl (*qv*) explosion. Explosions of the size that destroyed the Chernobyl containment would be more than enough to burst apart an AGR.

It is unlikely that the United Kingdom will build any more AGRs. The Central Electricity Generating Board has expressed its preference for the American Pressurized Water Reactor (PWR), and it has now obtained permission to build one at Sizewell (*qv*). The Sizewell PWR is intended as the first in a series of such reactors to be built in different parts of Britain. (See Also: *Nuclear accidents*Ⓝ, *Pressurized water reactor*Ⓝ.)

Aflotoxins

Food moulds that can damage the immune system and cause liver cancer. They were first identified in the 1960s, after 100,000 turkeys in a battery farm in the United Kingdom died of wasting diseases. Aflotoxins are also mutagenic (see *Mutagen*), inducing chromosomal damage in a wide range of animals. They have also been suspected to be a cause of Reye's Syndrome.

Modern agricultural techniques – particularly the use of artificial fertilizers (*qv*) that increase the water content of foods, thus encouraging the growth of moulds – have been suspected of being a major cause of aflotoxin contamination.

Aflotoxins pose a particular problem in the humid tropics and the sub-tropics. Surveys in the Sudan have revealed aflotoxins in 80 per cent of raw foods.

Acute aflotoxin poisoning causes immediate death, with cases being reported from India and East Africa. Children with severe malnutrition are thought to be particularly vulnerable, since they are less able to excrete aflotoxins than healthy children. (See also: *Cancer*Ⓛ, *Carcinogen*ⒻⓁ, *Post harvest losses*Ⓐ.)

Agent Orange

The code name given to a herbicide used extensively by US forces during the Vietnam War in order to defoliate the jungle and deprive the North Vietnamese of cover.

An equal mixture of 2,4,5-T (*qv*) and 2,4-D (*qv*), Agent Orange was heavily contaminated with dioxin (*qv*). Hospitals in North Vietnam reported that the rate of birth defects in sprayed areas had risen by 50 per cent. They also reported a significant increase in the rate of Down's Syndrome.

In the early 1980s, 2,500,000 Vietnam veterans (including veterans from Australia and New Zealand) sued both the US Government and the seven manufacturers of Agent Orange for alleged physical and psychological damages. The veterans reported illnesses ranging from chloracne – a disfiguring skin disease – to liver cancer, reproductive defects, genetic damage and mental disorders. The case was eventually settled out of court, with the manufacturers of Agent Orange agreeing to pay $180 million. The companies did not admit liability.

In Australia, where Vietnam veterans had also sued for injuries, a Royal Commission ruled that Agent Orange was not guilty of causing any adverse health effects among those who had served in Vietnam. (See also: *Biocide*ⒶⓅ, *Cancer*Ⓛ, *Phenoxy herbicides*ⒶⓅ.)

Aid

Transfers of capital – generally in the form of loans or grants – from governments and public institutions of the industrialized world to governments of the Third World. Where such capital transfers contain an element of "conditionality", they are recorded by the Development Assistance Committee (DAC) of the Organization for Economic Cooperation and Development (OECD). The same applies if the money is provided as grants rather than loans. Aid is also provided in the form of "technical assistance", notably by various specialized agencies of the United Nations.

Three-quarters of all "official aid" registered with DAC is "bilateral aid", provided directly from one government to another. Much of this takes the form of grants. The rest is channelled through the World Bank (*qv*), the International Monetary Fund (*qv*) and regional banks together with various funding institutions set up by the European Economic Community. Such transfers are referred to as "multilateral aid" in that the funds are derived from different countries. Of the capital sent out as bilateral aid, well over 50 per cent is provided to finance specific projects, ten per cent as food aid and two per cent as disaster relief.

Between 1970 and 1980, the

amount of official aid rose from $7,000 million to $35,600 million — an increase of 50 per cent when the figures are adjusted for inflation. Since then, aid has actually declined – Britain, for example, providing $2,700 million in 1979 but only $1,800 million in 1980 and $2,200 million in 1981.

Since 1982, the Third World has received less money in aid from the industrial world as a whole than it has paid out in interest on loans already contracted. In 1985, new loans totalled just over $15,000 million, whilst interest payments amounted to $55,000 million. The gap is even greater if all financial flows – including the repatriation of profits by Western companies operating in the Third World and capital flows, whether legal or illegal — are taken into account.

The net flow of funds from the Third World to the industrialized world has embarrassed the main aid agencies, in particular the World Bank. Various plans for reversing the trend have been formulated, notably by the former US Senator Howard Baker, but none has yet been adopted.

In recent years, the value of aid for fighting poverty and malnutrition in the Third World has increasingly been questioned, the principal argument being that aid has been used as a political and economic weapon in the interests not of the Third World but of donor countries.

Thus most aid is "tied aid" – that is, it is conditional on recipient countries buying goods and services from the donor countries. Eighty per cent of Canadian aid is tied aid, while two-thirds of British aid is said never to leave Britain (some put the figure as high as 98 per cent). Such aid is in effect little more than an export subsidy.

Similarly, much of the aid given to Third World countries is given in order to enable them to export foodstuffs and raw materials to the industrialized world. In effect, it acts as an import subsidy for donor countries. The loans incurred to permit the modernization of Third World agriculture under the green revolution (*qv*) come under this category, since they serve largely to permit the large-scale production of cash crops (*qv*) for the industrialized world.

Aid is also used as a blatant political tool. The bulk of US aid has gone to strategically important countries, while the world's ten poorest countries received less than five per cent of US bilateral aid.

Another reason for doubting the value of aid as a means of fighting poverty and malnutrition in the Third World is that much of it takes the form of arms. Military aid is now the biggest category of US aid, having increased from 22 per cent of the total budget in 1980 to 37 per cent in 1985. During the same period, US military aid to sub-Saharan countries rose by more than 40 per cent – this despite the region being ravaged by the greatest famine in human history.

In many cases, military aid is sent to blatantly oppressive governments – often dictatorships – whose lack of concern for the poor is well-documented.

Corruption ensures that as much as 20-30 per cent of aid never reaches the poor, but is instead siphoned off into private bank accounts and slush funds.

As presently constituted, aid projects rarely improve the welfare of the poor and the hungry. Instead, they serve largely to satisfy the economic, political and strategic interests of industrialized countries and the financial interests of the ruling elites of the Third World.

It is worth considering, too, that without aid the vast development projects that have led to so much environmental degradation in the Third World would never have been initiated. (See also: essay on *The Politics of Food Aid*, *Cash crops*Ⓐ③, *Dams*③, *Deforestation*③, *Development*③, *Third World debt*.)

Alaska

Since the territory received its statehood in 1959, Alaska's history has been characterized by a dramatic struggle between powerful, determined conservation and development lobbies. Many have regarded the state, an area one-sixth of all the other 49 US states put together, as America's last great wilderness, offering a last chance to achieve balance with the environment.

In 1968, 10,000 million barrels of oil and 736,000 million cubic metres (963,000 million cubic yards) of gas (25 per cent of proven US crude oil reserves and ten per cent of gas reserves) were discovered in Prudhoe Bay on Alaska's north shore. Extensive oilfields have been discovered elsewhere in the state.

The pressure to develop these, and the abundant resources of virtually all the major minerals, huge timber reserves and enormous fishstocks, has been great.

As yet, however, Alaska has been spared the unbridled development that many feared would be unleashed on the state. This has been due in part to state and federal legislation and in part to the economics of oil production in the permafrost.

Exploitation of the Prudhoe Bay field was halted until 1971 to enable the Alaska Native Claims Settlement Act to be drawn up and enacted. The legislation settled upon the state's 70,000 Eskimos, Aleuts and Indians 162,000 sq km (62,500sq miles) of land and $1,000 million in cash, vested in 12 Native Corporations in which all were equal shareholders.

In addition, the Secretary of the Interior was authorized to designate a further 324,000sq km (125,000sq miles) as national parks, national forests and wildlife refuges, subject to variable restrictions on development.

In 1980 the Senate passed a bill closing 429,000sq km (165,000sq miles), in varying degrees, to development. But early in 1987 the US Secretary for the Interior, Donald Hodel, recommended to Congress that oil development be permitted on 6,000sq km (2,300sq miles) of the Arctic National Wildlife Refuge.

A factor that has so far saved Alaska from depredation by the oil industry lobby is that seasonal limitations and environmental legislation have pushed up the price of Alaskan oil and gas compared to fuel produced under more favourable conditions. A fall in the price of Canadian gas was one of the factors behind the abandonment of the 7,700km (4,800 miles) Trans-Canada gas pipeline project. (See also: *Development*Ⓒ, *Extinction*Ⓒ, *Oil*Ⓒ.)

Aldicarb
A highly toxic insecticide of the carbamate (*qv*) group. Like other compounds in the group, aldicarb is a suspected carcinogen (*qv*). It is also a cholinesterase inhibitor, some people being especially sensitive to its effects.

Aldicarb is not selective and can harm non-target species. Gulls and stone curlews have died from aldicarb poisoning after consuming residues in earthworms.

Aldicarb is persistent in water. In 1983, the detection of aldicarb (known as Temik in the USA) in drinking water wells led to a year-long ban on its use in parts of Florida. In Long Island, New York, aldicarb in 1,000 wells was linked to a 46 per cent rise in spontaneous abortions.

In Britain, aldicarb is classed as a poison and is subject to special rules regarding protective clothing, which must be worn when it is used. In the Philippines, aldicarb is classified as too dangerous for general use.

The use of aldicarb is now restricted in the USA. Nonetheless, there is evidence that aldicarb has been sprayed illegally. Following a spate of unexplained illnesses in the mid-1980s, the public health authorities in California analyzed locally grown watermelons and found high levels of aldicarb residues – despite aldicarb being restricted for use on watermelons. Growers argued that the residues resulted from the previous spraying of aldicarb on cotton crops – but both the state authorities and the manufacturers deny that aldicarb persists in the soil longer than six months. A recent report by two researchers in Wisconsin has suggested that exposure to aldicarb may be linked to the spread of the AIDS virus. Aldicarb is known to interfere with the immune system and this, they argue, may leave people more vulnerable to AIDS. (See also: *Biocide*Ⓟ, *Cancer*Ⓛ Ⓟ, *Groundwater*Ⓟ, *Pesticides*Ⓐ Ⓟ.)

Aldrin
A toxic chlorinated hydrocarbon (*qv*) insecticide. A suspect carcinogen (*qv*), teratogen (*qv*) and mutagen (*qv*), it is classified by the World Health Organization as "highly hazardous". Once in the environment, aldrin is converted into dieldrin (*qv*).

Banned in Sweden, the use of aldrin has been severely restricted in the USA since 1976. In 1979, the EEC issued a directive to all member states banning the marketing and sale of aldrin in plant protection products. However, the commission agreed to permit certain "derogations" on uses. As a result, aldrin is still cleared for agricultural, horticultural and food storage uses. It is manufactured exclusively by Shell Chemicals. (See also: *Biocide*Ⓟ, *Cancer*Ⓛ, *Insecticides*Ⓟ, *Pesticides*Ⓐ Ⓟ.)

Alpha radiation
Positively charged particles released when one element transforms into another.

Compared with gamma (*qv*) and beta radiation (*qv*), alpha particles are heavy and slow moving with poor penetration. A sheet of paper or the outer layers of skin are enough to stop them.

Nonetheless, alpha particles can cause intense biological damage – possibly 10-20 times more than gamma radiation. The danger lies when alpha emitters are either inhaled or ingested with food, since they tend to accumulate in specific organs and tissues such as the lung and bone. (See also: *Cancer*Ⓛ, *Carcinogen*Ⓕ Ⓛ, *Plutonium*Ⓝ, *Radiation and health*Ⓛ Ⓝ.)

Alternative technology
Technology designed to use resources more efficiently and cause less environmental damage than technology in more widespread use but serving similar ends. Given this broad definition there are few technologies for which alternatives cannot be devised, and the pursuit of such alternatives has inspired many people within the ecological movement in many industrial countries.

The small-scale use of moving water in rivers, windmills, heat pumps, solar-energy collectors, and other devices to generate electricity for domestic use provide examples of alternative technologies. These alternatives have found application in many rural areas in industrial countries and are no longer considered strange. (See also: *Energy*, *Low energy housing*Ⓔ, *Photo-voltaic cells*Ⓔ, *Soft energy paths*Ⓔ, *Solar energy*Ⓔ, *Tidal power*Ⓔ, *Wave power*Ⓔ, *Wind power*Ⓔ.)

Amaranth
A widely used red coal-tar dye (*qv*), known in the USA as Red 2 and in the EEC as E123. It was first synthesized in 1878, and first produced on a large scale in the USA in 1914.

There are about 18 manufacturers in Western Europe, producing more than 300 tonnes a year. Amaranth is used not merely in food products, but also in drug preparations, cosmetics and inks.

It is one of the most widely used food dyes (*qv*) in the UK, and the most widely used red food dye. It can generally be found in soft drinks, sweets, and especially in products imitating cherries, strawberries, raspberries, or blackcurrants. In 1983, the British food industry incorporated about 50 tonnes of amaranth into their products. It can also be found in toothpastes and is available from British retailers as a food dye.

The doubts about the long-term safety of amaranth became particularly severe in 1970 when some work in the Soviet Union suggested that it was carcinogenic (see *Carcinogen*). This provoked an extensive discussion, and amaranth was banned in the USA in January 1976.

Amaranth is also banned in the USSR, Greece, Yugoslavia, Norway, Finland and Austria, but permitted and used in Britain and at least 63 other countries. (See also: *Additives*Ⓛ, *Cancer*Ⓛ.)

Amazonia
Stretching from the Atlantic coast of Brazil to the foothills of the Peruvian Andes, the Amazon Basin – better known as Amazonia – covers an area larger than the entire continent of Australia.

The world's largest tropical

Amazonia – the balance sheet of development and destruction

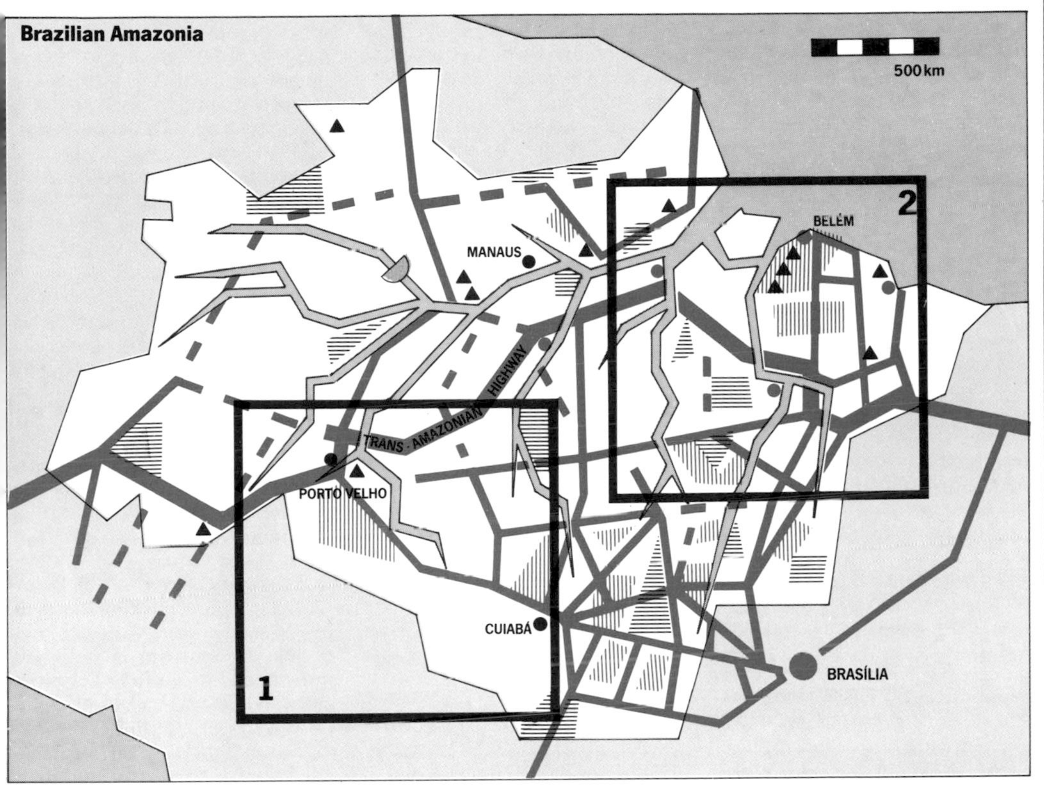

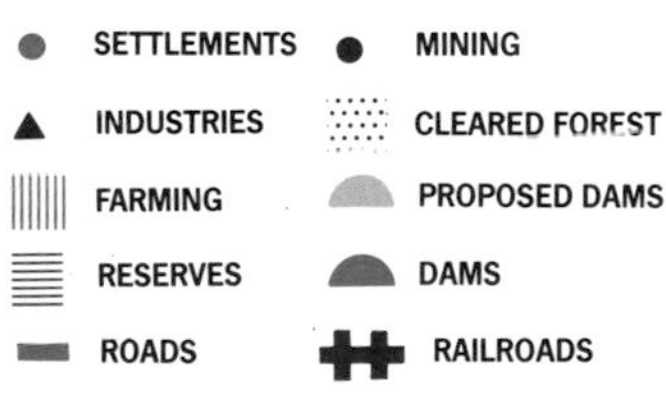

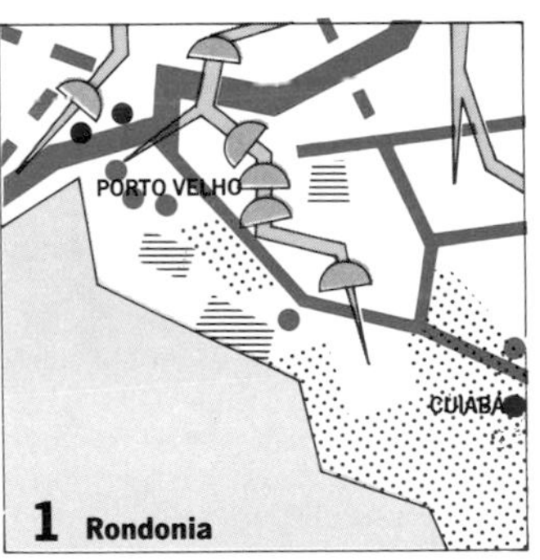

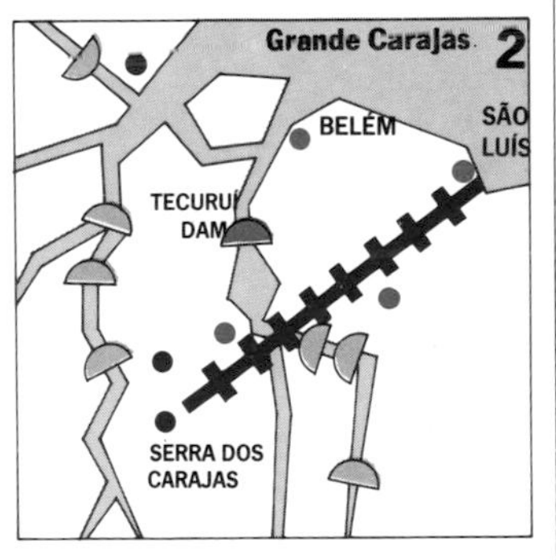

Most of Amazonia – some 5,200,000sq km – falls within Brazil, the remaining 1,800,000sq km being shared by Bolivia, Colombia, Ecuador, Guayana, Peru, Venezuela and Surinam. Since 1945 those nations have been intent on exploiting the region's vast natural wealth. In the late 1960s, Brazil launched "Operation Amazonia", an ambitious plan to promote a network of roads that would allow export-orientated industries to be set up in the heart of the forest.

The destruction caused by this and other operations has been immense, and it is continuing. As the map above shows, roads and railroads now invade the world's largest tropical rainforest; dams, mines, logging camps and settlements occupy the hundreds of thousands of square kilometres that have been cleared. And human beings have suffered with the forest. Five centuries ago there were 6-9 million Indians in Brazil: 200,000 survive.

The $1,600 million Polonoroeste road-building project is opening up the southern Brazilian state of Rondonia to thousands of settlers – an estimated 70-80,000 arrive each year. The state has already lost 30 per cent of its forest and, if clearance continues at the same rate, will have lost almost all of it within 10 years. Between 1981 and 1985 the World Bank provided $1,550 million for the project – but only 3 per cent of that figure was allocated to protecting the environment and the 28 different Indian tribal groups that will be affected by the development.

The giant Grande Carajas development in Brazil's north-east will affect some 900,000sq km and cost $62,000 million. The centrepiece of an assortment of mines, plantations, ranches and heavy industries, supported by roads, cities and ports, is the Serra dos Carajas open-cast mine, the world's largest. A 900km railroad will be pushed through the jungle to link the mine with São Luís on the Atlantic, where a deep sea port is to be built. Power for the development is to come from a series of dams. The first, the Tucuruí Dam, has already been built – and has flooded 2,160sq km of virgin forest.

rainforest, Amazonia has a vital role to play in regulating the Earth's climate. It is also a biological treasure trove, providing a habitat for an estimated one million animal and plant species including 2,500 species of trees, 1,800 species of birds and 2,000 species of fish. A hundred sq metres of forest can contain as many as 230 different tree species – as against a mere ten to 15 species in a typical temperate forest. In addition, Amazonia is the homeland for numerous tribal groups.

Since World War II, the Amazonian states have been intent on exploiting Amazonia's vast natural wealth and on "opening up" the forest to settlers. The destruction caused by development programmes is described in the accompanying illustration.

Satellite photographs reveal that the area of cleared land in Brazilian Amazonia more than doubled between 1975 and 1978 – from 28,600sq km (11,000sq miles) to 77,000sq km (30,000sq miles). Despite a 1977 resolution committing all eight Amazonian states to the "sustainable" development of the Amazon Basin, the destruction continues apace. Mining, dam projects, logging, ranching, road projects and colonization programmes pose the greatest threats. Ranching projects alone were responsible for the destruction of 38 per cent of the forest lost between 1966 and 1975, with road and highway operations accounting for 25 per cent.

Amazonia has immense mineral resources, Brazil earning some $9,000 million a year from mining operations. Gold, diamonds, uranium, titanium and tin have all been discovered on the Brazilian-Venezuelan border, threatening to devastate the homeland of 20,000 Yanomami Indians.

In north-western Amazonia, on the upper reaches of the Rio Negro, a gold rush has already led to the area being invaded by thousands of prospectors. For local indians, the experience has proved catastrophic. Recently, the Brazilian Government awarded mining concessions totalling 2,000sq km (770sq miles).

The environmental impact of the Grande Carajas (*qv*) project in north-eastern Brazil will be enormous. Some 15,000sq km (5,800sq miles) of forest will be cut simply to supply charcoal for the smelting of pig iron.

Some 30,000sq km (11,500sq miles) in Grande Carajas will be cleared for cattle ranching, all in pursuit of Brazil's ambition to become the world's largest producer of beef. Almost all of the meat will be exported.

Dam projects, too, are a major threat to the forests of Amazonia and its peoples. The dams planned for Brazilian Amazonia alone will flood an area the size of Montana.

Dam projects, mines, road projects and ranching have all brought encroaching cultivators (*qv*) in their wake. Though just two per cent of Amazonian soils are reckoned fertile enough to be farmed on a sustainable basis, the Brazilian Government has long encouraged settlers to colonize the rainforest under the slogan "Land without men for men without land."

The result has been catastrophic – both for the environment and for the landless peasants who seek a living in the forests. In Western Amazonia, the $1,600 million Polonoroeste road-building project is opening up the state of Rondonia to settlers, most of whom have been displaced from the fertile lands of southern Brazil in order to make way for vast government-backed plantations. The soils of Rondonia are so poor that many settlers have no option but to abandon their farms after the first harvest, forcing them to clear yet more forest in order to survive. An estimated 70,000 to 80,000 settlers arrive in Rondonia every year.

For the indigenous tribes of the Amazon, the development of Amazonia has proved nothing short of disastrous, 130 tribes having been wiped out in this century alone through disease, resettlement and the effects of culture shock. (See also: *Deforestation* ③, *Development* ③, *Greenhouse effect* Ⓕ, *Hamburger connection* ③, *Tribal peoples* ③, *Tropical forests* ③.)

Americium

A radioactive element that does not exist naturally on Earth, but is created in nuclear reactors.

Three different radioisotopes of americium can be formed. In terms of quantities generated, the most important is americium-241, itself a breakdown product of plutonium-241. Americium-241 has a half-life of 458 years. Once created, it will thus remain a potential pollutant for several thousand years.

Americium, an alpha emitter (see *Alpha radiation*), is at least one-third more radio-toxic than plutonium-239. The "safe" body burden is less than 0.04 microcuries.

Americium is a major constituent of the alpha wastes discharged from the Sellafield (*qv*) reprocessing plant in Cumbria.

In all, between 1968 and 1978 some 27,478 curies of alpha emitters were discharged into the Irish Sea via a pipeline from the Sellafield works. Forty-five per cent of that volume was made up of americium-241. Since October 1983, BNfl has embarked on a $16 million programme to reduce alpha discharges to no more than 200 curies a year – still ten times higher than the 20 curies of alpha wastes released annually into the English Channel from France's main reprocessing plant at Cap de la Hague (*qv*) in Normandy. Ultimately BNfl hopes to reduce its alpha discharges to the same level.

Contrary to expectations, both plutonium and americium are now coming ashore, thus contaminating the area around Sellafield. Hot spots of americium and plutonium have been found along the saltmarshes adjoining the Ravenglass estuary. Sheep grazing along the saltmarshes have been found with up to 100 times more plutonium in their livers than found in sheep from uncontaminated areas. (See also: *Plutonium* Ⓝ, *Radiation and health* Ⓝ.)

Antarctica

Antarctica is one of the last untouched wildernesses on Earth. The continent is the size of the United States and Mexico put

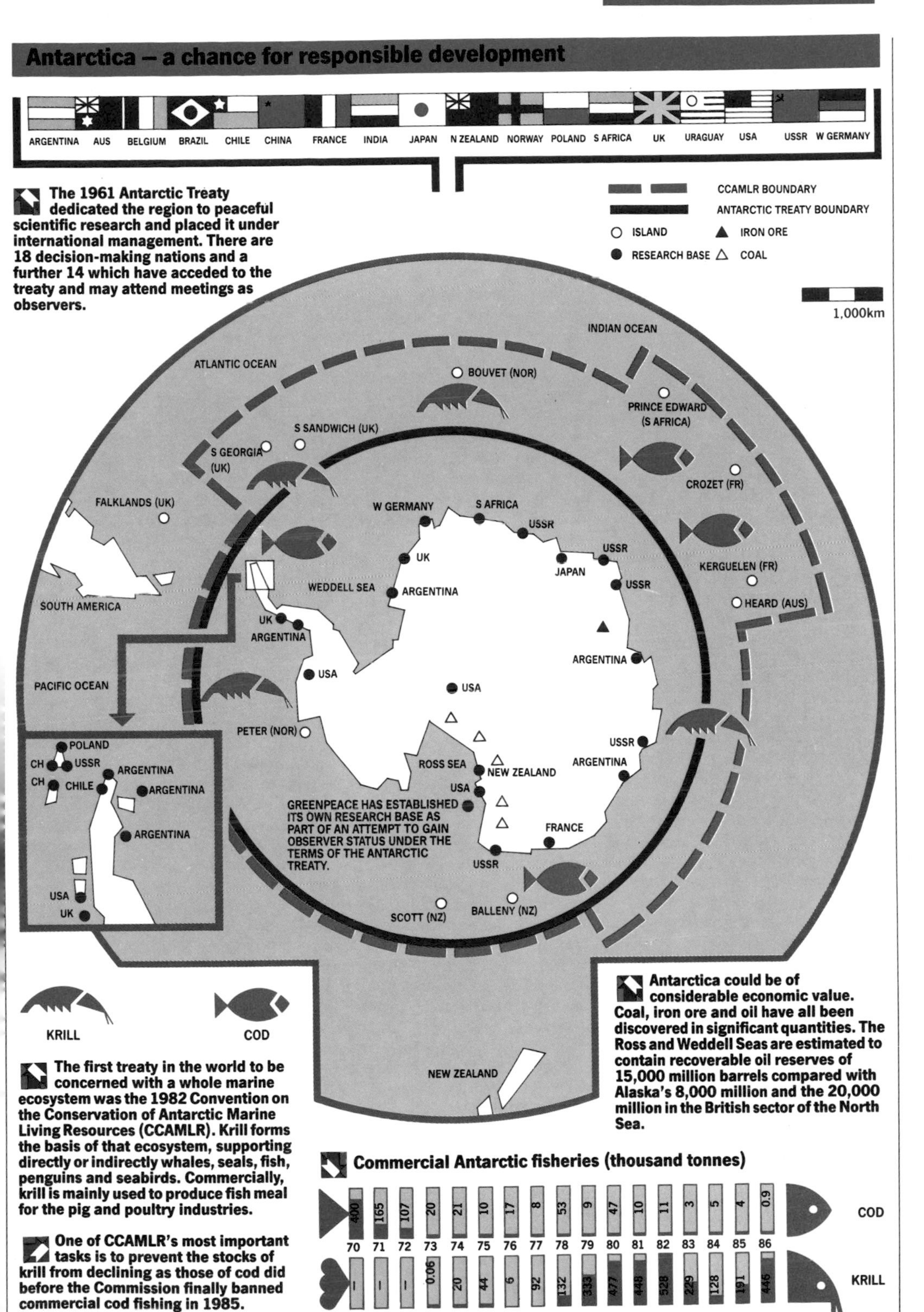
Antarctica – a chance for responsible development
ARGENTINA
AUS
BELGIUM
BRAZIL
CHILE
CHINA
FRANCE
INDIA
JAPAN
N ZEALAND
NORWAY
POLAND
S AFRICA
UK
URAGUAY
USA
USSR
W GERMANY
The 1961 Antarctic Treaty dedicated the region to peaceful scientific research and placed it under international management. There are 18 decision-making nations and a further 14 which have acceded to the treaty and may attend meetings as observers.
CCAMLR BOUNDARY
ANTARCTIC TREATY BOUNDARY
ISLAND
IRON ORE
RESEARCH BASE
COAL
1,000km
INDIAN OCEAN
ATLANTIC OCEAN
BOUVET (NOR)
PRINCE EDWARD (S AFRICA)
S SANDWICH (UK)
S GEORGIA (UK)
CROZET (FR)
FALKLANDS (UK)
W GERMANY
S AFRICA
USSR
USSR
UK
JAPAN
USSR
KERGUELEN (FR)
WEDDELL SEA
ARGENTINA
HEARD (AUS)
SOUTH AMERICA
UK
ARGENTINA
ARGENTINA
USA
PACIFIC OCEAN
USA
PETER (NOR)
USSR
POLAND
CH
USSR
ARGENTINA
CH
CHILE
ARGENTINA
ARGENTINA
USA
UK
ROSS SEA
NEW ZEALAND
ARGENTINA
USA
GREENPEACE HAS ESTABLISHED ITS OWN RESEARCH BASE AS PART OF AN ATTEMPT TO GAIN OBSERVER STATUS UNDER THE TERMS OF THE ANTARCTIC TREATY.
FRANCE
USSR
SCOTT (NZ)
BALLENY (NZ)
KRILL
COD
NEW ZEALAND
Antarctica could be of considerable economic value. Coal, iron ore and oil have all been discovered in significant quantities. The Ross and Weddell Seas are estimated to contain recoverable oil reserves of 15,000 million barrels compared with Alaska's 8,000 million and the 20,000 million in the British sector of the North Sea.
The first treaty in the world to be concerned with a whole marine ecosystem was the 1982 Convention on the Conservation of Antarctic Marine Living Resources (CCAMLR). Krill forms the basis of that ecosystem, supporting directly or indirectly whales, seals, fish, penguins and seabirds. Commercially, krill is mainly used to produce fish meal for the pig and poultry industries.
One of CCAMLR's most important tasks is to prevent the stocks of krill from declining as those of cod did before the Commission finally banned commercial cod fishing in 1985.
Commercial Antarctic fisheries (thousand tonnes)
COD: 400 165 107 20 21 10 17 8 53 9 47 10 11 3 5 4 0.9
70 71 72 73 74 75 76 77 78 79 80 81 82 83 84 85 86
KRILL: – – – 0.06 20 44 6 92 132 333 477 448 528 229 128 191 446

together, around 14 million sq km (5,400,000sq miles). Most of the land mass is covered in ice with an average thickness of 1,600m (5,250ft), though this can extend in some parts to a thickness of 4,000m (13,000ft). Though pack ice is present in the sea all year round, in the winter the ice expands to cover 200 million sq km (77 million sq miles) and prevents access to parts of the coastline normally accessible in the summer months.

Antarctica's environment is without doubt the world's harshest. The average temperature in the coldest months in the interior is –71°C (–160°F) and the lowest temperature ever recorded on Earth (–88°C, –190°F) was recorded there. Winds from the interior can sometimes exceed speeds of 300kph (186mph).

All this does not prevent Antarctica from being regarded as a continent of tremendous resource potential (see illustration). A large number of minerals have been found, though the logistics of recovering these under the extreme conditions of the continent could be prohibitively expensive. It is thought, however, that oil exploitation is a possibility.

Antarctica's enormous marine wealth has long been exploited. As early as 1824, British and North American sealers had almost exterminated the fur seal population and it was not until the mid-1970s that the seals recovered to any degree. Factory ships were already making a considerable impact on whales during World War I and by the 1930s catches had already started to decline.

Fin fish, such as the Antarctic cod, had been so overfished by the early 1970s that the catches declined to a fraction of their previous levels. Today, the Russian, and to a lesser extent Japanese, Polish and West German fishing fleets, are concentrating their efforts on fishing for krill (*qv*), a small shrimp-like creature that exists in enormous quantities in the Antarctic seas.

There is clearly tremendous interest in Antarctica's huge mineral and marine resource potential. However, the potential for irreversible environmental damage is very high. The pesticide DDT (*qv*) has already been found in penguin fat and eggs, though the nearest inhabited continent is 1,800km (1,100 miles) away. Biological processes operate only slowly and intermittently. The terrestrial ecosystem is particularly fragile and has little capacity to withstand physical, chemical or biological changes without itself being altered. Though there are more than 500 different species of lichen, 70 species of mosses, and two types of flowering plants, such vegetation is sparse and vulnerable. A human footprint in a bed of moss can remain for several years. Some damage has already occurred because of the research stations there.

The greatest threat, however, lies in commercial exploitation of Antarctica's resources. Oil is particularly menacing. The Antarctic seas are the roughest in the world and this, coupled with the extreme temperatures, would make the extraction and transportation of oil a hazardous enterprise at the best of times.

Oil pollution would cause a variety of disasters. It would threaten most of the animal life that depends on the marine environment, such as seals and penguins. An oil spill on ice would increase its capacity to absorb heat and would cause it to melt, with potentially disastrous consequences. Oil, in such a cold environment, would also take longer to decompose than in temperate areas.

Since 1961 a number of important protective conventions have been adopted by the member nations of the Antarctic Treaty. Most important among these is the Convention on the Conservation of Antarctica Marine Living Resources (CCAMLR), which is particularly interested in ensuring that the effects of krill fishing will not threaten other species in the ecosystem, though the establishment of krill quotas has been hampered by the fact that three of the parties to the convention, Japan, Poland, and the Soviet Union, are major krill-fishing nations.

In 1977, the consultative members of the treaty agreed not to proceed with any kind of commercial mineral exploitation until some kind of general agreement had been reached. Such an agreement has not yet been reached and it will probably not be until 1991 that any decision is made. The fate of Antarctica's marine and terrestrial environment still hangs in the balance. Conservationist interests have featured heavily in all the negotiations. However, the USA and Japan are pursuing a vigorous search for oil, in possible violation of a 1977 commercial moratorium, and that may yet tip the scales. (See also: *Mineral resources*©, *Oil*©.)

Antibiotics

Drugs used to combat bacterial infections. Mass-produced, synthesized antibiotics first became widely available after World War II and have undoubtedly saved many lives. But their abuse has led to many strains of bacteria becoming resistant to them, particularly to penicillin.

According to the World Health Organization (WHO), the number of intestinal bacteria showing signs of resistance tripled between 1976 and 1981. WHO warns that the general development of resistance could lead to a resurgence of such infectious diseases as pneumonia, meningitis, cholera, salmonella and venereal disease. Ironically, the problem of resistance is particularly severe in hospitals: 90 per cent of the staphylococcal infections found in hospitals are resistant to penicillin.

The use of antibiotics plays havoc with the body's own ability to fight infections. For that reason, practitioners of homeopathy and other forms of alternative medicines eschew the use of synthetic antibiotics, stressing the importance of a healthy diet and lifestyle as the first line of defence against infections.

Antioxidants

A group of chemicals that are added to oils and fats to inhibit oxidation and thereby slow down the rate at which foods become rancid and unpalatable. Oils and fats in food products oxidize from normal contact with atmospheric oxygen. In extreme cases they can become toxic, though they develop an unacceptable taste long before that happens. The effect of antioxidants is therefore to extend the shelf-lives of products to which they are added.

There are two broad groups of antioxidants, the natural ones and the synthetics. Some oils such as cold-pressed olive oil contain naturally occurring constituents that act as antioxidants, but most commercially used oils and fats do not enjoy such protection. Antioxidants are added to a wide range of fat-containing food products, such as fried snack foods, breakfast cereals, baked goods, cured meats, margarine and dessert products. In the EEC, all permitted antioxidants have E numbers, ranging from E300 to E321.

Unlike most groups of additives, the levels at which antioxidants can be used in particular types of products are stipulated in regulations, because their excessive use would constitute a health abuse. While some antioxidants such as ascorbic acid (E300) and the ascorbates (E302-E304) are safe, the safety of the synthetic antioxidants butylated hydroxyanisole (E320) and butylated hydroxytoluene (E321) is a matter of dispute. (See also: *Additives*Ⓛ.)

Appropriate technology

Technology that uses tools, machinery, and methods that can be applied readily by the people for whom they are intended, and can then be sustained and maintained by them without recourse to imported supplies that may be difficult to obtain. (See also: *Alternative technology*Ⓔ.)

Aquaculture

The farming of fish, mussels, oysters and other aquatic organisms. Aquaculture is now being actively promoted by the UN Food and Agriculture Organization (FAO) (*qv*) in many Third World countries as a new source of fish supplies.

At present, marine fisheries provide the bulk of world fish production, yielding 76.8 million tonnes in 1983. By the year 2000, however, FAO predicts that world demand for fish will have outstripped marine catches by between 20-30 million tonnes a year. In reality, the shortfall is likely to be even higher since fish stocks are increasingly threatened by pollution and overfishing.

Great hopes have been pinned on aquaculture as the solution to the expected shortfall in fish supplies. At present, it accounts for about ten per cent of world fish production. FAO would like to increase that by five to ten times by the end of the century – a goal that seems unrealistic.

The bulk of current production from aquaculture comes from smallholders using traditional methods of fish farming. But FAO is seeking to change this, promoting large-scale projects designed to maximize production. Unfortunately, FAO's plans are likely to make less (rather than more) fish available to the rural poor.

Indeed, traditional systems of aquaculture are highly sophisticated, reflecting the role that fish farming has played for centuries in the cultures of many peasant societies, particularly in South-East Asia. They are cheap to run, employ local technology and use existing sites, such as village ponds, paddy fields and irrigation canals. Aquaculture is thus complementary to other activities rather than competitive with them. In addition, the fish reared under traditional systems are readily available to local people.

By contrast, the prestige operations being promoted by FAO make use of expensive technology, require the clearance of new land and have high costs – not least the "opportunity cost" of the valuable land they take over.

Those high costs provide one reason why the "Blue Revolution", like the "Green Revolution" (*qv*) before it, cannot serve to feed the world's poor. To service the loans used to set up modern fish farms, their produce must necessarily be sold for urban consumption or for export.

FAO's plans to expand aquaculture have already hit problems. Attempts to set up modern fish-farms in the Nile Delta have failed, largely because no account was taken of local soil conditions. The sodic montmorillonite clay of the area is entirely unsuited to fish farming since, under water, it turns to a slurry and collapses.

Aquaculture requires a fertile clay soil and a reliable gravity-fed supply of uncontaminated water – itself an increasingly rare commodity. Such sites are scarce and, because they tend to be valuable for agriculture, are generally already in use. It is only in rare instances that converting them to aquaculture can help feed the hungry. (See also: *Development*③.)

Aquifer

A geological term that refers to those porous soils and rock formations through which water is able to percolate. Large amounts of water may be held in aquifers, which effectively retain water in the same way as a sponge.

By drilling wells or boreholes, engineers can tap the waters in an aquifer. More than 100 million Americans rely on aquifers for their drinking water and 40 per cent of the water used by US farmers for irrigation (*qv*) comes from groundwaters.

Aquifers can vary from being a few centimetres below the surface to being many hundreds of metres below. Their waters are recharged by the downward seepage of surface waters. They are therefore susceptible to pollution by contaminants within surface waters.

The pace at which groundwaters travel through an aquifer is snail-like, generally moving at less than a metre a month. The dark, lifeless environment of groundwaters makes the degradation of pollutants extremely slow. Once contaminated, therefore, pollutants can remain in an aquifer for hundreds of thousands of years. (See also: *Groundwater*Ⓟ, *Ogallala Aquifer*Ⓟ.)

Artificial fertilizer

Salts of nitrogen (N), phosphorus (P) and potassium (K) added to the soil to increase crop yield. Most artificial fertilizers used today consist of different proportions of N,P and K.

The use of artificial fertilizers had enormous impact on agriculture worldwide, especially after Haber in Germany developed his process for capturing nitrogen from the air and converting it into ammonia (NH_3). Farmers no longer had to leave their fields fallow to regain fertility or to follow crop rotations with the use of nitrogen-fixing legumes; nor did they need to keep animals to provide farmyard manure; instead they began growing the same crops year in and year out.

This, however, made it impossible to maintain a healthy soil structure, and farmers compensated for dwindling fertility and soil compaction by applying more fertilizers to crops that would respond specifically to such treatment.

Crops grown with artificial fertilizers tend to be poor in nutritional quality, to have a poor storage life, to be lacking in variety and to be contaminated with the residue of agrochemicals.

Research in France, particularly by Francis Chabousou of the National Institute for Agronomic Research, indicates that both artificial fertilizers and pesticides (*qv*) promote the proliferation of pests by upsetting the metabolism of the growing plants so that the cells contain higher than normal concentrations of soluble nutrients. But rather than abandon the unsound farming methods which encourage the pests, farmers have responded by using a battery of toxic herbicides (*qv*) and pesticides, thus causing widespread pollution.

An EEC Commission has announced plans to impose strict controls on the use of artificial fertilizers over the next few years. Denmark intends to cut its use of nitrogen fertilizers by 25 per cent. (See also: *Eutrophication* Ⓒ Ⓟ, *Green Revolution* Ⓐ ③, *Groundwater* Ⓛ Ⓟ, *Nitrate* Ⓛ Ⓟ, *Organic farming* Ⓐ Ⓛ, *Post harvest losses* Ⓐ ③, *Traditional agriculture* Ⓐ.)

Artificial sweeteners

Compounds that contribute sweetness, but no bulk, to food and drinks. They are to be distinguished from sugar and other sweeteners such as Mannitol, Sorbitol and Xylitol which provide both bulk and sweetness. Those taken voluntarily are most often used by diabetics or people wishing to lose weight.

In 1988 six artificial sweeteners are being sold in industrialized countries. The two main ones are saccharin (*qv*) and Aspartame (*qv*). Another group, cylamates, are permitted in 16 countries but banned in the UK and USA. Acesulfame-K is permitted in nine countries. Thaumat is permitted in Britain and Switzerland, while Belgium permits neohesperidien dihydrochalcone. There are at least seven other compounds that firms in the chemical and food industries wish to market.

In 1987, the British market for artificial sweeteners amounted to approximately $40 million, accounting for about 20 per cent of total European sales, while the American market is almost $150 million. Wordwide sales of Aspartame amounted to about $600 million in 1984 and have subsequently grown considerably.

It is almost invariably assumed that artificial sweeteners help consumers to lose weight because they contain fewer calories than sugar, despite the fact that there is no evidence to support that belief. In 1986, new research suggested that using them may actually stimulate the appetite for carbohydrates and thus even be counter-productive. Their usefulness to diabetics is less often questioned.

There have been, and continue to be, huge arguments over the safety and acceptability of artificial sweeteners. From an economic point of view, their increased use has generally contributed to the fall in the price of sugar, which, in turn, has severely damaged the economy of sugar-producing countries in the Third World. (See also: *Additives* Ⓛ, *Cancer* Ⓛ, *Cyclamates* Ⓛ.)

Asbestos

The generic name for a group of fibrous mineral silicates, consisting of crocidolite ("blue" asbestos), chrisotite ("white" asbestos) and amosite ("brown" asbestos).

Once mined, the asbestos fibres are separated out from the rock in which they are embedded. They are then spun into a cloth, which can be mixed with cement or resin to give it shape.

Because it is both highly resistant to heat and a poor conductor of electricity, asbestos has been used in a wide range of materials – from roofing slates to fire-fighting suits.

The dangers of asbestos lie in its fibres, which can range in length from one four-hundred-thousandth of a centimetre to more than 30cm (1ft). If inhaled, the fibres lodge in the lungs and the tissues of the bronchial tubes. The result is a crippling lung disease called asbestosis. In the final stages of the disease, victims can be so breathless that they are unable even to climb stairs. Asbestos also causes lung cancer, cancer of the gastro-intestinal tract and mesothelioma, a malignant cancer of the inner lining of the chest.

In the USA, official statistics put the potential death toll from exposure to all forms of asbestos at two million people. According to the US Environmental Protection Agency (EPA), 65,000 Americans currently suffer from asbestosis, a further 12,000 dying every year from asbestos-related cancers.

Many of those suffering from asbestosis have now sued the major US asbestos companies for damages. In 1982, inundated by claims, the Manville Corporation filed for bankruptcy, thus freezing all future suits against the company but allowing it to continue operating. At the time, Manville faced 100,000 suits.

In early 1987, Manville proposed a settlement whereby it would set up a trust fund to pay out $2,500 million over the next 30 years. The proposal has been opposed by both Manville's shareholders and those who have sued the company. The latter have demanded punitive damages on the grounds that Manville continued to sell asbestos without warning labels even after the dangers were fully appreciated.

The EPA has now proposed a total ban on the production of all asbestos products within the USA, arguing that "there is no level of exposure without risk." The ban would phase out all asbestos products within ten years.

In Europe, the Swiss Eternit group has agreed to eliminate asbestos from all cement products for sale in Switzerland and West Germany, except asbestos piping.

In Britain, both blue and brown asbestos have been officially banned since January 1986. White asbestos continues to be sold, despite numerous case studies linking its use to mesothelioma.

With tighter controls being imposed on the use and production of asbestos products in the industrialized world, many companies have simply "exported" their factories to the Third World – in particular to India, South Korea and Mexico – or to countries where regulations are less strict, such as South Africa.

A 1987 US study has suggested that most common asbestos substitutes, such as glass fibre, may also be a cause of lung cancer. (See also: *Cancer* Ⓛ, *Carcinogen* Ⓕ Ⓛ, *Export of pollution* Ⓟ ③, *Glass fibre* Ⓛ Ⓟ.)

Aspartame

Trade name for a synthetic sweetener made by combining two amino acids, L-phenylalanine and L-aspartic acid. It can be found in no fewer than 125 different products, but these are mainly soft drinks and soft drink mixes and low calorie yoghourts. It is also marketed as Nutrasweet and Candarel.

Since 1973, controversy has raged over the use of Aspartame. Because it is synthesized from a combination of two common, vital and naturally occurring amino acids, one might expect Aspartame to be one of the least problematic chemicals. But various studies have implicated the sweetener as a cause of brain damage – particularly when consumed in combination with monosodium glutamate (*qv*). It has also been suggested that Aspartame may disturb brain functions, provoking a variety of severe symptoms including epileptic fits.

Law suits against the marketing of Aspartame remain pending in US courts. In the meantime, Aspartame is permitted and used in the UK and USA. Though it is widely approved for use in table top sweeteners, its use is banned in foods and/or beverages in Austria, Belgium, France, Greece, Italy, Holland or Portugal. (See also: *Additives* Ⓛ, *Artificial sweeteners* Ⓛ.)

Aswan Dam

Opening Egypt's Aswan Dam in 1970, President Nasser promised that the dam would prove "a source of everlasting prosperity". Today, Aswan has become a symbol of all that can go wrong with big water development projects.

The dam has caused a dramatic increase in waterborne disease. Schistosomiasis (*qv*), a parasitic disease, has rocketed in villages close to Lake Nasser, behind Aswan, some communities suffering 100 per cent infection rates.

The perennial irrigation projects fed by Aswan have caused widespread salinization (*qv*). Prior to the dam, the accumulated salts had been flushed out of the land by the Nile's annual flood. Now 35 per cent of irrigated land is affected.

In addition, the fertility of downstream farmland has been badly affected by the reduced silt-load of the Nile. Before Aswan, the Nile deposited 100 million tonnes a year of organically rich sediment on nearly 10,000sq km (3,800sq miles) of land. Today, it deposits a few tonnes a year, the remainder being trapped behind the dam. To compensate for the loss of nutrients, Egypt must apply artificial fertilizers – at huge cost.

The reduction of silt has also led to coastal erosion and the loss of sardine shoals along the eastern Mediterranean. Before Aswan, sardine catches amounted to some 18,000 tonnes a year. After the dam opened, catches fell to 500 tonnes a year. This loss has been partly compensated for by the development of fishing in Lake Nasser and the setting up of an ocean-going fleet. (See also: *Dams* Ⓐ Ⓔ Ⓛ ③, *Development* ③, *Resettlement programmes* ③, *Waterborne diseases* Ⓛ.)

Atomic tests

The world's nuclear powers have carried out regular nuclear

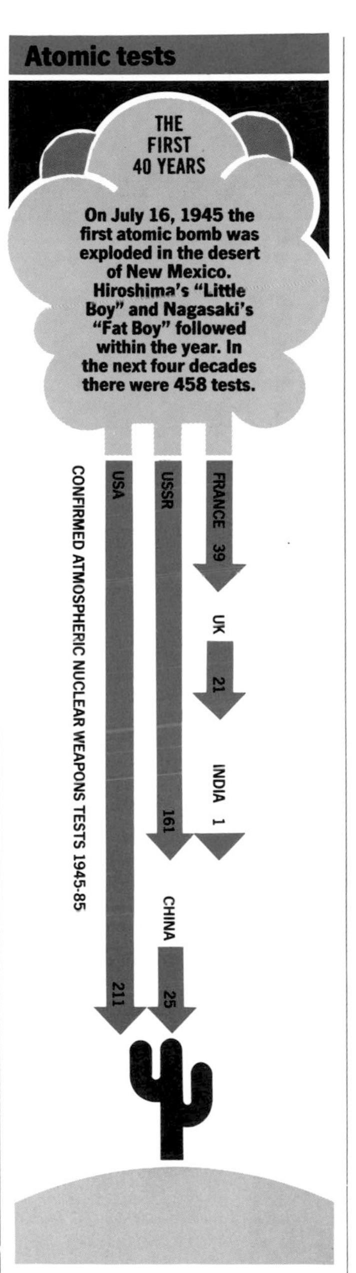

In 1963 atomic testing went underground when the US, UK and USSR signed the Limited Test Ban Treaty banning atmospheric tests. Over 100 other nations acceded to the treaty, but with two notable exceptions, France and China. France continued with atmospheric testing until 1974 with a further 37 tests, while China only began atmospheric testing after the treaty. By 1985 a total of 1,472 tests, both above and below ground, had been monitored from around the world.

weapons tests ever since 1945. These tests are detailed in the accompanying illustration.

In 1985 the USSR adopted a self-imposed moratorium on testing, but the initiative came to naught when the USA refused to reciprocate. The moratorium ended after 19 months.

Bitter controversy still surrounds the ecological and medical effects of the atmospheric tests conducted from 1945 to 1963. More than 250,000 US troops took part in the US atmospheric tests. Many were exposed to very high levels of radiation without any warning about the dangers. Some troops were positioned as close as 2,280m (2,500yd) from ground zero, the aim being to test their reactions in the wake of an atomic blast. The leukemia rate among veterans of one test – Operation Smokey – has since been found to be double the national average.

About 170,000 civilians living downwind of the main US test site in the Nevada desert were exposed to fallout from the tests. A 1979 study revealed the cancer rate among exposed children in southern Utah to be three times higher than normal: 500 adult and child cancer deaths are alleged.

Several tests sites – most notoriously Maralinga (*qv*) and Emu Fields in Australia and the Marshall Islands (*qv*) in the Pacific – are now known to have been throughly contaminated by radioactive fallout.

In the USA, Australia and the United Kingdom, servicemen and civilians exposed to fallout have sued their respective governments, alleging cancers and other health effects. In April 1987, a $2,600,000 award to nine civilian cancer victims of the US tests was overturned by the Federal Appeals Court.

Little information exists on the ecological and medical damage caused by the Soviet Union's atmospheric tests – or on those conducted by China. (See also: *Muraroa* Ⓝ, *Nuclear proliferation* Ⓝ.)

Balance of nature

A basic concept of ecology, regarded by Dr. Frank Egerton, a leading historian of ecological thought, as "the oldest ecological theory".

The idea greatly pre-dates the term "ecology", having been entertained by all known tribal peoples since time immemorial. In such societies, "natural" disasters – epidemics, floods, drought and the like – are traditionally regarded as divine retribution for the violation of basic cultural norms and, thus, the disruption of the natural order or "balance of nature". Only when that balance is restored by reverting to "the old ways" – ways that embody a respect for nature and that help to maintain ecological harmony – can order be restored.

In modern times, the concept of the "balance of nature" played a central role in the thinking of George Perkins Marsh, a 19th century American diplomat who was the first US minister to the kingdom of Italy and one of the most important early ecological thinkers. Marsh believed that, "All nature is linked together by invisible bonds, and every organic creature, however low, is necessary to the well-being of some other among the myriad forms of life with which the creator has peopled the Earth."

The notion of the balance of nature is very much part of the "organismic" or "holistic" view of the world promoted by such early academic ecologists as F.E. Clements, V.E. Shelford and E. Forbes at the beginning of this century. Clements and his colleagues stressed the equilibrium and stability to be found in nature, introducing such concepts as the "unity of nature", "mutualism" (*qv*), "stability" and "ecological climax".

But their ideas found no place in a society committed to development and "progress" (which Marsh described as a "war against the order of nature"). Indeed, other academic ecologists actively sought to discredit Clements. In an effort to gain scientific respectibility and to make "ecology" more acceptable to mainstream society, the old "holistic" concepts, including that of "the balance of nature", have been abandoned. Today, most academic ecologists reject the notion altogether, seeing perpetual flux and change where their predecessors saw balance and stability.

Nevertheless, the concept of the balance of nature has been revived within the ecological movement (*qv*) by deep ecologists (see *Deep ecology*), as indeed have its associated concepts of mutualism, ecological climax and so on. Their rehabilitation is playing a critical part in the development of a new ecological world view. (See also: essay on *Man and the Natural Order*, *Biosphere*Ⓕ, *Ecological succession*Ⓕ, *Gaia*Ⓕ, *Holism*Ⓕ.)

Baltic Sea
See: *Regional seas.*

Band Aid
The Band Aid Trust was established at the end of 1984 by the Irish pop singer Bob Geldof. It initially raised £9 million ($14 million) through the release of the record "Do They Know It's Christmas?", recorded by a large group of pop celebrities.

In the summer of 1985, Geldof organized Live Aid, a day-long concert event in London and New York, which was televized live and beamed round the world by satellite. As a result, the Band Aid Trust raised about $160 million.

Band Aid has no programmes of its own. Plans are submitted by other agencies and approved by a committee of ten unpaid professionals who include, among others, a social anthropologist, an epidemiologist, a tropical agriculturalist and an expert on nomads.

By February 1987, the trust had distributed money to 124 charities for a variety of different projects. These included, for example, an Oxfam seed distribution programme in the Sudan and a UNICEF vaccination programme in Mozambique.

The trust expects to run for another three years, after which the remaining $48 million will run out. This is in accordance with Geldof's original conception of an aid organization that would reach the Third World poor quickly, through its capacity to innovate, and thus avoid the bureaucratic troubles of established aid agencies. (See also: essay on *The Politics of Food Aid*, *Aid*③, *Famine*③.)

Basel accident
See: *Sandoz fire.*

Becquerel
Measure of radioactivity, named after A.H. Becquerel (1852-1908), who discovered the phenomenon in 1896. One becquerel (Bq) is one disintegration per second. The unit is replacing the curie (ci), which is equivalent to 37,000,000,000 bq. (See also: *Radiation and health*Ⓛ Ⓝ, *Radioactivity*Ⓕ.)

BEIR Committee
The Committee on the Biological Effects of Ionizing Radiation – or BEIR Committee – exists under the aegis of the US National Academy of Sciences (NAS).

Its first report, issued in 1972, proved a landmark in the debate on nuclear power. The report stated that there is a risk in any radiation dose (no matter how small) and that there is therefore no safe threshold below which radiation poses no risk.

However in 1980, the committee issued a third report that rejected the findings of the first report. BEIR III was extensively amended at the last moment before publication and many people, including the committee's chairman, who dissented from the report, felt that its conclusions had been modified to make them more acceptable to the nuclear industry.

The controversy centres on attempts to calculate the health effects of low doses of radiation. The main problem is that the amount of epidemiological data, though increasing, is small. Most data relate to the effects of high doses: much derives from information on Hiroshima and Nagasaki survivors. Thus the manner in which high doses are considered to relate to low doses is of great importance.

The 1972 BEIR report assumed a linear extrapolation from high to low doses: that is, the risks from low doses were taken to be proportional to the risks from high doses. In the BEIR III report, the committee assumed that the relationship was linear quadratic: that is, that the low-dose risk is less than proportional to the high-dose risk. (See also: *Radiation and health*Ⓛ Ⓝ.)

Benzene
A volatile and inflammable chemical, normally produced from petroleum but also obtainable from coal. It is used widely in industry as a solvent and as a raw material in the manufacture of chemicals. Benzene is also a component of petrol.

Benzene is highly poisonous with serious acute and chronic effects. Exposure to a high concentration of the vapour causes confusion, dizziness, unconsciousness and respiratory failure leading to death. If the affected person is moved to a clean air zone before serious damage occurs, recovery is possible. Because benzene is absorbed through the skin, this can add to the inhaled dose and aggravate the symptoms.

More serious and more widespread is chronic poisoning by benzene, whereby smaller quantities are absorbed frequently over a long period. Benzene has a cumulative effect and repeated exposure cause mounting damage. Its chronic effects are on the blood-forming tissues, progressively reducing the numbers of white and/or red blood cells present. Most serious is its ability to cause leukemia, a form of cancer, which may not be identified before it reaches an untreatable stage. For this reason, workers exposed to benzene should be medically examined on a regular basis. Industrial safety limits for benzene have recently been revised downward in many countries in view of its ability to cause leukemia. (See also: *Cancer*Ⓛ, *Carcinogen* Ⓛ Ⓟ.)

Beta radiation
Fast-moving energetic electrons, shed from the nucleus of atoms undergoing spontaneous radioactive transformation. The negatively-changed beta particles are able to penetrate several centimetres of living tissue, which

Bhopal

Events leading to the release of toxic gas over Bhopal

The Bhopal disaster in central India occurred when 30 tonnes of methyl isocyanate (MIC) escaped from its tank into a nearby slum area housing 200,000 people. A report drawn up eight months later, by two international trade union confederations, suggested that it was a combination of management mistakes, badly designed equipment and poor maintenance that caused the accident.

In the preceding four years there had been six accidents at Bhopal, some of which involved MIC, including one fatality in 1981 which led to an official investigation. The findings of that investigation were not acted upon until after the accident in 1984.

A safety inspection conducted by the parent company in 1981 found fault with the MIC tank control system, the reliability of some valves and the operator training programme. No subsequent check was made to establish whether or not appropriate changes had been introduced.

Union Carbide has been involved in three more important accidents including another leak, this time of chlorine, at Bhopal and a leak at Institute, West Virginia which put 134 people in hospital. Meanwhile, the battle for compensation from Union Carbide for the 1984 Bhopal accident is still being waged in the American courts.

DECEMBER 2 1984

1

THE REACTION

● MIC is a highly volatile and toxic chemical used in the production of certain pesticides. It will react with almost any impurity to generate heat and carbon dioxide. The heat causes the reaction to go faster, which in turn generates more heat, until a "runaway" reaction occurs. The Bhopal accident began when water entered the MIC storage tank through a leaky valve. The temperature and pressure in the tank started to increase...

2

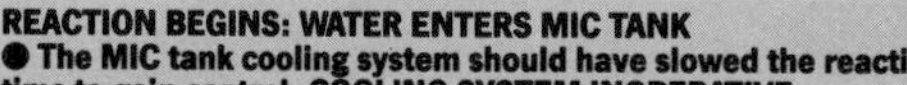

REACTION BEGINS: WATER ENTERS MIC TANK

● The MIC tank cooling system should have slowed the reaction giving more time to gain control: COOLING SYSTEM INOPERATIVE
● The incoming night shift operator should have been informed about the increasing pressure in the tank: NO ACTION
● The MIC tank temperature alarm, which would have detected the heat produced by the reaction, should have been reset: ALARM INOPERATIVE

3

RUNAWAY REACTION: TEMPERATURE AND PRESSURE INCREASE

● The MIC should have been transferred to the spare tank in order to reduce the pressure: NOT TRANSFERRED
● Sufficient gas masks should have been available for the operators in order to reduce the possibility of confusion: NOT SUPPLIED
● The vent gas scrubber should have been able to neutralize the MIC, but it was on standby and incorrectly activated: FAILED

4

SAFETY VALVE OPENS: GAS ESCAPES

● The flare tower which should have been used to burn off the escaping gas was under maintenance: INOPERATIVE
● The water curtain which should have been used to douse the escaping gas was designed to reach 15m into the air. By the time it was activated, one hour after the initial release, the MIC cloud had already risen to a height of 35m: FAILED

5

GAS DISPERSES INTO THE ATMOSPHERE

● The warning sirens should have been used immediately: ONE HOUR DELAY
● An evacuation plan should have been prepared beforehand: NO PLAN
● The civil authorities should have been informed about the toxicity of MIC and suitable emergency procedures: NOT INFORMED
● The population of Bhopal should have been informed about the movement of the gas cloud: NOT INFORMED

6

BHOPAL EXPOSED TO THE GAS

● Doctors should have been briefed on the most effective ways to reduce the risk of exposure; simply breathing through wet towels would have saved many lives: NOT INFORMED
● The medical authorities should have been clearly informed about the antidote for cyanide poisoning (a possible consequence of exposure to MIC), but this was withheld for fear of inducing panic: NOT INFORMED

Exactly how many people died as a result of the accident, no one will ever know. The official tally two years later registered 2,352 deaths, but this is quite clearly an underestimate. Some 8,000 bodies are cremated in an average year at Bhopal's three Hindi crematoria. The number rose in 1985 to 15,000 and this figure does not include the mass cremations that occurred immediately after the accident. The ecological and economic damage is even harder to estimate.

How many died?

DEATHS	2,352-10,000
DISABLED	17,000-20,000
EXPOSED	200,000
EVACUATED	70,000

LATE 1987 — STILL NO COMPENSATION

they damage by disrupting the electrons orbiting around the atoms making up the cells. Like other forms of radiation, beta particles form ions in the cell, which then interact with other atoms in aberrant ways. If the cell is not killed by such radiation, but continues to reproduce itself, it may form a malignant line of cells that ultimately grows into a cancer.

In sufficient intensity, beta radiation can burn the skin and other tissues. Marshall Islanders who were covered in fall-out from US atmospheric weapons testing suffered severe beta burns, as did a number of the firemen who fought the Chernobyl blaze during the accident of April 1986. (See also: *Radioactivity*Ⓝ, *Radiation and health*ⓁⓃ.)

Bhopal

The site of a major chemical accident in Central India, involving a pesticide plant belonging to Union Carbide. A chronology of the accident is given in the accompanying illustration.

Any accurate measure of the number of casualties is impossible for a number of reasons. Many of the bodies were cremated, destroying crucial evidence of the cause of death. Many died some time later and in a different locality, so their deaths were not registered with the Bhopal authorities. Many are still dying today.

Thousands are now suffering from a host of ailments including lung damage, vomiting blood, conjunctivitis, damaged eyesight, pains in the abdomen, lassitude and depression. Women are suffering from similar problems, and on top of these there seem to be gynaecological complications such as menstrual disorders, a white vaginal discharge and some evidence of reproductive disorders. The overwhelming impression is that the victims of Bhopal have lost their hold on health, already tenuous due to poverty and malnutrition, and that this, combined with the symptoms of lassitude and breathlessness, has seriously affected their capacity to work.

At the moment it is impossible to say in what way the victims of Bhopal will be compensated for their loss, but the prognostications are not good. Immediate emergency relief has been implemented but is plagued by lack of funds and by bureaucratic delays. One year after the disaster, less than 1,400 people had received the maximum government compensation grant of $830 for each life lost in the family. This means that the number of people compensated was below the official tally at the time, a gross underestimate anyway. The battle for compensation from Union Carbide, meanwhile, is being waged in the US courts and is likely to go on for years.

At a local and individual level, therefore, the Bhopal disaster is still very much a reality. Any alleviation of the pain and suffering caused by the accident is still in the distant future. At a more general level, the accident seems to have achieved little in the way of making the chemical industry either more accountable or safer.

Other companies have also suffered major accidents, the most notorious being the pollution of the Rhine by a spill from a Sandoz (see *Sandoz fire*) warehouse at Basel. Such events suggest that many more Bhopals lie in store. (See also: *Development*③, *Export of pollution*③, *Pesticides*Ⓟ.)

Bikini Atoll

See: *Atomic tests, Marshall Islands.*

Bilateral aid

See: *Aid.*

Bioaccumulation

See: *Bioconcentration.*

Biochemical oxygen demand

See: *BOD.*

Biocide

"Life-killing" – a term applied to those chemicals used to destroy living organisms that also interfere with, or threaten, human health and activities. They include herbicides (used against weeds), insecticides (used against insect pests), nematicides (used against eelworms and similar animals), acaricides (used against mites), fungicides (used against plant diseases and moulds), and rodenticides (used against rats and mice).

Chemicals such as antibiotics, used in medicine to kill harmful organisms, are not generally included in this group. Some biocides are selective, being most potent against a small number of species, while others are more generally toxic. A wide, spectrum biocide is formaldehyde (*qv*), poisonous to all forms of life and used to sterilize soil in greenhouses before crops are planted. The term pesticide (*qv*) is more generally used in place of biocide. (See also: *Cancer*Ⓛ, *Carbamates*ⓁⓅ, *Chlorinated hydrocarbons*ⓁⓅ, *Herbicides*ⒶⓅ, *Insecticides*ⒶⓅ, *Mercury* ⓁⓅ, *Organophosphates*ⓁⓅ.)

Bioconcentration

Once in the environment, many pollutants tend to accumulate within the tissues of living organisms. This process is known as bioconcentration and is described in the accompanying illustration.

Such bioconcentration of pollutants can have macabre consequences. The biologists Anne and Paul Ehrlich recall the experience of a village in Borneo:

"Some years ago, large quantities of DDT were used by the World Health Organization in a programme of mosquito control in Borneo. Soon the local people, spared a mosquito plague, began to suffer a plague of caterpillars, which devoured the thatched roofs of their houses, causing them to fall in. The habits of the caterpillars limited their exposure to DDT, but predatory wasps that had formerly controlled the caterpillars were devastated.

"Further spraying was done indoors to get rid of houseflies. The local gecko lizards, which previously had controlled the flies, continued to gobble their corpses – now full of DDT. As a result, the geckos were poisoned, and the dying geckos were caught and eaten by house cats. The cats received massive doses of DDT, which had been concentrated as it passed from fly to gecko to cat, and the cats died. This led to another plague, now of rats. They not only devoured the people's food but also threatened them

Bioconcentration – natural mechanisms that kill

Even though bone and fat capture pollutants, many living organisms would be able to tolerate present levels of pollution were it not for the food chain. This pyramid of predation – with plant eaters near the bottom and meat eaters at the top – ensures a steady, cumulative poisoning which peaks with the pyramid. As each creature eats its prey, it takes in almost all of the poison locked in the prey's bone and fat: this is the second phase of bioconcentration and in the North Sea can mean that a seal may carry 80 million times the PCB level of the seawater in which it swims. That level – 160 parts per million – is high enough to induce birth defects and declining fertility rates. In one of the earliest examples of bioconcentration recorded, levels of DDD sprayed on Clear Lake in California in the 1950s to rid it of gnats, ranged from 0.02ppm in the water to 1,600ppm in grebes: thousands of them died.

If pollutants were to pass naturally through living organisms, bioconcentration would not exist. But bone and fat are able to capture some of the most deadly pollutants, bone concentrating radionuclide strontium-90 and fat absorbing PCBs, DDT and other chlorinated hydrocarbons. The consequence is that the levels of pollutants in living organisms are far higher than the background levels in the environment.

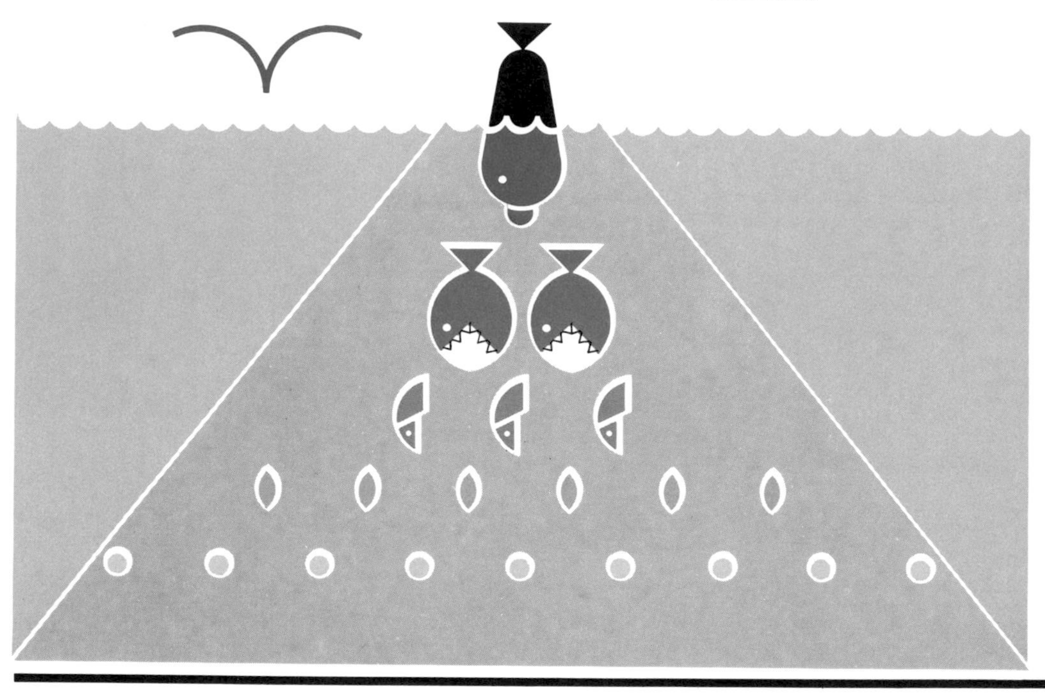

with yet another plague – this time the genuine article, bubonic plague. The Government of Borneo became so concerned that cats were parachuted into the area in an attempt to restore the balance." (See also: *Balance of nature* Ⓕ, *Dilution and dispersal* Ⓟ.)

Biological control

A general term describing a variety of biological techniques used in the control of pests.

In nature, potential pests are all controlled biologically. Thus birds, bats, fish and amphibians consume vast quantities of insects. Birds, reptiles and mammals, such as hedgehogs and moles, consume ground-dwelling invertebrates such as slugs.

Biological control by humans can take several forms. It may mean using a natural predator to control the pest. For example, red spider mite in greenhouses can be prevented from causing economic damage to crops by using a predatory mite. The pest is introduced first, deliberately, and allowed to breed. When its population density reaches an appropriate level, the predator is introduced. The two species maintain a balance so that the pest is always present but causes no harm.

Of course, this does not always work as planned. Introduced predators have often failed to reduce the populations of target species and have themselves proliferated, eventually becoming pests in their own right. Ferrets introduced into New Zealand to control rabbits, for example, proved more interested in native ground-nesting birds, whose numbers they decimated.

Another method of biological control involves disrupting the breeding of pest species. Where females mate only once in each breeding cycle, this may be achieved by breeding a captive population of the pest, separating the sexes, sterilizing the males chemically (or, more usually, by subjecting them to low doses of ionizing radiation), then releasing them at an appropriate time into the wild. There, males and females still mate, but a proportion of the matings are unproduc-

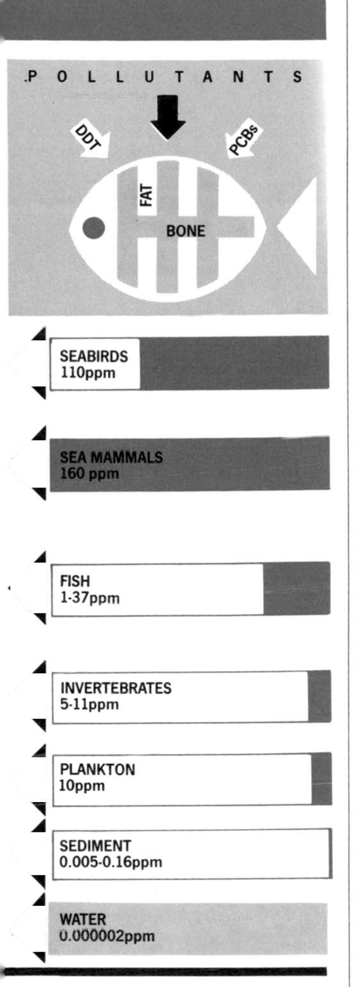

tive and an increase in population growth is prevented – or, at least, limited. However, this, too, can go wrong, the classic case being the release of supposedly sterile medflies in California – the result was a medfly epidemic (*qv*).

A third method of biological control involves synthesizing chemical substances called "pheromones", which many insects release to attract mates. The synthesized pheromones are then used to attract individuals of one or other sex to sterilize or kill them. (See also: *Balance of nature* Ⓒ Ⓕ, *Biocide* Ⓐ Ⓛ Ⓟ, *Pesticides* Ⓐ Ⓟ.)

Biological diversity
See: *Extinction* Ⓒ, *Genetic diversity* Ⓒ.)

Biosphere
The global ecosystem, whose constituent parts are all our planet's living things together with their inanimate (abiotic) environment.

The idea of the biosphere was first conceived by the great French naturalist Jean Lamarck. The term itself was coined in 1873 by the Austrian geologist Eduard Suess, and was later taken up by Edouard le Roy, a French mathematician, and by Henri Bergson, author of *Creative Evolution*.

The founder of the modern theory of the biosphere, however, is taken to be the Russian scientist Vladimir Vernadsky, who published a monograph entitled *The Biosphere* in 1929.

The concept has recently been considerably refined by Professor James Lovelock, who has accentuated the creative nature of Gaia (*qv*), as he refers to it, and its ability to maintain its own homeostasis in the face of potentially disruptive change.

Mainstream science shows little interest in the concept of the biosphere, whose structure and function, indeed whose very existence, is difficult to establish on the basis of accepted, reductionist scientific method. (See also: *Ecology* Ⓕ, *Ecosystem* Ⓕ, *Holism* Ⓕ, *Mutualism* Ⓕ.)

Biotechnology
Generic terms for those technologies that seek to use living organisms (or parts of living organisms) to modify existing forms of life or to create new forms of life.

As such, biotechnology presents both a benefit and a threat to mankind. At its most benign, it consists of nothing more harmful than, say, fermentation or the cross-breeding of plants and animals. At the other end of the spectrum lie genetically engineered organisms, potentially the most dangerous scientific development yet known to man. (See also: *Genetic engineering* Ⓟ.)

Bison
Also called the buffalo, the American bison is a large bovid that once existed in most of North America. When the first Europeans arrived, there were, on conservative estimates, 60 million buffalo. At that time they formed an important part of the economy of the Plains Indians, providing them with food, clothing and shelter.

Slaughter followed the European's colonization of North America. Initially killed for meat and hides, and then for hides alone, the buffalo was by 1900 on the verge of extinction. Only then were the remainder put in government reserves, where they have survived until today. Several thousand animals now exist in managed herds.
(See also: *Extinction* Ⓒ, *Genetic diversity* Ⓒ.)

BOD (Biochemical oxygen demand)
One of the official measures of how polluting a liquid is when it is discharged into a living watercourse. The way in which the BIOCHEMICAL OXYGEN DEMAND (BOD) of an effluent poses a threat to the environment is described in the accompanying illustration.

One of the main aims of sewage treatment is to reduce the BOD of the liquid so that it can be discharged into a river without causing serious pollution. Regulatory authorities concerned with water pollution normally set limits on the volume and BOD of an effluent before it is permitted to be discharged.

Organic matter is present in healthy streams and lakes and this will give the water a BOD of a few parts per million, a load which the system has no difficulty in handling.

Thanks to stricter pollution controls – principally on discharges of sewage – BOD levels in some of the major rivers in Western Europe and North America have been reduced dramatically in recent years. The most publicized success story is that of London's River Thames, which, in the late 1950s, was so polluted with sewage that no fish could survive in the 100-odd kilometres (60 miles) from Richmond to the sea. During the 1960s, new sewage works were built and many old ones were modernized. Gradually, the fish began to return to the river: by 1969, more than 50 species had been identified. Ten

Biochemical oxygen demand (BOD) – sewage that suffocates

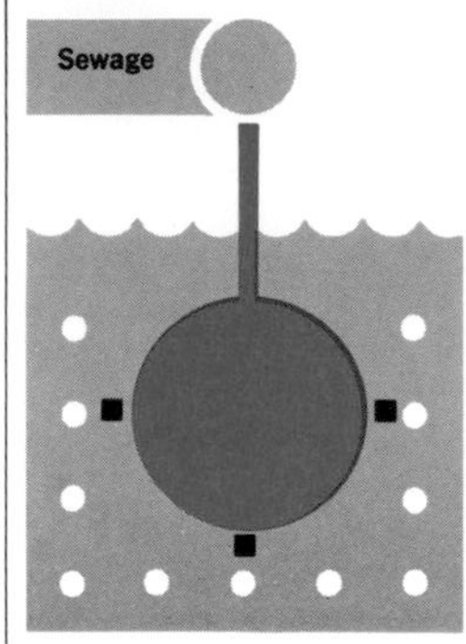

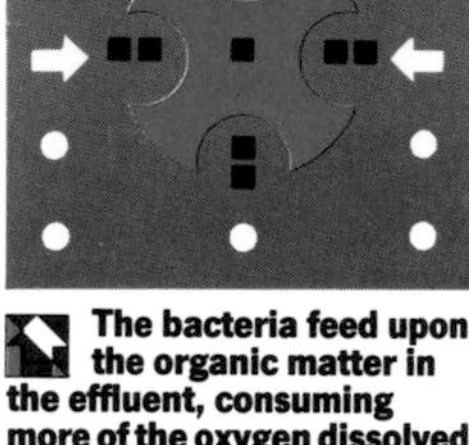

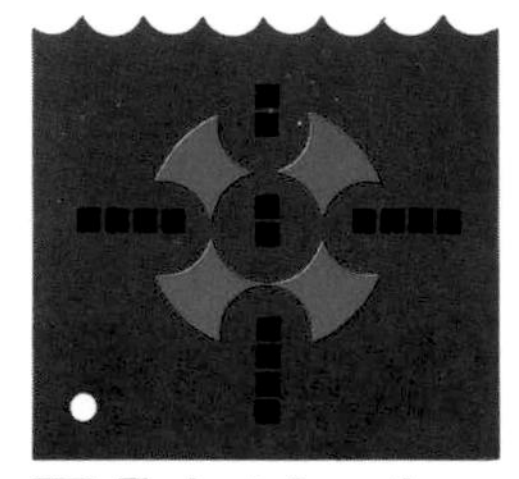

When an effluent such as sewage is discharged into a living waterway, it may provide the bacteria and other micro-organisms living there with such a rich food source that they rapidly reproduce, consuming all the oxygen in the water. Such an effluent is said to have a high BOD.

The bacteria feed upon the organic matter in the effluent, consuming more of the oxygen dissolved in the water. An effluent with a high BOD will do more harm in summer, when flow rates are low and the oxygen level is reduced; less oxygen will dissolve in warm water than in cold.

The bacteria continue to feed and begin to multiply, initially doubling their numbers each generation. As a result the level of dissolved oxygen in the water is further reduced. Warm weather accelerates the whole process by enabling the bacteria to reproduce even faster.

If the BOD of the effluent is too high, or the receiving waterway is unable to dilute it to a safe level, then the oxygen level will fall to such an extent that fish and other aquatic animals suffocate. One of the main aims of sewage treatment is to reduce the BOD of the sewage.

years later, the number was closer to 100. Birds, too, began to flourish – the numbers of swan, pochard, teal, shelduck, dunlin and tern have all risen since the 1960s.

But BOD levels are only part of the story. Organic matter is not the only form of pollution entering water courses: pesticides (*qv*) and nitrates (*qv*) from farmland, leachate (*qv*) from toxic waste dumps and chemicals discharged by industry are among others. As a measure of pollution, BOD can give no indication of the extent to which a waterway may be contaminated with such harmful chemicals – unless, of course, they raise the oxygen demand of the receiving waters. Nonetheless, those chemical pose a potent threat to public health and the environment – many of them being suspected or proven carcinogens (*qv*). (See also: *Eutrophication* Ⓛ Ⓟ, *Hazardous waste* Ⓟ, *Landfill* Ⓟ, *Micropollutants* Ⓛ Ⓟ.)

Browns Ferry

Site of a near-disastrous nuclear accident in 1975. At the time, Browns Ferry, in Alabama, was the world's largest nuclear power station, consisting of two 1,065 megawatt boiling water reactors.

The accident occurred when two electricians were checking the air flow through cable passages beneath the control room with a candle. The candle ignited some foam packing and the fire spread along the cables into the Unit 1 reactor building.

Both reactors were shut down, and the fire burned for seven hours. The station, which had supplied 15 per cent of the Tennessee Valley Authority's power, was out of action for 18 months.

Browns Ferry, rather than Three Mile Island (*qv*), marks the beginning of US disenchantment with nuclear power. The decline was sudden. 1974 was the peak year for US reactor order: the last order that has not subsequently been cancelled was placed in 1975. (See also: *Nuclear accidents* Ⓝ.)

Brundtland Report

Popular name for the report issued by the World Commission on Environment and Development, which was chaired by the Prime Minister of Norway, Mrs. Gro Harlem Brundtland. The commission, set up by the general assembly of the United Nations in 1983, was given the brief of formulating "a global agenda for change". In particular, it was charged with working out a strategy for achieving "sustainable development" (*qv*) for the year 2000 and beyond.

The report sets seven goals for the future: "Reviving economic growth; changing the quality of growth; meeting essential needs for jobs, food, energy, water and sanitation; ensuring a sustainable level of population; conserving and enhancing the resource base; reorientating technology and managing risk; and merging environment and economics in decision making."

On agriculture, it urges land reform, the substitution of "sustainable agriculture" for "chemically dependent" agriculture and a shift from export crops to food crops for local consumption. On energy, it recommends the adoption of soft energy (*qv*) systems, which it describes as "the best way towards a sustainable future." Industry is urged to "do more with less" and "to assess potential impacts of new technologies before they are widely used, in order that their production, use and disposal do not overstresss environmental resources." (See also: *Development* ③, *Sustainable development* ③.)

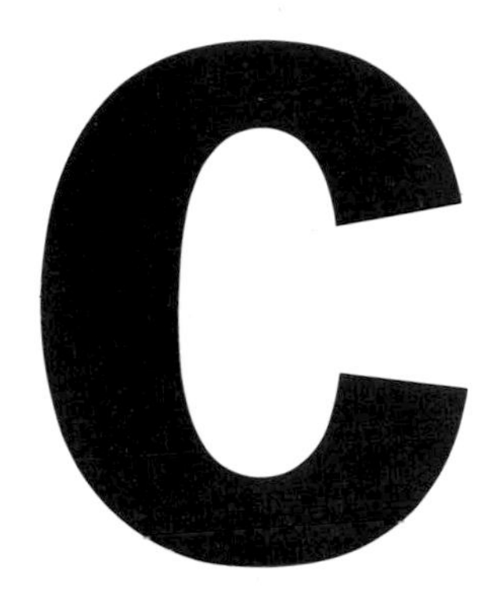

Cadmium

A toxic heavy metal that has had serious effects on human health in a number of incidents around the world. If cadmium compounds are swallowed in any quantity, their irritant and emetic action usually results in the material being expelled from the body before fatal poisoning occurs. If dust or vapours are inhaled, however, pulmonary oedema and death can occur.

Chronic cadmium poisoning causes damage to the kidneys and heart. Prolonged exposure also results in a loss of calcium from the bones, which then become brittle and break easily. In Japan, for example, a whole village was stricken by cadmium poisoning – known as "itai-itai" disease, meaning "it hurts, it hurts" – after industrial effluent contaminated rice supplies. One victim died as a result of 57 fractures, the last one occurring when she was covered by a sheet. Cadmium poisoning has also occurred when people have consumed food or drinks stored in cadmium-plated vessels.

Cadmium is used in metal plating, in certain plastics as a pigment and in some rechargeable batteries. It is naturally present in phosphate fertilizers and is discharged to the environment when they are refined. Zinc and lead (*qv*) smelting also results in cadmium pollution since the metal is usually present in zinc and lead ores. Because of its widespread use, cadmium frequently occurs at high levels in sewage sludge, as domestic wastes are rarely separated from industrial wastes that contain heavy metals. This limits its use as a fertilizer and can cause the metal to bioconcentrate (see *Bioconcentration*) in marine life if the sludge is dumped at sea. Incinerating refuse also disperses cadmium into the environment.

To reduce pollution by cadmium, restrictions on its use are being imposed in Sweden. West Germany has also taken action, banning the use of cadmium in pesticides. In Britain, a ban on cadmium plating in food utensils has been in force since 1956. Tough new controls on the use of cadmium are being contemplated by the EEC. (See also: *Heavy metals* Ⓕ Ⓟ.)

California condor

Once a common sight in the mountains of California during the 19th century, the California condor has since suffered a rapid decline. In the 1930s only one hundred birds remained, and by 1970 there were fewer than half that number. In January 1986 five remained in the wild and by April of the following year the last wild condor, a male, was captured and sent to join a captive breeding programme.

The demise of the condor can be blamed principally on hunting, their position at the top of the food chain and their lengthy reproductive cycle. Size and lethargy have made them tempting targets for the hunter's bullet and a signifiicant number have also died from ingesting lead bullets in animal carcasses. Agricultural poisons, known to cause fragile eggshells and deformed young, have adversely affected reproduction. Condors also cover up to 250km (150 miles) a day in search of carrion and man's inexorable expansion has encroached on the birds' hunting grounds.

Twenty-seven birds are now in captivity, of which 13 are female. The captive breeding programme has caused bitter controversy, though at the preseent rate of decline, it seems to have been the only option available. No birds have yet been raised in captivity, but there is room for optimism. See also: *Extinction* Ⓒ, *Genetic diversity* Ⓒ.

Cancer

A disease caused by a breakdown in the mechanisms governing cell division, organs and tissues with rapid cell division (such as the liver) being particularly susceptible.

Cancer is a "disease of civilization" (*qv*). Its incidence among tribal people living in their natural environment is usually extremely low, often indiscernible. Cancer is also rare in countries that are in the early stages of economic development. Thus, according to the World Health Organization (WHO), the crude rate of cancer in 1985 in West Africa was 94, and in Micronesia and Polynesia 151. By contrast, it was 638 in North America, 748 in Australia, New Zealand and Melanesia and more than 1,000 in Europe. The worldwide incidence of cancer in 1985 was 5,900,000 and the number of deaths was 4,300,000. In the USA, there were 900,000 new cases in 1986, with 450,000 deaths – an incidence of one in three, and a mortality rate of one in four.

Deaths from cancer continue to increase from year to year. According to WHO, between 1960 and 1980 they rose in 28 industrial countries by 55 per cent among men and 40 per cent among women. A five per cent decline in deaths from cervical cancer and a 12 per cent decline in stomach cancer among men (15 per cent among women) were more than made up by a 60 per cent increase in breast cancer and a 116 per cent increase in lung cancer among men (200 per cent among women).

In the USA, the incidence of cancer has increased steadily since the 1930s by about some two per cent a year and the mortality rate by some one per cent. For white males born in 1985, the probability of developing cancer is 36 per cent, and of dying from it, 23 per cent. For those born ten years earlier the probabilities are lower – 30 per cent and 19 per cent respectively.

The increased incidence of cancer is officially attributed to increased longevity, smoking and the consumption of fats and alcohol. Environmental pollution has been consistently played down as a cause of cancer. The chemical industry in Britain, for example, argues that only five per cent of cancers are caused by occupational exposure to carcinogens.

That view, however, is contested by a number of scientists, notably Professor Samuel Epstein of the School of Public Heath, University of Illinois. "Increased longevity is certainly a factor to take into account, but it must not be exaggerated since the cancer rate has seriously increased among children and is now an important cause of their deaths. So, too, diet and tobacco are important causes of cancer, but to single them out as the primary causes is to detract from the lethal impact of everyday exposure to carcinogenic chemicals."

Support for the view that pollution is a major cause of cancer comes from a variety of studies. In 1978, a US commission, appointed by the Secretary of Health, Education and Welfare, estimated that 38 per cent of all future cancers would reflect exposure to just six widely used occupational carcinogens. Some experts suggest that 80 per cent of cancers may have an environmental cause.

Studies show significant clusters of cancer in areas with a high density of chemical factories and toxic waste dumps. The Chesapeake Bay area is a case in point. In one locality, where the per capita level of toxic waste generated is 46 times greater than the national average, deaths by cancer are several times higher than the US average.

The cancer rate on the Chicago south side, a highly polluted area containing 22 chemical plants in addition to 31 operating or closed chemical waste dumps, was found by the Illinois Environmental Protection Agency in 1984 to be 20 per cent higher than in the rest of Chicago.

In the UK, the incidence of cancer of the lung and stomach is particularly high in northern industrial cities such as Liverpool, Southport, Manchester and Jarrow.

The carcinogenicity of chemicals and heavy metals such as radon gas, nickel ores, chromium, arsenic, mustard gas, vinyl chloride and asbestos has been well established by epidemiological studies among workers exposed to these carcinogens at their place of work. The carcinogenicity of a large number of the 9,000 or so synthetic organic chemicals, in which category most modern pesticides must be included, has also been established by the US Environmental Protection Agency (EPA) and other regulatory agencies.

A further consideration is that at least 20 per cent of the 100,000 people who die from lung cancer in the USA are non-smokers, among whom the lung cancer rate doubled between 1958 and 1969 and continues to increase.

Studies among animals also show pollution to be a cause of cancer. Extremely high rates of cancer have been found, for instance, among fish living in waters contaminated with agricultural and industrial pollutants. Thus, recent research shows that 40 per cent of the flat fish in certain parts of the North Sea now have cancer. Studies have shown that the incidence of cancer among tomcods more than two years old in the highly polluted Hudson River is 80-90 per cent, whereas among tomcods living in a clean environment it is no higher than 2.5 per cent.

The official view is that considerable progress has been made in the treatment of cancer. This, too, is contested by many critics. The overall cancer "cure" rate in the USA, as measured by survival for more than five years after diagnosis, is at present 50 per cent for whites and 38 per cent for blacks. It does not seem to have increased in the last decades, even though some progress has been made in the treatment of cancer of the cervix and a few rare cancers such as those of the testes, Hodgkin's disease and childhood leukaemia. The generally accepted treatment – chemotherapy – has serious side effects, and even when effective can increase the subsequent risk of developing a second cancer by up to 100 times. (See also: *Additives* Ⓛ Ⓟ, *Biocide* Ⓛ Ⓟ, *Carcinogen* Ⓟ, *Cancer research* Ⓛ Ⓟ, *Junk food* Ⓛ Ⓟ.)

Cancer research

In 1971, the US Government, by an Act of Congress, declared war on cancer. Researchers were confident that, with sufficient funds, they could establish the exact mechanism of carcinogenesis (see *Carcinogen*), and that with a budget of just $1,000 million a year for ten years, they could find a cure.

Funding for cancer research was thus increased from $230 million a year in 1971 to $815 million in 1977 (from £144 million to £509 million), and, at one time, some 7,000 scientists were employed on the programme.

However, despite 25 years of research, only moderate progress has been made either in understanding carcinogenesis or in the treatment of cancer. Meanwhile, the incidence of the disease, except in cases of stomach cancer and cancer of the cervix, has continued to increase.

This would suggest that the research has been misdirected. Much of it has been concentrated on the role played by viruses in carcinogenesis, on the development of a vaccine, and on chemotherapy.

On the other hand, little effort has been made to prevent cancer by reducing the public's exposure to carcinogenic chemicals and low-level radiation – the two principal known causes of cancer. Indeed, until the passing of the Toxic Substances Act of 1976, which was long delayed by chemical industry lobbying, there was no requirement in the USA for chemical companies to test any chemical before it was put on the market. Even today, the number of chemicals that have been seriously screened for their possible carcinogenic effects is a minute proportion of those that have been sold commercially.

Research on a cure is more glamorous and carries greater scientific kudos – Nobel prizes for example – than research on prevention. It also leads to the development of marketable products – such as anti-cancer drugs. Prevention does not. But most important of all, a policy of prevention, if actively pursued, would seriously compromise the viability of many industries by requiring a ban on the production, sale and use of carcinogenic substances. Among those that would be jeopardized are the chemical industry, the asbestos (*qv*) industry, the food-processing industry and the tobacco (*qv*) industry. Small wonder that they are most active in lobbying against such efforts at establishing a policy of cancer prevention. (See also: *Cancer*Ⓛ, *Delaney Amendment*Ⓛ, *Proposition 65*Ⓛ, *Radiation and health*ⓁⓃ.)

Candu reactor

A type of nuclear reactor developed by Canada. After World War II Canada decided not to manufacture atomic weapons, but continued work on power reactors using natural (unenriched) uranium, which it has in abundance, and heavy water as moderator (*qv*) and coolant. The result was the Candu reactor, the first units becoming operational in 1971 at Pickering, near Toronto. The Pickering power station will eventually comprise eight 500 megawatt reactors; another station at Bruce will have eight 750MW units. Candu reactors have been exported to India, Pakistan, South Korea and Argentina.

The design of the reactor is unusual in that fuel is contained in a large number of pressure tubes, rather than in a single pressure vessel. These are enclosed in a large steel tank, called a calandria. Heavy water under pressure passes through the tubes, and then to steam generators. Like Magnox reactors (*qv*), Candu reactors can be refuelled on load.

A British 100MW prototype pressure tube reactor, called a steam generating heavy water reactor (SGHWR), closely similar to Candu, except that it uses enriched uranium, has been operating at Winfrith since 1967. It almost became the basis for a modest programme of commercial heavy water reactors, because by 1974 it was apparent that the advanced gas-cooled reactors (AGRS) (*qv*) were proving difficult to build. The government, unwilling to order American pressurized water reactors (PWRS) (*qv*), decided on SGHWRS instead. In 1976 the decision was reversed, and two more AGRS were ordered instead.

Cap de la Hague

Nuclear reprocessing plant, situated on the coast of Normandy, and used to reprocess spent fuel from France's civil nuclear programme and from other countries, such as Japan.

Known officially as UP2, or Usine Plutonium 2 (UP1 is a military plant at Marcoule), Cap de la Hague opened in 1966, reprocessing Magnox (*qv*) fuel.

In 1969, France opted for pressurized water reactors (PWR) (*qv*) and built a new plant, known as the HAO plant, to enable Cap la Hague to handle the higher burn-up oxide fuel from PWRS. The HAO plant began operating in 1976.

Initially, it was intended that the plant should reprocess 800 tonnes a year, but this target was deemed impractical and the plant was finally designed for a throughput of 400 tonnes a year. Even so, the plant had considerable spare capacity. This allowed COGEMA, the company that runs Cap de la Hague, to bid for reprocessing contracts from abroad. In this they had some success because the only competition – the B204 Head End plant at Sellafield (*qv*) in the UK – had been closed following an accident in September 1973.

Within months of opening, the HAO plant was closed following a strike by process workers demanding better radiological protection. Ironically, workers at Cap de la Hague are exposed to far lower levels of radiation than workers at Sellafield – this despite the great difficulties of handling oxide fuel. Discharges to the sea from Cap de la Hague are much lower than from Sellafield – those for plutonium being 50 times lower for the same throughput.

The HAO plant has also been plagued by technical problems. By 1985, only 1,000 tonnes of PWR fuel had been reprocessed. Ignoring the first two years of operation (which suffered severe teething problems), an average of 140 tonnes a year were reprocessed from 1978 to 85.

That low throughput strongly suggests that the problems of reprocessing spent PWR fuel have yet to be solved. Indeed a 1982-83 report by the Castaign Commission (a government-appointed group of independent experts) came to the unequivocal conclusion that reprocessing in France is in a state of crisis.

Since then, little has happened to alter the commission's conclusions. Though two new plants, UP-3A and UP2-800, both with a planned throughput of 800 tonnes a year, are under construction, building work is apparently behind schedule, and it is unlikely that they will come into service until 1922–five years later than planned. A third new plant, UP3-B, was cancelled when the expected volumes of foreign business failed to materialize. With its current facilities, Cap de la Hague is expected to reprocess no more than 250 tonnes per year of PWR fuel between 1985-90.

The difficulties in reprocessing spent PWR fuel does not bode well for France's fast breeder reactor (*qv*) programme. Reprocessing fast reactor fuel is far more problematic than reprocessing PWR fuel. In particular, plutonium losses during reprocessing must be kept down to a few tenths of one per cent if there is to be a net gain of plutonium. This is ten times lower than has yet been achieved on a commercial scale. Yet, without efficient reprocessing, the idea of fast reactors "breeding" enough fuel for their replacements will prove a pipedream. (See also: *Reprocessing*Ⓝ.)

Captan

A fungicide used as a spray, dip or seed treatment against a variety of plant diseases, including those found on garden plants. Rooting compounds used for plant propagation sometimes contain it.

It is not an acute poison, but it is an irritant to the skin, lungs and eyes. More seriously, captan has been shown to cause mutations in bacterial and several other types of cells, including human cells. It has also been shown to be carcinogenic (see *Carcinogen*) in mice and teratogenic (see *Teratogen*) in rats, rabbits and hamsters. There are few specific controls on its use. (See also: *Biocide*ⓁⓅ, *Cancer*Ⓛ, *Mutagen*ⒻⓁⓅ, *Pesticides*ⓁⓅ.)

Car emissions

The internal combustion engine produces many different substances in its exhaust, several of which are serious pollutants. Carbon monoxide, an odourless and highly poisonous gas, can build up to dangerous concentrations if a car engine is run in confined spaces or if the car's windows are closed and a leak in its exhaust system allows fumes to enter it.

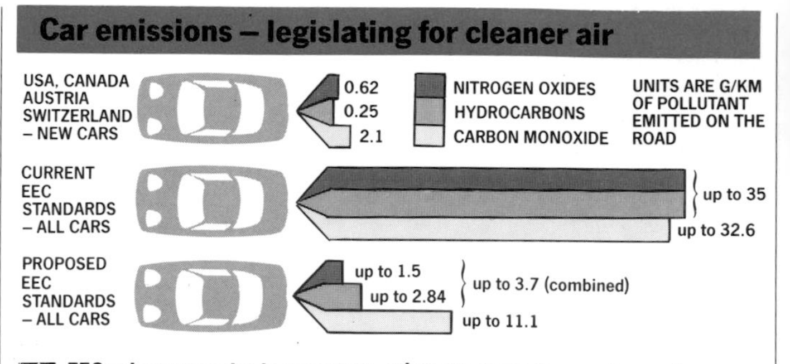

EEC exhaust standards are not as strict as the American ones, even allowing for the fact that they use a different test. A number of European countries, including Sweden from 1989, have opted to abide by the stricter US standards.

Under sunny conditions, nitrogen oxides and hydrocarbons combine with other atmospheric gases to produce ozone. In the upper atmosphere ozone forms a protective screen against harmful rays from the Sun, but in the lower atmosphere it contributes to the photochemical smogs that now pose a serious threat to human health in many large cities around the world from California to Greece.

Unburned hydrocarbons give motor exhausts their unpleasant smell and, with particles of incompletely burned fuel, form part of the visible smoke. Both these groups of substances are carcinogenic (see *Carcinogen*). Nitrogen oxides are emitted by all car engines, but those operating at higher temperatures produce most. They are irritant and poisonous gases.

The compounds listed above are known as primary pollutants, since they are polluting when produced. But hydrocarbons, nitrogen oxides and other components of the atmosphere combine, under the influence of sunlight, to form the secondary pollutant ozone (see *Ozone layer*). This is extremely poisonous and a powerful irritant in the lower atmosphere. Ozone, and the compounds it forms when it reacts with other pollutants, is one of the main ingredients of the damaging photochemical smogs afflicting heavily trafficked sunny cities such as Los Angeles, Athens and Sydney.

Many car engines still emit an extremely dangerous additional pollutant, lead. This is added to gasoline to improve its performance and is dispersed as fine particles in the exhaust. High levels of lead are now found in the general environment as a result of the use of leaded fuel. (See also: *Acid rain*Ⓟ, *Calatytic converter*Ⓟ, *Lead*Ⓟ.)

Carbamates

Organic nitrogen compounds, many of which are used as pesticides. The commonest types are the carbamate insecticides, which were developed in the 1960s and onward, providing new chemicals to be used against a wide variety of insect pests, including those that had become resistant to DDT (*qv*) and organophosphates (*qv*). Resistance to carbamates, however, has also now become a problem in some areas.

Carbamate insecticides range in hazard from the moderately toxic carbaryl to the highly poisonous aldicarb (*qv*). They resemble organophosphorus compounds in that they attack the nervous system by inhibiting the substance acetylcholinesterase.

Some carbamates are used as fungicides, a notable example being zineb–zinc ethylenebis(dithiocarbamate). This is a recognized carcinogen (*qv*) and, like many compounds in this group, is a powerful skin irritant. Several other chemicals in the carbamate group are suspected of causing lung cancer.

They are unselective and can readily kill non-target organisms, especially bees, but most do not leave long-lasting residues in the environment.

Carbaryl is used in insecticidal shampoos to kill head lice as well as in gardening and agriculture. (See also: *Biocide* Ⓟ, *Cancer* Ⓛ *Pesticides* ⓁⓅ.)

Carbon cycle
The circulation of carbon in the biosphere (*qv*). As a building block of the molecules that make up living cells, carbon is an essential element for life on Earth. Carbon is thus a basic ingredient of proteins, carbohydrates, hormones, bones, connective tissue and nerve cells. Equally important, as a primary component of carbohydrates such as sugar and starch, carbon is directly involved in photosynthesis and respiration, the processes by which energy from the sun in the form of light is used to build up energy-rich molecules. These molecules are then used at a later date to release their energy to the cell. The fossil fuels–coal, petroleum and natural gas–are all carbonaceous compounds which at a previous time in the Earth's history have been built up by living cells through photosynthesis.

Carbon, whether as a gas, liquid or solid, moves between the atmosphere, oceans, rivers and landmass through both geophysical and chemical processes and with living organisms acting as a powerful intermediary. Thus carbon in the atmosphere as carbon dioxide (*qv*) and carbon monoxide gas can get washed out as carbonic acid, which interacts with rocks to form bicarbonates, which flow into rivers and into the oceans. Equally, the oceans absorb carbon dioxide gas directly from the atmosphere. Living organisms affect the carbon cycle by accelerating the rate at which carbon dioxide interchanges between land, sea and the atmosphere. Marine algae such as coccolithophorids and other diatoms form elaborate limestone shells from bicarbonate, as do corals, and as solid limestone gradually sink to the ocean floor to form in time the limestone cliffs that are a prominent feature of both coastlines and inland hills. On land, the formation of peat, brown coal and black coal are also sinks for carbon dioxide. Meanwhile, if photosynthesis and respiration are in balance, there is no net change in the carbon dioxide balance in the atmosphere.

Geophysical processes such as volcanoes, spew out carbon dioxide as well as other gases, and tectonic plate movements, which can thrust limestone deposits on the ocean floor up onto land, are means by which carbon returns into circulation, as is the burning of fossil fuels. (See also: *Balance of nature* Ⓕ, *Carbon dioxide* Ⓟ, *Gaia* Ⓕ, *Greenhouse effect* Ⓕ Ⓟ.)

Carbon dioxide
Carbon dioxide is a natural component of air and a vital part of the carbon cycle (*qv*). It is produced when animals respire and when any material containing carbon is burned. It is also released by volcanoes. Plants use carbon dioxide as a food and, with the addition of water and sunlight, convert it into sugars during photosynthesis. Photosynthesis is ultimately the source of food for all but a few types of living organisms.

Carbon dioxide is not toxic but can suffocate since it cannot support life. It is used in fire extinguishers because it will not support combustion.

The build-up of carbon dioxide in the atmosphere now threatens to destabilize the world's climate, precipitating the so-called greenhouse effect (*qv*). The principal causes of that build-up are industrial emissions resulting from the combustion of fossil fuel and the cutting down of the Earth's tropical rainforests, a major sink of carbon dioxide. (See also: *Gaia* Ⓕ, *Greenhouse gases* Ⓟ, *Tropical forests* Ⓒ Ⓕ)

Carcinogen
Any agent that causes cancer (*qv*). Most carcinogens are also mutagenic (see *Mutagen*) and teratogenic (see *Teratogen*).

Even when a substance is supected of being carcinogenic, it is often hard to establish that it is indeed a carcinogen in the strict sense of the word. One reason is that many substances that are not carcinogenic in themselves can be transformed by living organisms into carcinogens – an example being the conversion of nitrite to nitrosamine in the human gut.

Equally, many substances act only as cancer promoters; that is, they do not cause cancers but instead promote the growth of tumours that have previously been triggered by other substances.

The long period – known as the latency period – between exposure to a carcinogen and the appearance of clinical symptoms of cancer also makes it hard to pin down a suspect substance as the cause of a cancer. For leukemia, the latency period is six to seven years, while for many other cancers it can be up to 40 years.

A further problem is that a carcinogen may cause clinical symptoms of cancer only in the offspring of those exposed. Thus, it was the daughters of women exposed during pregnancy to the hormone diethylstilbestrol (DES) (*qv*) who contracted vaginal cancer – not their mothers.

In addition, it is difficult to establish epidemiologically which substances have caused a cancer, when, in the highly polluted environment in which we live, the average person is exposed every day to a large number of known or suspected carcinogens.

The problem is compounded by synergic reactions between the different chemicals. It is known, for instance, that uranium miners (who already suffer from a high incidence of lung cancer) have an eight times higher chance of contracting the disease if they also smoke.

A further difficulty is that many carcinogens are harmful at levels so low they are very difficult to measure, or – like dioxin (*qv*), for example – can be measured only at a prohibitive cost.

Finally, efforts to identify carcinogens have been hampered by the low priority given by governments to research into the possible carcinogenicity of chemicals in the environment.

Even when the carcinogenicity of a chemical has been established, it is difficult to have it removed permanently from current use, often because of the extremely strong pressure exerted on government by industry. Thus, though the carcinogenicity of various coal-tar dyes (*qv*) has been well established, their use (in many varieties of soft drinks for instance) is still tolerated in most countries.

Since indisputable scientific evidence for the carcinogenicity of a chemical is so difficult to obtain, the only responsible course of action must be to ban all suspect carcinogens unless their safety can be demonstrated by exhaustive tests (including tests for additive and synergic (see *Synergy*) effects), carried out over appropriate periods of time, on sufficiently large samples. Where a suspect substance forms part of a specific chemical group – the phenoxy herbicides (*qv*), for example – the entire group should be regarded as suspect. (See also: *Cancer research* Ⓛ.)

Caribbean Sea
See: *Regional seas*.

Carrying capacity
The optimum population that a climax ecosystem can sustain without interfering with the ecosystem's basic structure, and hence undermining its stability.

Today, the impact of man's activities is often greater than the land can support. A recent study by UNESCO examined the impact of modern agriculture on the world's arid lands and concluded that in every region it looked at, the impact was greater than the land could support for long. It is not surprising, therefore, that these lands are everywhere subject to erosion (*qv*) and desertification (*qv*), an unequivocal sign that the carrying capacity of the land in question has been exceeded. This is not only true in the dry tropics – which are particularly vulnerable to degradation – but also to the much less vulnerable lands of the temperate areas.

Much emphasis is now placed on the need for "sustainable" development in the Third World. It is difficult to see how any development projects can be "sustainable" in areas where the carrying capacity of the land has been exceeded. In such areas, ecological rehabilitation can be achieved only by reducing the impact of man's activities.

Sustainability, in fact, implies de-development and not development, an inescapable conclusion but one that is not even suggested today because it is irreconcilable with present political and economic activities. (See also: *Development* ③.)

Cash crops
Crops grown for sale on the urban market or for export. Cash crops have long been promoted as essential to Third World development. The crops grown include such foods as coffee, cocoa, and vegetables, and other non-food items such as tobacco or cotton.

In those countries lacking mineral or oil wealth, cash crops are a major source of foreign exchange, without which it would be impossible to import the technology deemed necessary for development.

Critics argue, however, that by using land that could otherwise be used to grow food for local consumption, cash crops are a major cause of world hunger.

Vast areas of the Third World have been turned over to growing cash crops. In West Africa, for example, 70 per cent of Gambia's arable land and 55 per cent of Senegal's is used to grow peanuts. In Niger, the amount of land used for peanuts expanded fivefold – from 730sq km (290sq miles) to 4,320sq km (1,670sq miles) – in the 30-odd years between 1934 and 1968, the eve of the Sahel famine (see illustration).

In Africa as a whole, the production of coffee has quadrupled since the mid-1960s, while the output of sugar has tripled and that of cotton and cocoa doubled. The production of tobacco has risen by 60 per cent. Those increases were achieved by expanding the area under production rather than by increasing yields.

Significantly, a 1986 study reveals that despite food aid and cereal imports, the Third World as a whole is a net exporter of food to the industrialized world. The author of the report, Professor George Kent of the University of Hawaii, points out: "Of the food entering into international trade in 1976 ($123,650 million worth) 11.9 per cent went from richer to poorer countries, while 20.2 per cent went from poorer to richer." Moreover, the food exported from Third World countries was "on balance" of a higher quality than that imported by them.

Many cash crops are ruinous to the soil. Land used to grow peanuts, for example, typically requires a minimum of six years of lying fallow to recover its fertility. If that fallow period is not observed, the soil is quickly depleted of nutrients. One study in the Caramance region of Senegal found that land used to grow two successive crops of peanuts had lost 30 per cent of its organic matter by the end of the second year. Yields dropped by 20 to 40 per cent.

With the best agricultural land being used to grow cash crops, Third World peasants are increasingly forced to use marginal land to grow food for their own consumption. The result can be devastating to the environment. In the Sahel region of West Africa, the northward march of the peanut pushed local nomads onto desert margins, leading to widespread overgrazing.

Once locked into the cash economy, peasants are particularly vulnerable to the vagaries of the world commodities market. In Thailand, for example, thousands of square kilometres of forest were cleared between 1975 and 1985 to grow tapioca for export to the EEC as feed for pig and cattle. With beef and pork surpluses in Europe forcing a cut-back in production, the tapioca market collapsed, forcing many Thai peasants into debt bondage. Many have literally sold their children into child labour (*qv*).

Much of the foreign exchange earned from growing cash crops is spent on huge development projects – large dams for example – that bring few rewards for Third World peasants, benefiting instead the small urban and rural elites. Vast sums are also spent on buying Western consumer goods and armaments – $165,000 million being spent on defence by Third World countries between 1972 and 1982. Egypt alone spends $800 million a year servicing its military debt, while Argentina's debt eats up 61 per cent of its annual export earnings.

Many development activists in the Third World now argue that "delinking" from the cash eco-

Cash crops – the road the Sahel took to ruin

The apparent cause of the Sahel tragedy of the 1970s was overgrazing by nomadic herdsmen, but the real cause was dependence on a single cash crop – peanuts – at the expense of traditional husbandry. When drought struck, 100,000 people died.

In essence, income from peanuts persuaded farmers to interfere with an age-old system that was designed to provide for hard times. They were encouraged to do so by private interests seeking profit and by the French government which, through the 1960s and 1970s, led European resistance to massive competition from US soya bean imports. Niger, one of the most severely affected of the Sahelian countries, had 1,420sq km under peanuts in 1954; by 1968, the year the drought began, it had 4,320sq km.

By the 1960s, farmers were heavily dependent on peanuts for large portions of their income. Food imports became necessary, but the terms of trade were declining. France offered improved seed, and the fertilizer it needed, but it was not used, partly because farmers were already over-committed in payments for the new seed and new technology, partly because of poor publicity, and partly because of bad distribution. And yet the soil on which the peanuts were grown needed fertilizer: the crop so exhausts the soil that after three years the land must be left fallow for six.

A situation in which the farmers faced declining prices, increasing debt and declining soil fertility was made worse when, in 1965, France was obliged, by Niger's associate membership of the EEC, to begin reducing its guaranteed price. Between 1967 and 1969 the price fell 22 per cent. To maintain reasonable living standards, peanut production had to be increased. The farmers were now on a treadmill.

Their solution was to bring into production the land that had traditionally lain fallow as a buffer against crop failure; and to push north toward the desert occupied by the nomadic herdsmen. The traditional interplay between farmer and herdsman (see below) was already breaking down; when drought came the expansion was to prove a recipe for disaster.

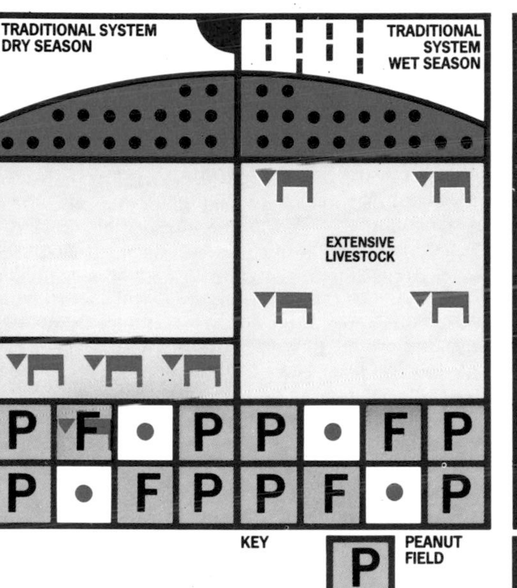

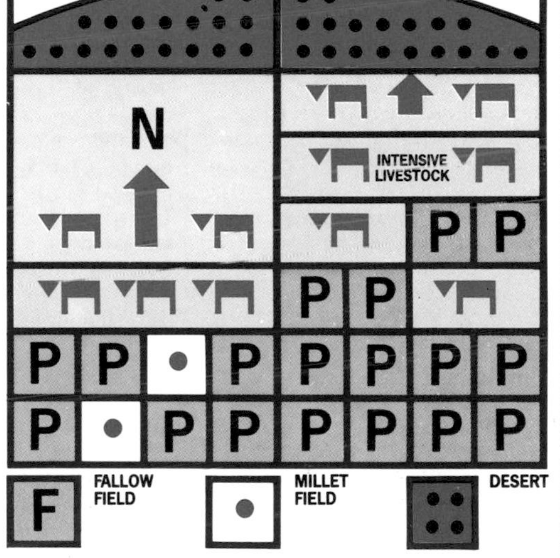

KEY: P PEANUT FIELD; F FALLOW FIELD; MILLET FIELD; DESERT

Traditional husbandry in the Sahel hinged on interaction between southern pastoralists and northern nomadic herdsmen. The pastoralists grew millet and sorghum as staple foods and bartered them for animal products, and fertilization of their land by the herds. As an insurance against crop failure, certain zones of land between villages were left fallow. It was these fallow zones which, in the 1950s and 1960s, were increasingly used for peanut cultivation.

The immediate effects of this use of fallow land were to deprive herdsmen of grazing for their cattle during the dry season and to deprive the pastoralists of natural fertilizer. Artificial fertilizer was available, but – for reasons given above – was not used. The farmers' only choice was to exploit even more fallow land.

With the decline of the caravan salt trade at the end of World War I, the herdsmen's buffer against bad times had become large herds. They had also been encouraged to increase the size of their herds by French-sponsored attempts to develop a fresh meat industry. The herds needed water, and wells were dug; but along the trek routes to them vegetation was destroyed and soil erosion increased. Overgrazing began in the 1950s.

nomy is the only hope for improving the status of Third World peasants. A recent study from Tanzania reveals that the health of villagers improves dramatically when the national economy is depressed. The reason is clear enough: with the government having less money to spend on roads, the national distribution system breaks down. As a result, villagers can no longer sell their cash crops and therefore begin to grow food for their own consumption. (See also: *Dams* Ⓐ ③, *Development* ③, *Green Revolution* Ⓐ ③, *Irrigation* Ⓐ ③, *International Monetary Fund* ③, *World Bank* ③.)

Catalytic converter

A device fitted to motor vehicle exhausts to destroy some harmful gases produced by the engine. Expensive metals such as platinum are used to bring about this process, which takes place in two stages. First, nitrogen oxides are reduced to nitrogen (a normal constituent of air) and then carbon monoxide and unburned hydrocarbons are oxidized to harmless products. For the system to work efficiently the proportion of air in the exhaust must be carefully controlled and electronic devices are used to ensure this. Some catalytic converters – known as oxidation catalysts – simply destroy carbon monoxide and unburnt hydrocarbons so that other techniques, such as lean-burn engines, are required to control nitrogen oxides. Catalytic converters are poisoned by lead, so vehicles fitted with them must run on unleaded fuel. They are also fairly expensive and need to be used properly, since poor maintenance reduces their efficiency. This is a serious problem as, in practice, necessary maintenance is not always done. Catalytic converters must be fitted to all new Japanese and US cars by law, and West Germany is also introducing them. (See also: *Acid rain* Ⓟ, *Lead* Ⓟ.)

CFCs

See: *Chlorofluorocarbons*.

Chernobyl

Soviet nuclear power plant, located some 80km (50 miles) from Kiev, in the Ukraine, where a major nuclear accident occurred on April 26 1986.

The accident, caused by an experiment involving the deliberate switching off of safety systems, resulted in the deaths of at least 31 people. More than 200 others suffered acute radiation sickness, 135,000 had to be evacuated from the area, and radioactivity fell across Europe from Scandinavia to Greece. Estimates of the number of cancer deaths likely to occur as a result of the accident, vary. According to Britain's National Radiological Protection Board, only 2,000 people in the EEC are likely to develop cancer over the next 50 years. Others put the figure far higher. A study by the US Department of Energy estimated that there will be 39,000 extra cancer deaths, 21,000 of them in Europe, while Professor John Gofman of the University of California calculated that there will be a million or more victims.

Prior to Chernobyl, the nuclear industry had claimed that the chances of a serious accident occurring at a nuclear power plant was no more than one in a million "reactor years" – or even one in ten million reactor years. That claim is now seen to be false. Indeed, since Chernobyl, the US Regulatory Commission has estimated that, before the end of the century, there is one chance in two that a serious nuclear accident will occur in the USA. Since the USA possesses about a quarter of the world's nuclear reactors, we can expect another "Chernobyl" every eight years.

In July 1987 six former senior officials and technicians at the plant were sentenced to terms of imprisonment in labour camps for blatant violation of the safety regulations. (See also: essay on *Nuclear Energy after Chernobyl*, *Nuclear accidents* Ⓝ, *Radiation and health* Ⓛ Ⓝ.)

Child labour

According to the International Labour Organization (ILO), some 75 million children between the ages of eight and 15 are forced to work for a living. In India alone, there are more than 16 million child labourers. In Thailand, where the figure reaches more than one million, the number grew by six per cent between 1983 and 1984.

Many children are forced to work, even as prostitutes, to help pay off their family debts. In most cases, their wages are paid in advance, so they are in effective "debt bondage" to their employes. Working conditions are generally atrocious and pay is minimal. Many children are treated as virtual slaves.

The problem of child labour has been exacerbated by the increasing indebtedness of Third World peasants – in many cases, the direct result of the modernization of agriculture through the Green Revolution (*qv*). The high costs of such agricultural inputs as fertilizers and pesticides, and the low prices commanded by their produce, have left peasants in debt to money-lenders for several generations. Environmental degradation – particularly deforestation – has compounded the problem, with droughts and floods causing crop losses and thus increasing indebtedness. (See also: *Cash crops* Ⓐ ③, *Development* ③.)

China syndrome

Nickname for an uncontrolled meltdown in a nuclear reactor, in which the loss of reactor coolant and the failure of emergency systems would be followed by the molten nuclear reactor core, still generating immense heat, falling through the floor of the reactor building and burning its way down – "toward China". (See also: *Nuclear accidents* Ⓝ, *Three Mile Island* Ⓝ.)

Chipko movement

A grassroots campaign opposing government deforestation (*qv*), which originated in the Himalayan foothills of the Indian state of Uttar Pradesh in the early 1970s. "Chipko" means "to embrace" in Hindi, and the movement is so called because women embrace trees to prevent their being felled. The movement draws on a long history of village-led opposition to commercial felling of both virgin and communal forest and to restrictions on villagers' access to these. It also owes a great deal to Mahatma Gandhi's philosophy, both in terms of its commitment to non-violent resistance and in terms of Gandhi's village-oriented approach to economic development.

The Chipko campaign resulted in some early success, culminating, in 1980, in a meeting between one of the movement's two leaders, Sunderlal Bahuguna, and the late Prime Minister, Indira Gandhi, after which a 15-year ban was imposed on commercial green felling in the Himalayan forests of Uttar Pradesh. This gave the movement time to extend its support base. Bahuguna undertook a 4,780km (3,000 miles) march to carry the Chipko message to villagers along the whole length of the Himalayan range. At the same time, Chandi Prasad Bhatt, the other leader, initiated a massive tree-planting effort involving local people.

Initially, the Chipko movement was concerned with the just allocation of rights to exploit the forests. As time has passed, however, it has matured into a fully-fledged environmental movement. Underpinning the Chipko philosophy today is the recognition that forests are a fundamental resource, ensuring the viability of both human life and wildlife in the Himalayas. Their importance to humans lies in the fact that they prevent soil erosion, thus preserving precious agricultural land. They retain water, reducing fluctuations in the supply of water and preventing flash floods. They are also a crucial source of fuel and fodder for the villagers. In the words of one of the Chipko movement's basic slogans, "What do the forests bear? Soil, water and pure air." This important message is spreading to other parts of the Indian subcontinent, such as Karnataka, Rajasthan, and Orissa. Such success gives some substance to the hope that the rapid deforestation of India might also be checked and even reversed. (See also: *Eucalyptus* Ⓒ ③.)

Chlordane

A toxic and persistent chlorinated hydrocarbon (*qv*) insecticide. It is readily absorbed through the skin and has caused fatal poisoning in humans as a result. The nervous system is the principal target, with respiratory failure being the cause of death. It is a suspect carcinogen (*qv*) and teratogen (*qv*).

Chronic chlordane poisoning causes damage to the liver and kidney and the material has caused cancer in laboratory animals. An association between chlordane exposure and leukemia in humans has been demonstrated and there is evidence suggesting that chlordane, in a mixture with the similar compound heptachlor, may cause brain tumours in children exposed to it before and after birth.

In a few developed countries – notably France and West Germany – chlordane is banned altogether. In several others – Israel and the USA, for example – it is severely restricted, usually to uses such as termite control. Even so, many millions of people in the USA have been exposed to chlordane as a result of termite spraying. Tens of millions of homes across the nation have been treated.

Chlordane is used in agriculture in the Third World, which is contrary to the advice of the World Bank. Though chlordane is nominally proscribed in the EEC, it is still on sale to the general public in Britain as a worm killer.

Non-target organisms are harmed by chlordane, which accumulates in food chains. Groundwater contamination has also been reported. Heptachlor, another chlorinated hydrocarbon insecticide, resembles chlordane in its toxicological and environmental effects. (See also: *Biocide* Ⓟ, *Cancer* Ⓛ, *Export of pollution* Ⓟ ③, *Groundwater* Ⓟ.)

Chlorinated hydrocarbons

Hydrocarbons are organic chemicals containing only carbon and hydrogen in their molecules. It is possible to replace one or more of the hydrogen atoms with atoms of chlorine to produce a chlorinated hydrocarbon. Some chlorinated hydrocarbons may also contain oxygen.

The simpler chlorinated hydrocarbons are almost entirely synthetic, rarely occurring in any quantity in nature. The bacteria involved in decomposing waste material, as part of the natural cycles of decay and recycling, cannot cope easily with these new synthetic substances. As a result, chlorinated hydrocarbons tend to persist in the environment.

The most notorious chlorinated hydrocarbon is the insecticide DDT (*qv*) which, together with other similar compounds such as dieldrin (*qv*) and hexachlorcyclohexane (HCH) (*qv*), has caused massive damage to wildlife due to its ability to persist in food chains. Chlorinated hydrocarbon insecticides are hardly soluble in water but dissolve readily in fat. This means that they are stored in the fatty tissues of the body. Consequently, small amounts eaten regularly build up to high levels in the tissues – a process known as bioconcentration (*qv*). If a predator eats several small animals with large amounts of chlorinated hydrocarbon insecticide in their tissues, it may absorb a fatal dose.

Insecticides are not the only chlorinated hydrocarbons capable of causing pollution. The contamination of water supplies by chlorinated solvents (*qv*) is a growing problem, as many of these materials are toxic and suspected carcinogens (*qv*). Where fluorine is included in the hydrocarbon molecule as well as chlorine, gases known as chlorofluorocarbons (*qv*) are formed: these are used as propellants in aerosol cans and may be having a disastrous effect on the protective ozone layer (*qv*).

If chlorinated compounds are burned at low temperatures – for instance around 500°C (930°F) – significant quantities of highly toxic dioxins (*qv*) are produced. When plastics such as PVC or any items containing chlorinated solvents are burned in waste incinerators, a health risk may thus be created. A particular problem arises when PCBS (*qv*) are burned because they form dioxins easily, and serious dioxin contamination has resulted from fires involving electrical transformers containing PCBS.

Because of the environmental problems they have caused, chlorinated hydrocarbon insecticides are banned or restricted for many uses or are being phased out in most developed countries of the world. In the Third World they are still sold, often by companies based in countries where these chemicals are banned. (See also: *Cancer* Ⓛ, *Circle of poison* Ⓟ ③, *Groundwater* Ⓛ Ⓟ, *Hazardous waste* Ⓛ Ⓟ.)

Chlorinated solvents

Solvents are liquids used to dissolve materials for a wide range of purposes. Chlorinated solvents – chlorinated hydrocarbons (*qv*), widely used in industry and a growing number of domestic products – are one group of solvents that are causing increasing environmental problems.

Some, such as perchlorethylene and trichlorethylene, are used for grease removal in dry cleaning establishments or industrial processes such as microchip manufacture. Chloroform has also been used industrially as a de-greaser, and the closely related carbon tetrachloride has been used as a fire extinguisher as well as a spot remover for clothing. Paint removers often contain methylene chloride, while thinners for typewriter correcting fluid are frequently based on trichlorethylene.

Most chlorinated solvents are poisonous to some degree. Inhaling them can produce drowsiness and unconsciousness, possibly leading to collapse. Placing recently dry-cleaned clothing in a warm, sealed car can lead to a dangerous build-up of solvent fumes, while the abuse of correcting fluid by glue-sniffers has had fatal results. Prolonged exposure to these materials can cause damage to the kidneys, liver and, in some cases, the heart.

Several chlorinated solvents are proven or suspected carcinogens (*qv*). Chloroform has been withdrawn as an ingredient of medicines in North America and parts of Europe for this reason and question marks hang over methylene chloride and trichlorethylene.

Chlorinated solvents persist in the environment as they are not easily broken down by micro-organisms. If dumped in landfills (*qv*) or lagoons, they may evaporate – which may lead to damage to the ozone layer (*qv*) – or percolate through the soil to groundwater (*qv*). Many wells in the USA have been badly polluted by carelessly dumped chlorinated solvents, and in some places taking a shower can cause an unacceptably high exposure to solvent fumes as the material evaporates from the heated water. Water supplies have been lost in Britain and the Netherlands through similar pollution. Other groundwater sources may be at risk since it takes a long time for the solvents to reach the water table from the surface.

In Britain, the Department of the Environment estimates that ten per cent of potable aquifers could be contaminated with concentrations of chlorinated solvents above the limits recommended as safe by the World Health Organization. The two solvents causing most concern are trichloroethane and tetrachloroethylene.

Chlorinated solvents cannot easily be burned since they form toxic partial combustion products. If their vapours are exposed to a naked flame, or drawn through a cigarette, the poisonous gas phosgene is formed. Combustion in an incinerator at too low a temperature leads to the formation of dioxins (*qv*).

Until such time as the carcinogenic pollutants can be phased out, waste chlorinated solvents are best recycled where possible or disposed of in specialist high-temperature incinerators with stringent pollution controls. (See also: *Hazardous waste* Ⓛ Ⓟ.)

Chlorination

Chlorination simply means "adding chlorine" and has two environmental contexts. In industrial chemistry, chlorination means adding chlorine atoms to other molecules, often hydrocarbons, to produce a wide range of new chemicals, including the chlorinated hydrocarbons (*qv*). Chlorination reactions can be hazardous, but the most dangerous aspect of this process is the bulk storage of chlorine gas on the factory site. Chlorine is highly toxic – it was used as a weapon in World War I – and a major leak from an industrial plant could have devastating consequences.

In gaseous form, chlorine is also used to disinfect water. It is a powerful germicide, killing most micro-organisms. Concern has been expressed at the presence of chlorinated organic chemicals in treated water. These substances, known as micropollutants (*qv*), are formed when the chlorine reacts with traces of other chemicals – some of natural origin and some from industrial discharges – present in the raw water. Some of these substances are known or suspected carcinogens (*qv*). In laboratory tests, chlorinated drinking water samples have been shown to cause mutations in bacteria.

Further investigations are continuing into this problem, but some scientists are recommending the use of ozone for water disinfection, especially where levels of trace organic compounds in the raw water are high. Many swimming pools now use ozone for disinfection since it does not lead to irritant chlorination products being formed in the water and does not involve the storage of toxic chemicals.

Chlorofluorocarbons (CFCs)

Chlorine-based compounds used as aerosol propellants, refrigerants, coolants, sterilants, solvents and in the production of insulating foam packaging used by the fast food industry. CFCs have long residence times – of 100 years or more in some instances – in the lower atmosphere. There are no natural mechanisms for their removal and CFCs break down only under the action of ultraviolet radiation when they enter the upper atmosphere. Consequently, atmospheric levels have risen sharply over the past ten to 20 years and will remain high even if CFC releases cease immediately.

Concern has arisen because CFCs are implicated in two major threats to the global environment posed by air pollution. They are greenhouse gases (*qv*) – that is, they allow incoming solar radiation to pass relatively unhindered but trap outgoing terrestrial radiation before it can escape to space, heating the planet through the "greenhouse effect" (*qv*). CFCs are fifty thousand times more effective as greenhouse gases than carbon dioxide, and rising levels could be responsible for the expected global warming of up to 1°C (1.8°F) over the next 30 to 40 years.

CFCs may also be responsible for depletion of the stratospheric ozone layer (*qv*) as the chlorine released as CFCs break down in the upper atmosphere reacts with ozone. Concern over the threat of ozone depletion led to a selective ban on CFC use in certain countries during the 1970s and to the Vienna Convention of 1985, which represents the first step towards global control of CFC production and release. As drafted, the convention would bind signatories to freezing production of CFCs at 1986 levels by 1990, followed by a reduction of 20 per cent by 1992. After further discussion, cuts of 30 per cent would be introduced either in 1994 or 1996 – the aim being a 50 per cent reduction by the year 2000. Environmentalists had hoped for a total ban by then.

Whether or not the agreed reduction will be sufficient to avoid severe damage to the ozone layer remains to be seen. Releases would have to be cut immediately by 85 per cent if no further rise in atmospheric concentration is to occur.

West Germany has stated that it intends to press ahead with a unilateral ban on CFCs by the end of the century. The USA, France and Britain are opposed to stricter regulations.

Circle of poison

The term "circle of poison" was coined by two American researchers, David Weir and Mark Shapiro, to describe the international trade in pesticides banned or restricted in the West. The complete circle is described in the accompanying illustration.

The victims of the circle of poison begin with Western chemical workers and end with West-

Circle of poison – the ban that rebounded

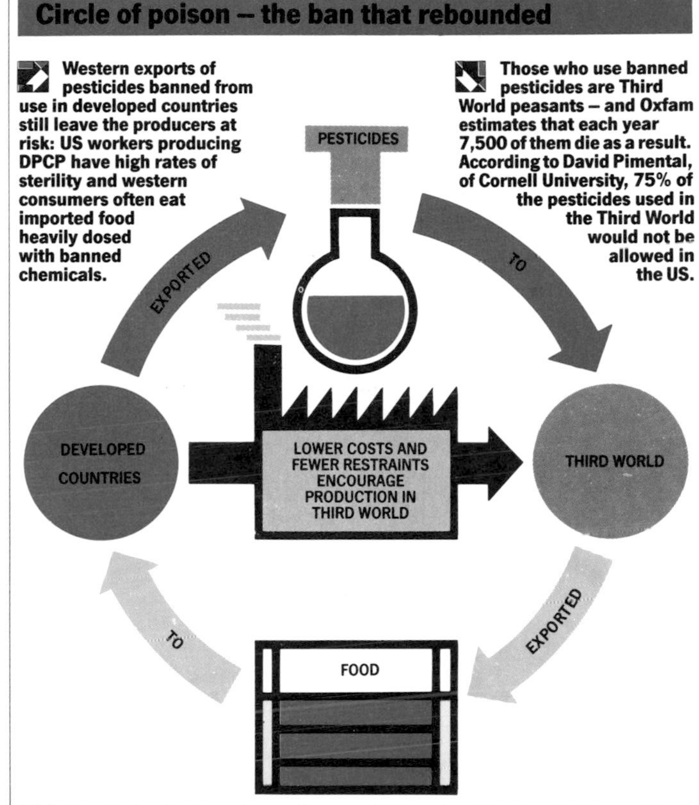

Western exports of pesticides banned from use in developed countries still leave the producers at risk: US workers producing DPCP have high rates of sterility and western consumers often eat imported food heavily dosed with banned chemicals.

Those who use banned pesticides are Third World peasants – and Oxfam estimates that each year 7,500 of them die as a result. According to David Pimental, of Cornell University, 75% of the pesticides used in the Third World would not be allowed in the US.

The circle of poison closes when Western consumers eat Third World produce: one test of Guatemalan milk showed a DDT level 90 times higher than US safety limits permitted. Coffee beans have also shown evidence of banned pesticides. But only a fraction of imported foods is tested.

FRANCE 10%
8%
SWITZERLAND 6%
BELGIUM 5%
JAPAN
UK 13%
W GERMANY 20%
USA 30%

The United States leads the world in the export of pesticides, with West Germany and the United Kingdom in second and third places. President Carter tried to curb exports; President Reagan removed the restraint. Lobbying by exporters has ensured that a 1985 FAO code has no legal force.

A clear example of the shift from home consumption to export is offered by British export figures for DDT. In 1980, exports were 1,840kg; in 1984, after the pesticide was banned in Britain, they rose to 125,503kg. The United States exports more than 18 million kg a year. Producers have also set up pesticide factories in less developed countries to avoid controls at home: El Salvador produces nearly a quarter of the world's output of parathion; and Union Carbide set up its plant in Bhopal to manufacture Sevin and other carbaryl pesticides.

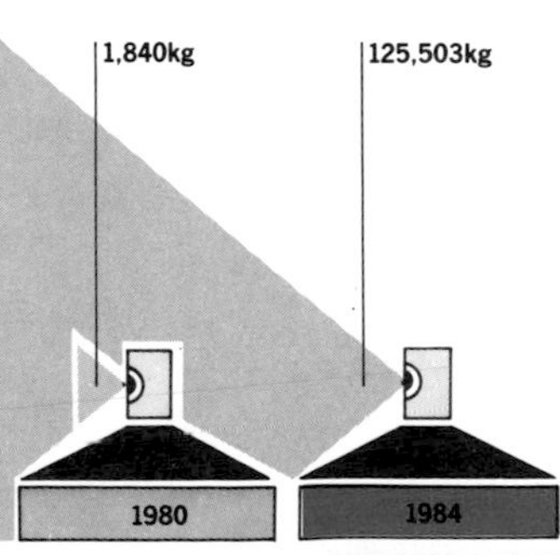

ern consumers, who frequently eat imported food heavily contaminated with banned pesticides. In between lie the Third World peasants who use the chemicals. Examples of the chemicals involved include DDT (*qv*), hexachlorocyclohexane (HCH) (*qv*), and dieldrin (*qv*).

In 1981, Oxfam estimated that 750,000 people are poisoned every year by pesticides (*qv*) and 10,000 die – 50 per cent of all poisonings occurring in the Third World and most involving banned chemicals. In Costa Rica, for example, farmhands using the chemical DPCP on banana plantations have been found to have a high rate of sterility.

Studies undertaken in America have consistently shown imported foods to be contaminated with pesticides. In 1983, the National Resources Defense Council (NRDC), an independent group, tested imported coffee beans for pesticides and found "multiple illegal chemical residues on every sample taken." One specimen was contaminated with DDT, BHC, lindane, aldrin and chlordane.

The true extent of the problem can only be guessed at since the USA and other authorities test only a minute fraction of imported foods. Of the 2,000 million oranges imported into the USA in 1982, only 14 were tested for pesticide residues by the Food and Drug Administration (FDA). Moreoever, says the NRDC, only a limited number of pesticides are tested. No tests, for example, are made for DBCP, a known carcinogen (*qv*) banned in the USA but widely used in the Third World.

In 1985, the UN Food and Agriculture Organization (FAO) issued a code of practice aimed at controlling the export of banned or restricted pesticides. The code requires prior notification to be given to the government of the importing country, which must give its explicit consent to the import. However, lobbying by Britain and other major exporters has ensured that the code has no legal force. (See also: *Development* ③, *Export of pollution* Ⓟ ③.)

Climate change
See: essay on *Man and Gaia*.

Co-disposal
The process by which industrial wastes, usually in a liquid form, are disposed of on the same landfill (*qv*) site as domestic wastes. The reasoning is that the domestic wastes are capable of absorbing a reasonable volume of liquid, thereby preventing the run-off of effluent into streams or the excessive generation of leachate (*qv*) which could be a problem if the wastes were dumped into a non-absorbent material. In addition, the biological processes that occur within a landfill site may help to break down some of the organic materials in the waste and render them harmless. A further advantage to the site operator is that it can boost the income available from a site since higher charges can be made for disposing of liquid wastes, which will not take up very much space in the dump.

At a co-disposal site, trenches are normally dug in the surface of the deposited domestic waste and the liquids are pumped into them. Great care must be taken to ensure that the capacity of the site to absorb liquid is not exceeded and incompatible wastes, which may release toxic or inflammable gases if mixed, must not be allowed to come into contact with each other. Co-disposal sites must be carefully monitored by means of boreholes to ensure that leachates containing dangerous materials do not escape. Should they do so, they may contaminate domestic water supplies.

Some research on co-disposal site leachates suggests that the leachate produced from such sites is not very different from that emanating from a purely domestic waste dump. Such research has been interpreted by supporters of co-disposal as evidence that it renders pollutants harmless. An alternative interpretation, now finding increasing support, is that the toxicity of domestic leachate has previously been underestimated. Co-disposal is not permitted in the USA. (See also: *Groundwater* Ⓛ Ⓟ, *Hazardous waste* Ⓟ.)

Co-generation
See: *Combined heat and power*.

Coal
A hydrocarbon "fossil" fuel formed when plant remains are buried under anoxic conditions. Peat forms the lowest rank, with most water and volatiles and least carbon per unit weight, followed by lignite ("brown coal"), then "hard coal" and ending with the highest rank, anthracite. In terms of the energy they yield when burned, one tonne of hard coal is equal to about three tonnes of lignite, these being the two coals in greatest use worldwide.

In 1984, world production of hard coal was a little more than 3,000 million tonnes, and of lignite just over 1,000 million tonnes. World reserves are vast.

Today, there is concern about the inevitable carbon dioxide (*qv*) emissions from the combustion of coal, which are increasing the carbon dioxide content of the atmosphere and threatening to cause climatic changes through the greenhouse effect (*qv*).

There is further concern about emission of sulphur dioxide and oxides of nitrogen from coal-fired power stations, as these are involved in the production of acid rain (*qv*). These emissions, however, can be greatly reduced by the use of scrubbers (*qv*) and also by fluidized bed (*qv*) technology.

They could be further reduced by switching to combined heat and power (*qv*) energy systems that would drastically reduce the amount of coal required to generate a given amount of energy, by making use of the waste heat for district heating programmes. The contention that coal-fired power stations are more expensive to build and to operate than nuclear power stations, with which they have recently been replaced in a number of countries, cannot be substantiated.

Coal-tar dyes
A group of synthetic food dyes (*qv*) used widely in food and drinks. First discovered during the late 19th century, they were used to dye fabrics. From the technical point of view, they are preferable to many natural vegetable dyes in that they are very strong, fast and chemically stable. At least 17 of them are in wide-

Combined heat and power (CHP)

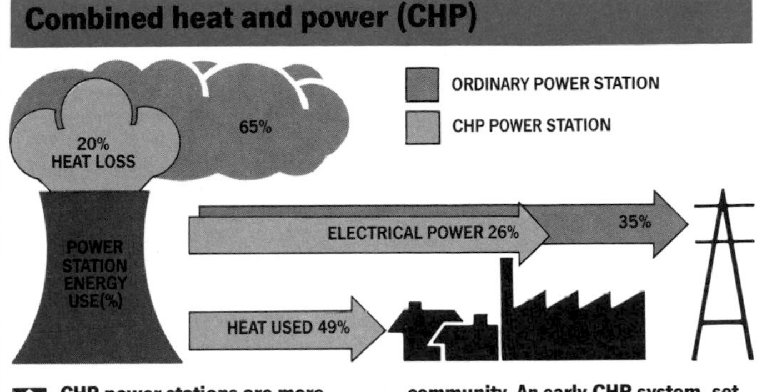

CHP power stations are more efficient at converting fuel into usable energy than ordinary power stations. They achieve this by using the waste heat, which is normally just lost into the environment, to provide hot water and space heating for the local community. An early CHP system, set up in London in 1950, ran under the Thames from Battersea power station to a number of prestigious apartment blocks on the opposite bank, including Dolphin Square. Battersea was closed down in 1984.

spread use worldwide, though none is permitted in Norway. The best known of the dyes are amaranth (*qv*) and Tartrazine (*qv*).

There has been a long-standing controversy about the safety of coal-tar dyes. During this century, at least a dozen of them have been banned or withdrawn because they have been shown to cause cancer in laboratory animals. They have also been implicated in provoking acute intolerance among a small but indeterminate number of people. In particular, they have been shown to provoke hyperactivity, asthma, eczema, urticaria and migraine.

Nonetheless, their use continues to be approved by the World Health Organization, the Food and Agriculture Organization of the United Nations, the EEC and numerous national authorities, including the US Food and Drugs Administration and Britain's Food Advisory Committee. (See also: *Additives*Ⓛ, *Cancer*Ⓛ.)

Colourings

See: *Food dyes*.

Combined heat and power (CHP)

Also known as co-generation. Most power stations, particularly the gigawatt-size ones used by the Central Generating Board in the UK, produce electricity with an efficiency of around 35 per cent. The remainder of the heat energy is discharged in to the environment as waste. Combined heat and power systems provide a means by which the "waste" heat is put to good purpose in providing hot water and space heating in domestic and industrial buildings. Through CHP schemes the efficiency with which energy is used can be as high as 75 per cent or more. Furthermore, the use of waste heat will in itself tend to reduce electricity demand, particularly when the electricity is used for heating purposes.

Many European cities – Copenhagen and Hamburg, for instance – have CHP schemes, and worldwide there are at present some 114 gigawatts of heat and power capacity. By improving energy use, CHP leads to fuel saving and reduced energy costs. It also tends to reduce the monopoly hold of the electricity supply industry, since CHP schemes can often best be operated by municipal authorities. (See also: *Energy conservation*Ⓔ, *Soft energy paths*Ⓒ.)

Containment

All nuclear reactors have a layer of concrete 2 to 5m (6 to 16ft) thick surrounding the pressure vessel. This is the primary containment, sometimes called the biological shield.

The pressurized water reactor (*qv*) primary containment encloses the entire primary circuit – steam generators, main pumps, etc. – as well as the core within its pressure vessel. The reactor building itself is designed to act as a further secondary containment, within which is the secondary circuit and the turbo-generators. Advanced gas-cooled reactor (*qv*) stations have a similar arrangement, but the steam generators and pumps of the Magnox reactors (*qv*) are outside the primary containment. Some Magnox stations do not have adequate secondary containment, which is a characteristic they share with the Soviet RBMK reactors, as was brought to light in the aftermath of the Chernobyl accident.

However, well-designed containment vessels do not provide fail-safe protection in the event of any nuclear accidents. Some experts dispute that the containment vessel at Three Mile Island (*qv*) could have withstood an explosion on the scale of Chernobyl. Even where the containment does not rupture, the radioactive material it "traps" must eventually be released into the environment. In the case of Three Mile Island, trapped radioactive gas was vented directly into the atmosphere, despite public protests. (See also: essay on *Nuclear Energy after Chernobyl*, *Nuclear accidents*Ⓝ.)

Contraception

Contraception is perceived as an essential component of modern life and, in particular, as a solution to the population explosion in the Third World. Of all the contraceptive methods available in the West, two have become particularly popular: the pill and the coil (intra-uterine device or IUD). In Britain in 1983, 2,900,000 women used the pill and 600,000 the IUD. The figures for the USA in 1982 are 8,400,000 and 2,200,000 respectively. Unfortunately, both forms of contraception also appear to have serious health effects.

In 1983, two studies published in *The Lancet* established the dangers of using male and female hormones to block ovulation. One found large increases in cervical cancer in women using the pill and the other found significant increases in breast cancer in women who had been taking the pill before they reached 25 years of age. In general, it seems that the pill can be a contributory factor in the pathogenesis of a number of diseases ranging from headaches and migraines to cardiovascular diseases and cancer. There is also strong evidence to suggest that taking the pill during the first few weeks of pregnancy (which occurs in ten per cent of all pregnancies in the USA) has teratogenic (see *Teratogen*) effects on foetuses and may increase vulnerability to cancer.

The dangers of the coil have not been investigated as thoroughly as those of the pill, though it is clear that they exist. It is thought to work by causing an inflammation that either prevents the egg from implanting itself or interferes with spermatozoa in the uterus. The uterus is well-protected, so it is not surprising that the insertion of a foreign body into it should lead to serious infections and other complications, such as miscarriages and perforation of the uterine wall. Some measure of the unpleasant side-effects of IUD use can be obtained by looking at the rate of removal during the first year of use. This varies from 30 to 40 per cent, according to the type of coil.

Both these methods of contraception are clearly unsatisfactory from the point of view of health. Yet there is tremendous official and unofficial pressure to adopt them at an early age. Many consider it deplorable that received opinion on sexual liberation should so endanger the health of women of all ages. (See also: *Cancer* Ⓛ, *Demographic transition* ③, *Family planning* Ⓛ, *Population explosion* ③.)

Convenience food
See: *Junk food.*

Conversion products
Chemicals entering the environment rarely remain in the same chemical form for very long. A wide range of processes acts on them, many involving living organisms, and the conversion products that result may be radically different from the original.

In many cases, a complicated substance is broken down into a range of simpler ones, a process known as biodegradation. Many organophosphorus pesticides, for instance, are broken down by the action of water, bacteria and other chemicals in the environment into, ultimately, carbon dioxide, phosphate and water – all of which are much less harmful than the original insecticide. Organic materials in domestic waste – such as proteins and cellulose – are broken down to nitrogen, carbon dioxide, fatty acids, water and methane.

In some cases, the conversion products may be more harmful than the original material. In certain conditions mercury (*qv*) compounds can be converted by bacteria into methyl mercury, a much more hazardous form. The pesticide heptachlor is converted first into heptachlor epoxide and then into heptachlor epoxide ketone, each of which is more toxic than the preceding form. Nitrogen oxides and unburned hydrocarbons from vehicle exhausts – themselves unpleasant air pollutants – are involved in the production of the toxic gas ozone and, when exposed to sunlight, a range of powerful irritants called PANS. (See also: *Bioconcentration* Ⓕ Ⓟ, *Cancer* Ⓛ Ⓟ, *Synergy* Ⓕ.)

Coral reef
Among the most diverse ecosystems in the world, coral reefs are produced in shallow warm seas by colonies of small marine organisms – known as polyps – that secrete limestone to form tubes within which they shelter and from whose open ends they feed. The myriad tubular skeletons left behind by the polyps build up into the reef. The polyps feed on tiny brown algae, known as *Zooxanthelae*, which colonize the inner surface of the polyps' tubes. The population is thus mainly fed from the reef itself rather than from the water, enabling reefs to survive in the marine equivalent of deserts. The relationship between the polyps and the algae is a perfect example of symbiosis.

The largest reef in the world is the Great Barrier Reef (*qv*), off the north-east coast of Australia. It extends for more than 3,000km (1,240 miles) and in terms of the number of species it supports is the richest ecosystem on Earth.

Coral reefs are among the most productive ecosystems in the world, their potential yield in fish being estimated at nine million tonnes – about 12 per cent of the present world fish catch. Overfishing is threatening many reefs, particularly those where dynamite is used to kill the fish.

Reefs are extremely vulnerable to changes in their environment. In particular, they cannot tolerate pollution that enriches the water but increases its opacity; they must have light to permit photosynthesis by the algae and thus require clear water, containing

Coral reefs – degrading a diverse ecosystem

	■	▲	●	▼	◆	□	△	○	▽	◇
INDO-PACIFIC	■	▲	●	▼	◆	□	△	○	▽	◇
SOUTH ASIA	■		●	▼	◆		△	○		
SOUTHEAST ASIA	■	▲	●	▼	◆	□	△	○		◇
EAST ASIA	■	▲	●	▼	◆					
EAST AFRICA	■	▲	●	▼	◆			○		
MIDDLE EAST	■	▲	●		◆					◇
WIDER CARIBBEAN	■	▲			◆	□	△	○	▽	◇

Coral reefs are the most productive and diverse of all marine ecosystems. They support one third of all fish species. But today they are under threat, all around the world, from a wide variety of activities and pollutants, ten of which are listed in the key on the right.

■ SILTATION
▲ FISHING
● SAND DREDGING
▼ MINING
◆ CURIO TRADE
□ TOURISM
△ CHEMICALS
○ DOMESTIC WASTE
▽ NUCLEAR WASTE
◇ OIL

little nutrient or organic matter. They do not occur, for example, close to the mouths of large rivers because of the cloudiness of water carrying sediment. Many reefs are being destroyed by sewage pollution, which encourages the excessive growth of the algae, eventually smothering the polyps. Silt run-off into the sea – the result of deforestation (*qv*) and misguided agricultural practices – also poses a threat.

Coral reefs are subject to other forms of degradation. Many are threatened by the proliferation of the predatory Crown of Thorns starfish, which annihilate the polyp population, a phenomenon that is possibly related to pollution. Building booms in tropical islands have also led to the mining of reefs to obtain limestone, leading to coastal erosion and increased storm damage. Tourism has resulted in many reefs being depleted of their coral for sale in souvenir shops or to decorate aquaria; the sale of coral has grown enormously in recent years. (See also: *Development* ③, *Balance of nature* Ⓑ.)

Coronary heart disease

A disease of the blood vessels of the heart. It is now one of the two principal causes of death among industrial populations, the other being cancer. About four out of ten men and two out of ten women alive today will die of it. As such, it has achieved the status of an epidemic.

In England and Wales alone, half a million people are diagnosed as suffering from heart disease. For 100,000 of them, it will prove fatal. In Britain as a whole, the disease affects 12 per cent of men between the ages of 50 and 59. In the USA, the incidence among men of the same age group is 21 per cent.

There is every reason to suppose that the disease begins to affect its victims early in life. One study found 15 per cent of a sample of US soldiers killed in the Korean war had damage to their coronary arteries.

Coronary heart disease is a true "disease of civilization" (*qv*), being almost unknown among primitive peoples, as is established in a number of well-documented studies.

The disease also seems to have been rare at the beginning of this century. Paul White, a famous heart physician, saw his first case of angina in 1921, during his second year of private practice.

Features associated with our modern way of life that have been pinpointed as major causes of heart disease are overeating; high consumption of saturated fats, sucrose and refined carbohydrates; smoking; and lack of exercise. (See also: *Bread* Ⓛ, *Junk food* Ⓛ.)

Creys Malville

Site, 40km (25 miles) from Lyons, France, of the world's first commercial-scale fast breeder reactor FBR (*qv*). Called "Super-Phénix", the reactor is 51 per cent owned by Electricite de France (EDF), with other European countries (including the UK) having different shares of the remainder. Work began on the reactor in 1975. In July 1976, violent anti-nuclear demonstrations, brutally suppressed by anti-riot police, took place there. One death and many serious injuries occurred.

Construction proved difficult and costs tripled. It is now estimated that the Super-Phénix has cost between $2,500 million and $3,000 million – about two to three times more than France's light water reactors. France's atomic energy authority – the CEA – regards this escalation in costs as being of no significance, arguing that Super-Phénix is only a prototype and that future breeder reactors will be much cheaper. Electricity from Super-Phénix will be twice as expensive as that from France's conventional nuclear reactors. EDF is now in serious financial trouble, owing a total of $35,000 million by 1986. France had originally intended to build six fast reactor stations after Super-Phénix, but this plan was quickly dropped. There would, in any case, be a huge surplus of generating capacity in France by the end of the 1980s.

Another difficulty arose when France made it clear that it considered the plutonium bred in Super-Phénix as being available to its weapons programme. Objections were raised not only in France, but also in the other countries involved in the project. As a FBR, Super-Phénix is intended to "breed" its own fuel by transforming non-fissile uranium-238 into fissile plutonium-239. However, "breeding" depends critically on mastering the reprocessing (*qv*) of high-burn-up fuel and ensuring that the plutonium loss during reprocessing is kept very low, possibly as low as a few tenths of one per cent. This is ten times lower than achieved to date.

Super-Phénix uses liquid sodium – which burns vigorously when in contact with air – as a coolant. In view of the massive amount of sodium it contains – some 5,000 tonnes – a fire would be very serious indeed. It has also been suggested that, unlike conventional reactors, FBRS can explode like an atom bomb.

The core of the Super-Phénix contains 5.5 tonnes of plutonium (*qv*) – 55 times more than was contained in the Nagasaki bomb. An accident on the same scale as Chernobyl would release between five per cent and 15 per cent of that plutonium into the atmosphere. With the cities of Geneva and Lyons just 50km (31 miles) away, the consequences would be too awful to contemplate.

In April 1987, Creys Malville was temporarily shut down after a serious leak of liquid sodium coolant in a tank used to store fuel rods. Given that Super-Phénix is only a prototype – and hence an experimental reactor – its construction and operation probably ranks as one of the most dangerous experiments ever conducted. (See also: *Nuclear accidents* Ⓝ.)

Cyclamates

A class of artificial sweeteners (*qv*) that has been the focus of a long-standing controversy.

There are actually three closely related compounds: cyclamic acid, calcium cyclamate and sodium cyclamate. They were discovered in 1937, but first introduced into the American market in the early 1950s, when they were allowed in preparations intended for diabetic or obese patients. They came into more general use at the end of that decade.

Cyclamates are cheap to produce, approximately 30 times sweeter than sugar, though not as sweet as saccharin (*qv*). They are, however, more stable than Aspartame (*qv*) and can be incorporated into products that are to be cooked.

In 1968, cyclamates were implicated as having adverse effects on the liver, kidney and chromosomes. A year later, they were linked to the development of cancer in rats. The US government responded by banning cyclamates in October 1969, and the UK followed suit a month later.

The use of cyclamates continues, however, to be permitted in almost all other European countries, though they are often restricted to table-top sweeteners.

Efforts have continued to try to get cyclamates reinstated in the USA and the UK. In the mid-1980s American experts were saying that cyclamates were probably not carcinogenic (see *Carcinogen*), or at worst only weakly so, but they could not dismiss the possibility that they might promote the growth of cancers that had been initiated by other agents. Other possible toxic effects, including testicular atrophy, have also not yet been ruled out.

Throughout these lengthy debates, the Joint Expert Committee on Food Additives, the UN Food and Agriculture Organization and the World Health Organization continued to approve the use of cyclamates. In 1982, the UK Government's advisory committee recommended that the use of cyclamates could be accepted provisionally, pending the receipt of more detailed data. The British government did not, however, accept this recommendation and cyclamates continue to be banned in the UK. (See also: *Additives*Ⓛ, *Cancer*Ⓛ.)

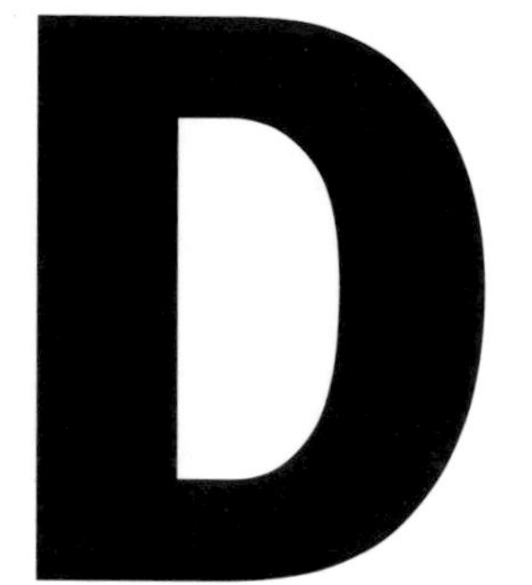

Daintree

Australia's wet tropical rainforests, and the struggle to preserve them, have become symbolized by environmentalists' efforts to stop a track being bulldozed through Daintree, a rainforest between Cape Tribulation and the south of Bloomfield on Queensland's far north coast. The battle to stop the Daintree track rallied public opinion in Australia in support of the preservation of rainforests. As a consequence the name "Daintree" is now synonomous with the much larger issue of tropical forest preservation in Australia.

Australia's tropical forests occur along a 700km (435 miles) coastal strip between Townsville and Cooktown in north Queensland and comprise an area of 8,000sq km (3,088sq miles), or less than 0.1 per cent of the Australian land surface. The rainforests are unique as they are a virtual living museum of plant evolution, containing 13 out of the 19 known primitive flowering genera in the world today. As such, the forests are closely linked to the origins of the world's flowering plants. Indeed, British botanist David Bellamy has described the area as the birthplace of flowering plants.

The major focus of the campaign to save these rainforests is to have the Australian Government nominate the region for the World Heritage List. This would enable Federal laws to be used to override the State Government of Queensland and stop the area being destroyed by logging, real estate subdivision, road construction, mining and agricultural clearing. Environmentalists have recently succeeded in gaining assurances from the Federal Government that the region will be nominated.

With more than 75 per cent of Australia's rainforest having been cleared for agriculture and urban development in the last 200 years, it is imperative that what remains is quickly secured within national parks. (See also: *Deforestation*Ⓒ.)

Dams

Despite their benefits, dams carry heavy ecological and social costs. The bigger the dam, the more extensive the likely damage. The effects of large dams are described in the accompanying illustration.

Worldwide, there are more than 13,000 "large" dams – that is, dams more than 15 metres (50ft) in height – already operating. Almost one-third of those dams were built between 1961 and 1981 and, at the last count, in 1981, a further 1,300 new dams were either under construction or at the planning stage. By 1990, the worldwide total of dams more than 150 metres (490ft) high is expected to reach 113, of which 49 will have been built in the 1980s.

On the face of it, hydroelectricity (*qv*) is extremely cheap, one kilowatt of installed capacity costing approximately $1,000 at 1984 prices. Moreover, being both renewable and non-polluting, it is one of the more benign sources of energy available to man, and is thus seen as a major potential source of energy.

Dams are also seen as having an essential role to play in the battle against world hunger, by providing water for irrigation projects.

Vast areas of land have been, and are expected to be, flooded as a result of dam projects. In 1985, dams still under construction or at the planning stage were projected to flood 32,753sq km (12,645sq miles). Land lost to flooding is often extremely fertile. In Sri Lanka, the Victoria Dam (constructed with money from Britain) flooded 28sq km (11sq miles) of agricultural land. In India, the Srisailam hydroelectric project near Hyderabad flooded some 430sq km (166sq miles). Thousands of square kilometres of valuable tropical forest have also been drowned by dams.

Millions of people have been uprooted to make way for the reservoirs of large dams. Seventy-eight thousand people were displaced by the Volta Dam alone: 120,000 by the Aswan Dam (*qv*) in Egypt; and 93,000 by the Damodar in India. Still more will be moved in the future. If it goes ahead, China's massive Three Gorges Dam (*qv*) project will alone submerge ten cities and partially flood another eight.

Dams – the side effects of an apparent blessing

In the 20 years between 1961 and 1981 enough concrete and fill was used in the construction of dams to cover the island of Cyprus with a layer a metre thick; and as much again will have been used by the time the new generation of dams come into service in the 1990s. From being a farming aid since the dawn of agriculture, large dams have become a growth industry, providing irrigation, water storage, flood control and power. But are they unreservedly good?

9,000 MILLION CUBIC METRES

8,000 MILLION

1961-81

1981-90s

GOOD

Irrigation

● According to the FAO, 100,000sq km of land need to be brought under irrigation by the year 2000 if there is to be any hope of food supplies keeping pace with demand. Much of that land is in arid and semi-arid areas. With groundwaters already severely depleted in many places, and rainwater too diffuse and unpredictable, dams are seen as having a vital role to play, their reservoirs providing water both where and when it is needed.

Power

● More than 20 per cent of the world's electricity is already supplied by dams. Hydro-electricity is cheap, renewable and non-polluting and is thus seen as a major potential source of energy. Installed hydro-electric capacity stood at 541,976 megawatts in 1984, but, even so, much of the potential of the world's rivers remains untapped. The World Energy Conference estimates that 19,000 terawatt-hours could be tapped without technical difficulty, against the 1,300 terawatt-hours produced today. That potential output is equal to that of about 3,000 nuclear reactors.

BAD

Resettlement

● Millions of people have been uprooted to make way for dams. China's massive Three Gorges plan will involve the resettlement of 3,300,000.

Loss of land

● Vast areas of land have been flooded by dam projects. In 1985, the International Commission on Large Dams estimated that up to 1981 308,423sq km had been flooded by dams – an area equivalent to the whole of Italy. Amazonia's planned dams will flood an area as big as Montana.

Disease

● Dam reservoirs are a major source of waterborne disease. As a result of the Aswan Dam, some villages near Lake Nasser have a 100 per cent infection rate for bilharzia. In Ghana, before the Volta Dam was built, the rate was 2 per cent; now it is 80 per cent.

Disruption of ecology

● Downstream ecosystems are severely disrupted by dams. They trap silt and thus hold back valuable nutrients, with a consequent effect upon fisheries that can far outweigh the benefits that follow the setting up of fisheries in a dam's reservoir. Before the Aswan Dam was built, sardine catches in the eastern Mediterranean totalled some 18,000 tonnes a year: by 1969 the catch was down to 500 tonnes a year. The 100 million tonnes of sediment deposited on farmland fell to just a few tonnes, and to compensate for the loss of nutrients, Egypt must apply artificial fertilizer at a cost of about $100 million a year. Reduced silt deposition below dams also leads to coastal erosion: the homes of 10,000 people living in Keto on the Togo coast, below the Volta Dam, fell into the sea as a result of erosion. And, finally, dams contribute to earthquakes, the weight of water in them increasing the likelihood of fault movements.

All too often, past resettlement (*qv*) projects have been characterized by inadequate compensation, low-grade land and inappropriate housing. For many – particular those from isolated tribal cultures – the move proves disastrous. With their land, their sacred shrines, their historic monuments, their homes and their villages drowned beneath the reservoir, their whole life falls apart. Many simply drift towards the nearest city, there to join the burgeoning number of slum dwellers. The reservoirs of dams, together with their associated irrigation works, provide ideal breeding grounds for the vectors of waterborne diseases, in particular schistosomiasis (*qv*) which is better known as bilharzia, a parasitic disease. Mosquitoes carrying malaria (*qv*) also thrive in irrigated areas.

Because they trap silt which would otherwise be carried downstream, dams deprive downstream ecosystems of vital nutrients. The impact on fisheries can be severe – far outweighing the often temporary benefits which follow the setting up of fisheries in a dam's reservoir. The reduced silt-load of rivers below dams also leads to severe coastal erosion. Land downstream can also suffer from salt water intrusion (see *Farakka Barrage*), caused by the reduced flow of impounded rivers.

The sheer weight of the water impounded in the reservoirs of large dams has triggered earthquakes. Such earthquakes have occurred at a number of sites throughout the world, in China, India, Zimbabwe and Greece. In 1981, the area around the Aswan Dam suffered an earthquake of magnitude 5.6 on the Richter scale. The Volta Dam and the Bratsk Dam have also experienced induced earthquakes despite being in "low-risk" areas.

Fears over the severe ecological and social costs of large dams have led to considerable pressure being placed on the World Bank (*qv*) and other development agencies to cease the funding of large-scale water development projects.

In some instances, that pressure has yielded results. Funding for the Bakun Dam in Malaysia and several other projects around the world has been suspended – at least for a short period of time – pending action to "safeguard" the environment and ensure "adequate" resettlement plans. Following the overthrow of President Marcos, the controversial five-dam Chico River Project has also been abandoned. (See also: *Development* ③, *Environmental refugees* ③ Ⓛ, *Irrigation* Ⓐ ③, *Salinization* Ⓐ.)

DDT
Known technically as Dichloro-Diphenol-Trichloroethane, DDT is one of the most widely used pesticides (*qv*) in the world. First synthesized in the 19th century, DDT was introduced into US agriculture on a wide scale in the 1930s. Since then it has been used in many other fields throughout the world, notably for malaria eradication.

DDT is highly persistent in the environment. The bioconcentration (*qv*) of DDT has led to the thinning of egg-shells in many bird species and the consequent failure to reproduce. The most famous examples are the osprey, the peregrine falcon and the bald-headed eagle, symbol of the USA. In 1945, Dr V.B.Wigglesworth of the British Agriculture Research Council, described DDT as "a blunderbuss discharging shot in a manner so haphazard that friend and foe alike are killed."

Research in the 1960s revealed DDT to be potent carcinogen (*qv*). As a result, DDT was banned in the USA in 1971 – but only after an intense court battle with the major producers. It has also been banned in West Germany, France and many other industrialized countries. In Britain, use of DDT was subject to a voluntary ban from 1974 to 1984, when it was finally withdrawn from the market. There is evidence, however, that it is still widely available on the black market.

Despite the ban on its use in most industrialized countries, exports to the Third World continue unabated – though Kenya has announced its intention to ban DDT altogether. It seems clear that many companies see the Third World as a dump for unused stocks of DDT. High levels of DDT have been found in the milk of many nursing mothers. (See also: *Cancer* Ⓛ Ⓟ, *Circle of poison* ③ ③, *Malaria* Ⓛ ③.)

Debt crisis See: *Third World debt.*

Decay products
See: *Conversion products* Ⓛ Ⓟ, *Radioactivity* Ⓛ Ⓝ.

Decommissioning
Term used to describe the dismantling and disposal of old nuclear reactors. To date, experience of decommissioning is limited. The original pressurized water reactor (PWR) (*qv*) at Shippingport is presently being decommissioned, as is the small prototype advanced gas-cooled reactor (AGR) at Sellafield. It is intended to decommission the latter, and all UK commercial reactors, in three stages, the last of which may be delayed for 100 years and would involve cutting up the reactor pressure vessel by remote control, possibly under water. Decommissioning of commercial reactors will involve large volumes of radioactive wastes (see *Nuclear waste*) (but no high-level wastes) and considerable costs, officially estimated at $480 million for the Sizewell B PWR, or 20 per cent of its capital cost. However, the Washington-based Worldwatch Institute calculates that decommissioning will cost about $1 million a megawatt. This would work out at 40 per cent of the capital costs of a PWR.

The costs of decommissioning are not fully taken into account when estimating the total costs of operating reactors, nor are sufficient funds set aside for financing decommissioning. In Britain, for example, the application of the government-determined five per cent per annum discount rate gives a small present value to costs that are incurred far into the future. Thus most of the costs of decommissioning, like reprocessing (*qv*) and long-term fuel cycle costs will have to be met by future generations. (See also: essay on *Nuclear Energy after Chernobyl.*)

Deep ecology
A movement inspired by the

Norwegian philosopher Arne Naess, who was the first to use the expression. His ideas have been further developed by Bill Devall, Georg Sessions, Warwick Fox and others.

The movement arose as a reaction both to the academic ecology taught in universities and the environmentalism of the ecological movement, neither of which questions the "dominant metaphysics" underlying our modern industrial society; hence their classification as "shallow ecology".

Warwick Fox sees deep and shallow ecology as differing in three main respects. The first is that shallow ecology views humans as separate from their environment, and favours environmental preservation only because it is ultimately in the human interest. By contrast, deep ecology sees man as part of nature, indeed, in Fox's words, as "just one particular strand in the web of life" or "just one kind of knot in the biospheric net."

Secondly, deep ecology rejects the "mechanical materialism" that characterizes the industrial worldview and reflects outdated 19th century notions of physics. Instead, deep ecology sees the world, as does the physicist David Bohm, as "unbroken wholeness which denies the classical idea of the analyzability of the world into separately and independently existing parts." Significantly, this "unified field" view of the new physics is also, as Fox notes, that of eastern mysticism.

Thirdly, shallow ecology accepts, at least by default, the dominant ideology of economic growth. Its adherents are therefore seen as willing, in Fox's words "to accept the reduction of all values to economic terms for the purposes of decision making." Deep ecologists, however, see economic considerations as firmly subordinate to ecological ones.

The movement undoubtedly represents a serious attempt to free environmental and ecological thinking from the fetters of the "dominant metaphysics" – that of our industrial society – within which, they are, indeed, still largely imprisoned. (See also: *Conservation* Ⓕ, *Ecological movement* Ⓛ, *Green politics* Ⓛ, *Holism* Ⓕ.)

Deep-well injection

Method of waste disposal used to dispose of both municipal and hazardous wastes. The wastes are injected under pressure into porous rock strata, generally at depths of 600 to 1,800m (2,000 to 6,000ft). The wastes displace any liquids or gas in the porous rock and are then trapped within the strata by the pressure of overlying rocks. In 1981, an estimated 3,600 million gallons of waste were injected underground in the USA. Estimates put the number of deep wells at 70,000, with wastes coming from the oil and gas industries, the pharmaceutical industry and the chemical and steel industries.

The principal danger of deep-well injection lies in the threat of wastes migrating into deep aquifers, thus contaminating water supplies. Such migration can result from the improper plugging of wells or from the fracturing of rocks. Overpressurized wells can also erupt, as happened in Eries, Pennsylvania, in 1968. Deep-well injection has also be linked to earthquakes. (See also: *Hazardous waste* Ⓟ, *Land disposal* Ⓟ.)

Deforestation

Trees are essential to the ecological wellbeing of the Earth. They clean the air, conserve the soil, maintain its fertility, store water, provide a habitat for wildlife and play a vital role in regulating the climate. In addition, they provide a source of fodder for animals, and food fruits, fibres, firewood and timber for man.

Today, the world's forests – both temperate and tropical – are under siege. In the industrial world, trees are increasingly falling victim to disease and the insidious effects of air pollution – most notably acid rain (*qv*); 14 per cent of all Europe's forests are now suffering from acid damage. In the tropics, the threat is more direct – millions of square kilometres of forest being cut down, flooded or burned each year as forests are felled for timber, cleared for agriculture, or converted into enormous reservoirs through the damming of rivers.

Since the turn of the last century, the Third World has lost nearly half of its forests. The figures for individual countries illustrate the extent of the crisis (see illustration).

Poverty, overpopulation and ignorance are commonly held to be the principal causes of deforestation. In support of that view, the UN Food and Agriculture Organization (FAO) (*qv*) points to the damage done by the 250 million peasant colonists who have moved into the forest in search of land. Describing their "slash-and-burn" (*qv*) methods of cultivation as "environmentally destructive" and "inefficient." FAO accuses these "encroaching cultivators" (*qv*) of being the prime agents of forest destruction, destroying 50,000sq km (19,000sq miles) of forest a year and degrading a further ten million.

The charge is grossly misleading. First, blaming encroaching cultivators for deforestation obscures the massive damage caused by plantation and ranching, road, logging and operations and dam projects. In peninsular Malaysia, where encroaching cultivators are few and far between, the logging industry has destroyed half the country's rainforests since 1960. In Ethiopia, massive plantations have stripped the Awash Valley of its forests, 60 per cent of the land now being under cotton with another 22 per cent devoted to sugar plantations. In central America, cattle ranching is responsible for the clearance of almost two-thirds of the forests. In Brazil, official government figures reveal that 60 per cent of forest destruction between 1966 and 1975 was caused by large-scale cattle ranching projects (38,650sq km/15,000sq miles) and road building (3,075sq km/1,190sq miles). Less than 17.6 per cent was lost to state colonization projects.

Secondly, the problem of encroaching cultivation is itself the direct result of policies intended to open up the forest to development. In Indonesia, more than 3,600,000 peasants have already

Deforestation – global destruction of a vital resource

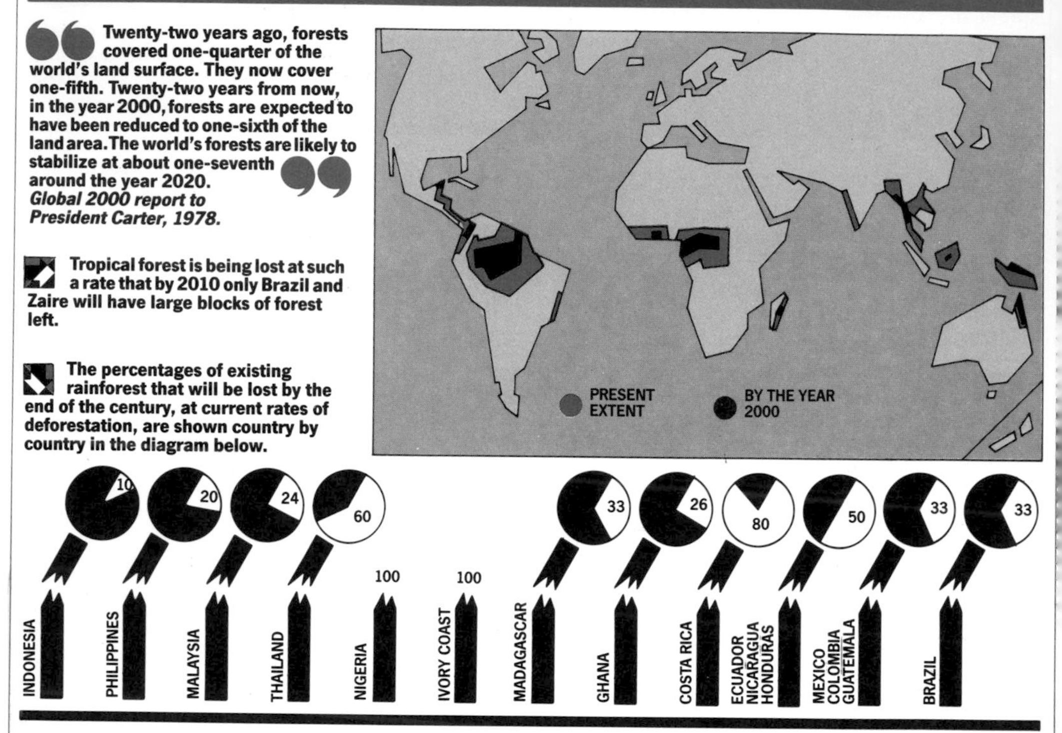

been moved into the densely forested outer islands of the archipelago as part of the country's transmigration programme (*qv*). More than 33,000sq km (12,700sq miles) of rainforest are at risk from the project. In Amazonia (*qv*) thousands of colonists have been sent into the jungle. Every road built brings settlers who spread out into the forest, every dam brings new settlements on the shores of its reservoirs, every mining project attracts scores of migrant workers.

Thirdly, the issue of encroaching cultivation cannot be separated from the problem of landlessness in the Third World. At present, land holdings are concentrated in the hands of very few people – 93 per cent of arable lands in Latin America being held by a mere seven per cent of land owners. Much of that land is used for plantation agriculture or ranching – thus denying its use to landless farmers. In the absence of land reforms, those farmers then have little choice but to invade the forests. Almost all those peasants who have invaded Rondonia in Amazonia have done so because their lands in the State of Rio Grande do Sul have been taken over by export-orientated plantations and ranches.

Undoubtedly encroaching cultivators do cause extensive forest destruction, but to blame them for the deforestation is quite unjustified.

In destroying their forests, the countries of the Third World are following a trail already blazed by the industrialized world. When Columbus first discovered America, the country was covered by an estimated 3,200,000sq km (1,235,000sq miles) of forest. By 1930, less than 400,000sq km (154,000sq miles) were left – and, today, the figure is down to 220,000sq km (85,000sq miles).

But while the forests of the world's temperate areas can be stripped with relative impunity, the effects of deforestation in the tropics are environmentally disastrous.

Stripped of its forest cover, the land rapidly succumbs to erosion (*qv*), leading to the silting up of rivers and the loss of fertility. Eventually, if the land is not reforested (see *Reforestation*) it becomes victim to desertification (*qv*). Meanwhile, because the forests are no longer available to soak up rain and store it in the soil, a drought-flood cycle (*qv*) is initiated, with streams and rivers drying up during the hot, dry season only to become raging torrents during the wet. The loss of biological diversity as a result of deforestation is also a cause for increasing alarm.

Many environmental groups in the Third World now argue that there is no solution to the problem of deforestation as long as Third World countries remain within the market economy. They urge "delinking" as the solution to the problem. (See also: essay on *Acid Rain and Forest Decline*, *Development*③, *Greenhouse effect*Ⓟ③, *Sedimentation*Ⓕ③, *Tropical forests*③.)

Delaney Amendment

The Delaney Amendment,

enacted as a US federal law on September 6 1958, prohibits the intentional use of any carcinogenic (see *Carcinogen*) chemical as a food additive under any condition. This was based on the principle that, as was stated at Congressional hearings in 1960, "scientifically, there is no way to determine a safe level for a substance known to produce cancer in animals."

The amendment has come under concerted attack ever since its enactment. Most of the criticisms have been levelled by industrial groups, trade associations and their consultants. The argument has hinged mainly on the scientific methodology of testing for the toxicity of chemicals. The anti-Delaney lobby argue that it is possible to establish a threshold dose below which exposure to a chemical is safe. They also argue that chemicals can be made to appear carcinogenic by, for example, testing at higher doses thus calling into question the value of carcinogenicity tests in general. However, it is quite clear that the scientific basis of the Delaney Amendment was, and remains, totally valid. To date, it is scientifically impossible to determine a safe level of exposure to carcinogenic chemicals, let alone predict any synergism (see *Synergy*) between a carcinogen and another chemical. This was the message of an ad hoc committee report submitted to the Surgeon General on April 22 1970.

Despite these attacks, the Delaney Amendment remains in force today and is the main legal plank protecting the American citizen from food adulterated with carcinogenic chemicals. However, the US Food and Drug Administration is currently trying to change the amendment to permit the addition of carcinogenic food additives at levels described to be "devoid of significant risk." (See also: *Additives*Ⓛ, *Cancer*Ⓛ.)

Demographic transition

The theory that the rate of the population growth (though not overall numbers) slows down and eventually declines as material standards of living rise. It is based on population trends in Europe following the industrial revolution. Population levels that initially escalated later slowed down. Now several industrialized countries have reached, or are approaching, "zero" growth rates.

The same demographic transition, it is confidently asserted, will also now occur in the Third World as development proceeds, first through family planning (*qv*) and, secondly, by providing people with financial security that will supposedly reduce their desire to have so many children.

There are two flaws to the theory. First, it is contradicted on the ground. Secondly, it is conveniently forgotten that a growing population is not intolerable as such, but because of its impact on the natural environment. That impact, however, is not simply a function of the number of people in the world but also of their level of material consumption. To reduce the rate of population growth by increasing material consumption is thereby self-defeating, since the impact on the environment is likely to increase.

Moreover, though the rate of population growth in the West has indeed fallen with rising affluence, this does not mean that it will necessarily do the same in the Third World. The theory claims that people have more children because they are financially insecure. But while economic development reduces financial insecurity for a few, it increases social insecurity for the mass of people by destroying the extended family. It also destroys the traditional community by eroding traditional cultural patterns, by degrading the physical environment, and by removing tribesmen and peasants from their land to accommodate large export-oriented farms and other development projects. Under such conditions of social alienation, people will tend to have more rather than fewer children.

Recent reductions in fertility in isolated countries, in particular Brazil, Colombia and Mexico, tend, if anything, to confirm that people have fewer children when they are financially insecure. Indeed, the research shows that the decline not only occurred during a period of economic depression, but also among poor rather than rich communities.

Lester Brown, Director of the Worldwatch Institute considers that though demographic transition may have marked "the advance of all developed countries," in the Third World it "may be reversed for the first time in modern history." (See also: *Population*③ ,*Population explosion*③ .)

Desalinization

The process by which drinking water is obtained from seawater or other sources of water containing high salt levels. The simplest process involves evaporating the water and condensing the vapour produced, the dissolved salts being left behind in the evaporation vessel. Large amounts of heat are required for this process and it is practised on a large scale only in the oil-rich desert countries where cheap energy is available. Desalinization programmes based on nuclear reactors have been suggested, but none is currently operating. A newer process, called reverse osmosis, has been developed in which a membrane separates the water from the dissolved salts. It, too, is expensive. The most promising means for desalinization under development is the solar-powered still using natural sunlight.

In the USA, drainage water from large-scale irrigation schemes has raised the salinity of the Colorado River to such levels that a massive desalinization plant has had to be built at Yuma on the US-Mexico border to reduce salinity levels before the river crosses into Mexico. Under an international treaty, the US guarantees to protect the quality of the Colorado's water. (See also: *Irrigation*Ⓐ, *Salinization*Ⓐ.)

Desertification

Final stage in a slow and insidious process of land degradation, which starts with the loss of vegetative cover and ends with the destruction of the soil's fertility and its transformation into barren desert. The most vulnerable soils are to be found in the dry tropics.

Nomadic pastoralism (*qv*) and

traditional peasant agricultures are frequently, but quite unjustifiably, blamed as the principal "cause" of desertification. In fact, traditional farming systems in the Third World place minimum pressure on the land. By contrast, modern mechanized farming and the introduction of vast livestock rearing projects have led to overgrazing and unsustainable agricultural intensification. The problem has been compounded by the pushing of peasants onto more and more marginal lands.

Significantly, the 1977 UN Conference on Desertification concluded that nomadic pastoralism represented the only sustainable method of husbandry in the fragile lands bordering the Sahara.

Following the conference, UNEP was charged with implementing an action plan on desertification. The plan has yielded minimal results, most of its projects being directed at monitoring the extent of desertification rather than combating its spread. Political indifference has also dogged the project, with governments from both the Third World and the industrialized world refusing to provide more than a fraction of the funds required. (See also: *Cash crop* Ⓐ③, *Deforestation* Ⓐ©③, *Erosion* ©③, *Overgrazing* Ⓐ③, *Salinization* Ⓐ©③)

Development

It is generally believed that poverty and malnutrition in the Third World can be combated only by economic development – hence the subordination, during the last 40 years, of all other considerations to the achievement of this goal.

Economic development has certainly led to the emergence of a middle class consisting largely of government officials, but also of traders and manufacturers, some of whom enjoy considerable affluence. But that class represents only a fraction of the total population of the Third World. And specific policies designed to encourage economic development have caused increasingly serious social and ecological problems that have undoubtedly worsened the plight of most Third World people.

Even the institutions that have spearheaded development (such as the World Bank, the UN Food and Agriculture Organization and USAID), admit that their policies have been a failure and propose to replace them with more sensitive types of development that have been variously called "rural development," "appropriate development," "ecodevelopment," "alternative development," "integrated development," and "sustainable development". Unfortunately, none of these has been adequately defined, still less applied. Indeed, the accent is still on precisely the same sort of vast capital-intensive projects – such as large water development projects, livestock rearing programmes, vast plantations, etc – that are known to have caused the current problems. These policies seem best to satisfy the short-term political and economic requirements of those who formulate them.

This becomes still clearer if we consider that the policies that go under the name of development are very similar to, indeed scarcely distinguishable from, those that were pursued under the aegis of the colonial system.

The prime object of colonialism was to obtain access to cheap raw materials, cheap labour and a captive market for Europe's manufactured goods. Cecil Rhodes expressed it clearly. " We must find new lands," he wrote, "from which we can easily obtain raw materials and at the same time exploit the cheap slave labour that is available from the natives of the colonies. The colonies would also provide a dumping ground for the surplus goods produced in our factories."

The economic development of the Third World achieves precisely the same goals. It consists in bringing Third World countries within the orbit of international trade by influencing them to eliminate subsistence agriculture and artisan modes of production catering for a largely local market and to replace them with capital-intensive plantations and factories geared to the international market.

If development has caused so many problems in the Third World it is for the following principal reasons.

1. Development inevitably causes a population explosion by destroying culturally in-built population-control strategies. It also leads to the inevitable displacement of tribesmen and peasants from their traditional lands to accommodate various development projects, and to the destruction of their highly cohesive social groupings. The resulting social alienation encourages people to have more children for the security with which such children may eventually provide them.

2. Development, encourages the import of manufactured goods. This has the effect of putting local artisans out of business. Many of them are forced into urban slums.

In addition, all imported goods must eventually be paid for. Third World countries cannot do so from the sale of expensive high technology devices as do most of the industrial countries of the West. They can do so only by selling the produce of their soil.

3. Historically this has entailed, first, the selling of new trees, if they have them. Indeed, development of such countries as Indonesia and Malaysia has been financed largely from the sale of tropical hardwoods, with disastrous ecological consequences. The deforestation (*qv*) involved has contributed significantly to the present escalation of malnutrition and famine.

Once the forests have gone, Third World countries must resort to selling cash crops, such as sugar, tea, coffee, jute, cotton, rubber and tobacco, as well as tropical fruits and vegetables. The large-scale mechanized cultivation of such crops on vulnerable tropical soils can only lead to erosion and desertification (*qv*). It also means vast livestock rearing projects which are equally destructive to the soil.

The large-scale export of food, which may be tolerable in thinly populated countries such as the USA, Canada, Argentina, Australia and New Zealand, cannot be tolerated in grossly overpopulated countries of the Third World. In such countries every square kilometre of land made available for producing export crops is an square kilometre of land made unavailable for feeding local people. And when local people are already undernourished, as is increasingly so throughout the Third World, the export of agricultural produce is only possible at the cost of increasing malnutrition (*qv*) and famine (*qv*).

Setting up vast enterprises geared to the export trade also means displacing the tribesmen and peasants who happen to derive their livelihood from the land. This has occurred on a massive scale, since in many countries as much as 50 per cent of the agricultural land is used for export crops. This gives rise to rural poverty and to vast slums to which the environmental refugees (*qv*) inevitably gravitate.

Tribal people and peasants are also displaced to accommodate the physical infrastructure of the growing conurbations and their industrial base. Still more agricultural land must therefore be diverted from feeding rural people, who once again end up being forced off the land and into the cities. The vast teeming slums of the Third World can be seen as the necessary concomitants of economic development.

There is now increasing pressure on Third World countries to maximize foreign exchange earnings in order to pay interest on the massive debts they have contracted – partly, at least, to pay for uneconomic development projects such as large dams. As a result, the development process has been stepped up, with social and ecological considerations increasingly being subordinated to that overall goal.

Ironically, all this misery may well be in vain. The Third World cannot conceivably attain the sort of affluence that we know today in the industrial world. Neither its vulnerable and, indeed, fast deteriorating soils, nor the adverse climatic conditions of tropical areas, render feasible the agricultural surplus required to finance development. Nor are world economic conditions today anything like as favourable to economic development as they were when western countries first developed. In the next decades, de-development rather than development may well be the rule.

For these and other associated reasons, the very principle of development as a panacea to the problems of the Third World should be seriously questioned. (See also: *Aid* ③, *Cash crops* ③, *Dams* ③, *Demographic transition* Ⓐ ③, *Hamburger connection* Ⓐ ③, *Population explosion* ③, *Third world debt* ③, *Tropical agriculture* Ⓐ ③.)

Dichlorvos

An organophosphorus insecticide, also known as DDVP, used as a liquid in agriculture, but more familiar as the active ingredient of slow-release fly strips for domestic use. In the USA such strips are marketed under the brand name "The No Pest Strip", and in Britain as Vapona.

Dichlorvos is a toxic material that affects the nervous system. Like organophospates, very small quantities can badly affect sensitive people. It has been found to be mutagenic (see *Mutagen*) in laboratory experiments and this suggests that it may also cause cancer. American researchers suggest that women of child-bearing age should handle dichlorvos with caution, if at all. (See also: *Cancer* Ⓛ Ⓟ, *Carcinogen* Ⓑ Ⓟ, *Insecticides* Ⓟ, *Organophosphates* Ⓟ, *Pesticides* Ⓛ Ⓟ.)

Dieldrin

A highly toxic insecticide that has caused serious harm to wildlife in many parts of the world. It has been used in agriculture against a wide variety of insect pests, around the home to kill moths, and in the garden to destroy unwanted insects. Carpets are sometimes impregnated with dieldrin to protect them from moth attack, and it has been widely used as a timber treatment.

Like other chlorinated hydrocarbons (*qv*) dieldrin is persistent in the environment and accumulates in food chains. Dieldrin pollution is known to have played a major role in the decline of the British otter, now extinct in England and Wales. The decline of the Sandwich tern population of the Dutch Wadden Sea – the number of breeding pairs fell from more than 40,000 in the 1940s to just 1,500 in 1964 – was also blamed on pollution by dieldrin and closely related compounds, endrin (*qv*) and telodrin.

Studies conducted in the USA as early as 1962 showed dieldrin to be carcinogenic (see *Carcinogen*) in mice. Seven years later, the newly-formed Environmental Protection Agency proposed that the use of dieldrin, together with aldrin, an insecticide that is converted to dieldrin in the environment, be severely restricted. Dieldrin and aldrin were banned for most uses in the USA in 1976.

In Europe, the regulation of dieldrin and aldrin has varied considerably from country to country. In Britain, the Pesticide Advisory Committee of the Pesticide Safety Precautions Scheme concluded that a ban was unnecessary, though it did recommend voluntary restrictions on use. The same data led the French Government to impose a ban in 1972. In West Germany, on the other hand, the authorities decided that it was enough to restrict the use of the two insecticides, restrictions first being imposed in 1971 and then being strengthened in 1974 and 1979. In 1981 the European Community banned the marketing of aldrin and dieldrin in plant protection products. Even so, exports are still permitted. In 1985, clearance for the use of dieldrin in wood preservation was withdrawn in Britain.

Collectively, dieldrin, aldrin, endrin and telodrin are known as the "drins". In Europe, their discharge into waterways and their dumping at sea are now subject to increasingly stringent controls. Even so, as much as half a tonne of drins may be entering the North Sea each year.

Shell Chemicals continue to manufacture dieldrin and aldrin at their Pernis plant in Holland. In the Third World, where restrictions are less effective (and sometimes non-existent), dieldrin and the other drins are still widely used. Though Shell will not divulge its production figure, in 1983 it was estimated that 99 per cent of the drins manufactured at Pernis were exported, mostly to the Third World. That would have amounted to almost 5,000 tonnes. (See also: *Aldrin* Ⓛ Ⓟ, *Cancer* Ⓛ, *Circle of poison* Ⓟ, *Insecticides* Ⓟ, *Pesticides* Ⓟ.)

Diethylstilbesterol (DES)
Synthetic hormone, commonly known as DES, which until the mid-1970s was widely prescribed to pregnant mothers to prevent miscarriages.

First developed in 1938, DES was banned in the USA in 1971 after high rates of a rare vaginal cancer were found among a number of the daughters of those who had taken the drug during pregnancy. Sons of DES mothers are reported to be more prone to infertility, while the mothers suffer higher-than-normal cancer rates. In the USA alone, between four million and six million mothers and children were exposed to DES during pregnancy.

The reputation of DES as a drug that prevented miscarriage had little factual basis. According to Valerie Miké of the New York Hospital, Cornell Medical Centre, "Of the six clinical trials carried out in the early 1950s with concurrent controls, not one found DES to be effective in preventing miscarriages."

Many hundreds of thousands of people have also been exposed to DES through its use as a growth promoter in livestock. Studies from Italy and Puerto Rico have incriminated DES residues in meat as a cause of premature sexual development of children.

In the USA, the use of DES in agriculture was finally banned in 1979, after a protracted legal and political battle dating back to the late 1950s. In Europe, the EEC imposed a ban in 1981, though DES had been outlawed in many member states many years previously. The ban was vigorously opposed by Britain. (See also: *Cancer* Ⓛ, *Carcinogen* Ⓕ Ⓛ, *Hormones in meat* Ⓐ Ⓛ Ⓟ.)

Dilution and dispersal
A philosophy that is fundamental to pollution control policies in most industrialized countries. Based on the assumption that, once in the environment, pollutants will be diluted and dispersed to harmless levels, it has been used to justify the dumping of wastes at sea, the discharge of wastes to rivers, and the emission of air pollutants via chimney stacks.

Given their vastness, it is easy to assume that the oceans or the atmosphere have an infinite capacity for assimilating pollutants. Time and again, however, pollutants that have been released on the asumption that dilution and dispersal will render them harmless have ended up bioconcentrated (see *Bioconcentration*) within specific species or, as with SO_2 emission, causing damage to ecosystems far away from the point of release – in the form of acid rain (*qv*). (See also: *Pollution.*)

Dioxins
A family of 75 closely related compounds, the most notorious of which is tetrachlorodibenzo-para-dioxin (or TCDD for short). It first achieved notoriety as the chemical released during the Seveso (*qv*) accident and as a contaminant of Agent Orange (*qv*). TCDD pollution was also responsible for the evacuation of Times Beach and is now thought to pose a major threat to drinking water supplies in the Niagara Falls, New York, area.

The dioxins have never been manufactured deliberately, having no uses in their own right. Nonetheless, they occur as inevitable contaminants of a number of products and manufacturing processes involving chlorinated phenols. Such products include the phenoxy herbicide 2,4,5-T (*qv*) (which contains TCDD) and 2,4-D (*qv*) which is contaminated by octadioxin.

In both France and the USA, buildings have been heavily contaminated by TCDD following fires involving the burning of electrical insulating equipment ontaining PCBS (qv). Dioxins – notably TCDD – have also been produced by the incomplete incineration of chlorinated wastes and domestic rubbish. Crematoria have also been found to release dioxins when plastics in coffins are burned. Recent work has suggested that the burning of gas from waste dumps may produce a dioxin hazard.

The health effects of the dioxins – particularly TCDD – are highly controversial. In 1984, the Carcinogen Assessment Group of the US Environmental Protection Agency (EPA) reviewed the published data on the health effects of TCDD and concluded that it was the " most potent animal carcinogen" it had ever evaluated. The agency estimated that TCDD was 50 times more carcinogenic (*qv*) than such well-established carcinogens as vinyl chloride (*qv*).

Industry does not dispute that TCDD is carcinogenic in animal tests, but disputes that it is harmful to humans in low doses – insisting that its worst effect is a severe skin disease known as chlorachne.

That view is contradicted by several major epidemiological studies, which show TCDD to be a cause of cancer, birth defects and other adverse health effects. According to the EPA, TCDD should be regarded as "probably carcinogenic in humans". Significantly, many experts now consider that there is no "safe" level of exposure to TCDD. (See also: *Cancer* Ⓛ, *Incineration* Ⓟ.)

Direct action
The name given to Greenpeace's own brand of environmental protest (see illustration). It is derived from the Gandhian technique of "Satyagrtaha" or "non-violent passive resistance". The method's success relies on the fact that it forces an opponent to defend his position. Pacifism and clear argumentation also allow the protester to retain a moral advantage. Undoubtedly, too, one of the reasons for the success of direct action has lain in its ability to capture the attention of the media. Photographs of the Greenpeace inflatable rafts steering themselves in between whalers' harpoons and their targets have a great deal of appeal, and they bring seemingly remote issues directly to the public's attention. Through direct action, Greenpeace has played a

Direct action – Greenpeace confronts the vested interests

71

Greenpeace ship pickets Amchitka, Alaska, atomic test site

Greenpeace is formed when a group of Canadians and Americans charter a vessel to take them into the area around Amchitka island, in the Aleutians, where the US had proposed to carry out several nuclear tests. "We want peace and we want to make it a green one" is the philosophy of the new organization, which quickly becomes international.

Result
Only one test of the proposed series is carried out. Today, the island of Amchitka is a bird sanctuary.

72/74

Vega rammed, crew beaten, at French nuclear test site

Mururoa atoll in the Pacific becomes a Greenpeace target as the French begin testing nuclear weapons. *Vega*, a 38ft sloop, sails into the 285,000sq km "forbidden zone". It is rammed by a French minesweeper and the crew is arrested. *Vega* returns to Mururoa: this time the crew are savagely beaten, one being partially blinded by a blow from a rifle butt.

Result
Film of the attack is smuggled to the world's press and an outraged public demands that the testing be stopped. The French halt atmospheric tests at Mururoa.

76/77

Seal slaughter target of protest by Greenpeace

The Canadian seal slaughter sees Greenpeace in Newfoundland: the Canadian government produces the Seal Protection Act to deny it access. Protesters are thrown into sub-zero temperature waters. In 1977 Greenpeace blocks the path of the sealing vessels and uses a harmless but indelible dye to spoil the commercial value of pelts.

Result
Continued action by Greenpeace contributes in 1983 to an EEC ban on imports of seal pelts and products. The seal slaughter is reduced to a sixth of its previous level.

78/79

Greenpeace keeps up pressure on whaling nations

"Pirate" whaling operations in Peru and Chile, and by Spain, are exposed by Greenpeace. A major offensive begins in Iceland to defend the threatened fin whales. Opposition to legal whaling leads to the arrest of *Rainbow Warrior*. In 1979 *Rainbow Warrior* returns to Iceland and is arrested again. An illicit trade to Japan in whale meat is exposed.

Result
Greenpeace's campaign and public opinion leads in 1981 to reduced whaling quotas. By 1987 all nations except Iceland, Japan, South Korea and Norway have stopped whaling.

82/83

Radioactive waste dumping delayed by direct action

Three members of a Greenpeace ship's crew narrowly escape injury from falling barrels of waste; occupation of dumping platforms and cranes delays dumping. Attempts are made to block the Windscale outfall pipe. The London Dumping Convention votes by 19 to 6 to suspend dumping of radioactive waste for two years.

84/85

Rainbow Warrior sunk in Auckland harbour

86/87

Greenpeace gets observer status at EEC conference

In the Antarctic, Greenpeace raises its World Park Antarctica flag in an attempt to prevent exploitation of the Continent. The *Beluga* analyzes the major pollution sources entering the North Sea and, as a result, Greenpeace is awarded observer status at the 1987 North Sea Conference of EEC Ministers.

STOP NUCLEAR TESTING

STOP WHALING

STOP SEAL SLAUGHTER

STOP WASTE DUMPING

major role in ending commercial whaling, nuclear dumping at sea, and the hunting of seal pups. However, use of the technique is not restricted to Greenpeace alone. Similar actions have been taken, less flamboyantly perhaps, all over the world – notably by the Chipko movement (*qv*) in India – and it would be no exaggeration to say that Satyagraha and its variants represent one of the main forms of public protest in the 20th century. (See also: *Green politics*.)

Diseases of civilization
Term currently used to denote a constellation of diseases whose incidence seems to increase with per capita economic growth. Included in this category are most forms of cancer (*qv*), coronary heart disease (*qv*), diabetes, tooth caries, peptic ulcer, appendicitis and diverticulitis (a disease of the bowel).

Though the incidence of these diseases seems to vary in different countries in direct proportion to per capita GNP, in the rich industrial countries it is generally highest among those on low incomes.

Many factors help to explain the association between the diseases of civilization and economic development. The principal factors would appear to be increased stress; increased exposure to chemicals and radiation (see *Radiation and health*); and the dramatic changes in diet that occur as society adopts a modern way of life, with natural, uncontaminated high-fibre wholefoods giving way to refined, processed foods that are often devitalized and contaminated with agricultural chemicals and food additives (*qv*).

The role played by diet and lifestyle as causes of the diseases of civilization has been stressed in numerous studies. One of the first to make the connection was a naval doctor, Surgeon Captain Cleave, who found that the diseases afflicting the industrialized world were largely absent among Africans living on traditional diets, but were present among Africans fed on a modern diet. Research carried out by D.P. Burkett (discoverer of Burkett's lymphoma), also in Africa, confirmed this thesis, as did that of Sir Robert McCarrison in India.

More recently, Dr. Albert Darmon of the Peabody Museum at Harvard University has established that the diseases of civilization are largely absent among Solomon Islanders, who still lead a traditional lifestyle.

Dr. Ian Prior of Victoria University in New Zealand has also shown this to be true of the inhabitants of Puka Puka, an isolated island of the Cooks group, though the diseases of civilization were much more in evidence on the nearby island of Rarotonga, which has undergone considerable modernization.

Prior's highly documented 30-year study of the health of the Tokelau Islanders also reveals that the diseases of civilization are largely absent among islanders who remain on isolated atolls, but develop very rapidly among those who migrate to New Zealand, where they are exposed to a modern diet and way of life.

If the poor in the industrial countries have a higher incidence of the diseases of civilization than the better off, as has been shown to be true in repeated studies in the UK, then, as Dr Richard Wilkinson of Sussex University points out, this can be largely attributed to the tendency of the poor to eat cheap, devitalized, processed foods and to their very low consumption of fresh fruit and vegetables. Though this thesis is becoming increasingly accepted, efforts to improve the diet of the working man in the industrial world have failed, largely because of commercial pressures.

In general, it would seem that the diseases of civilization are inextricably liked with our modern way of life and to combat them effectively would require socio-economic change. (See also: *Cancer* Ⓛ, *Junk food* Ⓛ, *Development* ③.)

Dounreay
Located on the northern coast of Scotland, Dounreay has been the centre of British fast reactor development since 1954. The first experimental reactor started up in 1959 and operated until 1981, and the larger 250MW prototype fast breeder reactor (*qv*) was ordered in 1966 and began operating in 1974. Development to commercial-sized reactors has been much more difficult than first envisioned, despite the large amounts of money spent. Fast reactor research is the largest single energy research item funded on the British Government, and the UK Atomic Energy Authority's (AEA) major project. Spending since 1963 has been at an annual rate of £90-140 million ($144-224 million) at present values, but the AEA has estimated that another 20 to 25 years' work, and £1,300 ($2,080) million (exclusive of £2,000 ($3,200) million construction costs), will be needed before a commercial station is feasible.

A collaborative agreement was signed with France, West Germany, Italy, Belgium and the UK in January 1984 to build three commercial demonstration fast reactors. The third of these will be in the UK, but the reprocessing plant to service all three is to be at Dounreay, built and operated by British Nuclear Fuels Limited and the UK Atomic Energy Authority. A public inquiry was held into the proposal between April and November 1986. In July 1987 a draft report gave the go-ahead for the plan. The proposed reprocessing plant has been opposed by environmentalists, including local fishermen who fear that fish stocks will be contaminated. (See also: *Creys Malville* Ⓝ, *Reprocessing* Ⓝ.)

Drinking water
See: essay on *Water Fit to Drink?*.

Drought-flood cycle
In recent years countries in the dry tropics have been subjected to increasingly severe droughts, alternating with equally devastating floods. Both phenomena have a common cause: environmental degradation, in particular deforestation.

Among the many ecological functions fulfilled by forests is the control of run-off to rivers, typically 95 per cent of the annual rainfall in the forested watershed of rivers in the tropics being trapped in the elborate sponge-like network of roots that underlies the forest floor. Only five per cent of the rainwater is released immediately to the rivers.

When the forest is removed, however, there is no longer any "sponge" to absorb the water. As a result, the situation is, in effect, reversed, five per cent of the rainfall being trapped in what remains of the soil and 95 per cent being released, often over a very short period, in local rivers. The result is massive flooding. Where settlements have been built on the floodplains, as is common, the effects of flooding can be devastating. (See also: *Development* ③, *Deforestation* Ⓒ ③, *Famine* Ⓐ ③.)

Dutch elm disease

A disease caused by a fatal partnership between a fungus *Ceratosysisulmi*, and two species of beetles, the greater and lesser elm bark beetle. The bark of trees killed by the fungus is used as breeding sites for the beetles. When the beetle larvae emerge as adults, they fly to healthy trees to feed, carrying with them spores of the fungus, which in turn infect new trees. Once the fungus enters a tree it can travel through it at as fast as a metre (3ft) an hour. It can also spread through the natural root grafts between trees.

Very little is known about where Dutch elm disease originated. The first recorded outbreak of the disease dates from 1918, though it may have been the cause behind the catastrophic decline in the European elm population 5,000 years ago. The first epidemic in historical times reached its height in Britain and the Netherlands from 1931 to 1937, killing up to 20 per cent of the trees in southern England. By 1929 the disease had crossed the Atlantic to the United States. During the 1960s, a new virulent strain of the fungus developed in the northern USA and Canada, killing many elms. In 1964, it arrived in Britain again through infected timber. By 1976, nine million out of a population of 23 million elms were dead. The disease effectively eliminated all elms in southern England, though some survived in Brighton. (See also: *Tree diseases* Ⓒ Ⓟ.)

Ecological community

See: *Ecosystem.*

Ecological movement

The ecological movement developed in the United States in the late 19th century. Ecologists distinguished themselves from conservationists and preservationists, with whom they nevertheless had much in common. Their response to the perceived ecological crisis of the times was a more fundamental one. It was our whole attitude to our relationship with the natural world that had to change. Man was an integral part of nature, and if he destroyed nature then he destroyed himself too. It followed that our policies should subordinate man's apparent interests to those of nature.

Today, as in the 1890s, ecologists differ from conservationists and environmentalists in that they see the ecological crisis as a fundamentally social one. Setting up national parks and adopting stringent pollution control legislation, they argue, is not enough to solve the ecological crisis; more fundamental changes are required, changes in the way we view our basic relationship with the natural world, changes in the way our society is organized in its basic social and economic goals. What is required is a new ecological world view – and a new ecological society geared to the maintenance of a stable or "steady state" relationship with the natural world. Such ideas underlie deep ecology (*qv*) and are reflected in the platform of green parties in different European countries. (See also: *Ecology*, *Green politics*.)

Ecological succession

An important principle of ecology is that ecosystems develop in a series of stages that must occur in the right order – a process referred to as "succession". This continues until a "climax" is reached – a situation from which there is then little change, as it is the most stable achievable in the circumstances.

The idea is an old one. However, in recent years, it has come under attack, the first critic being the biologist Professor Herbert

Gleason in the 1920s, though his views were not, at the time, taken seriously. Later, with the development of the New Ecology after World War II, Gleason's views were resuscitated and expanded, and are now generally accepted among academic ecologists.

Modern ecologists have abandoned the old concept of succession for three reasons. First, it implied that an ecosystem was a sort of "superorganism" whose development took the form of a strategy, rather than being the result of random changes, and that this strategy was coordinated by the ecosystem itself. These concepts are irreconcilable with the tenets of modern science, which sees the world as atomised and random. It also implied that the goal of ecological development is the achievement of stability (defined as a state in which discontinuities are minimized) whereas science, in line with our modern industrial society, is committed to perpetual change in a single direction. In the natural world, such change can occur only by reversing the successional process and by maintaining artificially the ecosystem at its most productive pioneering stage. This is ecologically, at least, the least advanced stage in the process of succession – that most marked by discontinuities such as floods, droughts, epidemics, and population explosions among animals, plants and microorganisms.

The principle of ecological succession to a climax is thus contrary to modern notions of nature being in a state of continual flux. Succession implies stability: hence actions that interfere with succession encourage instability. From the ecological point of view, therefore, modern economic development, which inevitably disrupts the process of succession, cannot be identified with progress.

Many ecologists, recognizing, nevertheless, that ecological succession is a fact, have offered theories intended to reconcile ecological considerations and industrial growth. Detailing them is beyond the scope of this book, and the reader is referred to the specialist literature.

Ecology

The term "ecology" has been defined in different ways by different ecologists, according to their particular world view. Some have taken ecology to be "scientific natural history", others have seen it as the science of how living things relate to their natural environment or as "the science of communities".

Many early ecologists saw ecology as an approach rather than a discipline. Barrington Moore, the first President of the American Ecological Society, declared at a meeting of his society in 1919 that the science of synthesis, and such a science was clearly essential if we were to understand the functioning of the world as a whole rather than isolated bits of it.

Unfortunately, the full development of such a science was not allowed to occur. Ecology, of the type taught in our universities, soon lost its holistic (see *Holism*) character and became yet another reductionistic and highly specialized discipline.

However, the term was taken up in the late 1960s by environmentalists who sought to develop philosophical and, indeed, scientific foundations for their efforts to preserve what remained of nature from destructive economic development. Such basic ecological principles as that of the balance of nature (*qv*), the importance of diversity as a means of increasing stability, the holistic principle that the whole is more than the sum of its parts, and the principle of ecological succession (*qv*) to a climax have been taken up by the ecological movement (*qv*), while paradoxically they have been largely abandoned by academic ecologists. (See also: *Deep ecology* Ⓕ.)

Economics and ecology

Economists are having increasing difficulty in predicting economic developments, which can no longer be understood simply by perusing the value of "leading indicators". *Business Week* has gone so far as to state that "when all forecasts miss the mark, it suggests that the entire body of economic thinking – accumulated in the 200 years since Adam Smith laid the basis for modern theory with his inquiry into the Wealth of Nations – is inadequate to describe and analyze the problems of our times".

What has gone wrong? The answer seems to be that modern economics is studied in almost complete isolation.

The economic process, however, does not occur within a closed system but within the world of living things, which it affects dramatically and which is, in turn, affected by it. As Lester Brown of the Worldwatch Institute writes. "Four biological systems – fisheries, forests, grasslands and croplands – form the foundation of the global economic system. In addition to supplying all our food, these four systems provide virtually all the raw materials for industry except minerals and petroleum-derived synthetics. The condition of the economy and of these biological systems cannot be separated".

In the last decades, the impact of our economic activities on fisheries, forests, grasslands and croplands has escalated, causing increasingly serious ecological degradation. In some cases, those "ecological costs" are now being translated into economic costs.

Current economic theory postulates that as ecosystems are degraded, so the demand for their produce increases – until it becomes economic to restore their proper functioning. The argument is flawed on a number of counts – not least because ecological degradation is rarely reversible, at least on a historic time-scale. However much people are willing to pay for the dodo, it can never be brought back, nor can the tropical forests we are systematically annihilating today, nor the arable land we are transforming into deserts – except on a purely symbolic scale.

For these and other reasons, a new economics is required, one that takes into account biological, ecological and indeed social factors. Such an economics would have practically nothing in common with the "closed-system" economics which are still allowed to guide present policies throughout much of the world.

Ecosystem

Term coined by the Oxford ecologist A.G. Tansley in 1934, though the concept is much older, to describe the basic functional unit in ecology. The ecosystem includes the "ecological community" – which is made up of closely interrelated living things – and its physical environment. These are inseparable and interact upon each other. (See also: *Ecological succession*Ⓕ, *Ecology*Ⓕ.)

El Niño

A recurring climatic phenomenon affecting most severely the Pacific coast of South America, but with possible effects over a much wider area. It has been suggested that a similar phenomenon may also affect the South Atlantic. The effects are first felt in December (the Spanish name means "the Christ child").

El Niño is not fully understood, but begins with a reduction in the trade winds in the tropical South Pacific. This reduces the "pushing" effect on surface waters, allows the layer of warm surface water in the eastern Pacific to accumulate, and prevents the upwelling cold water reaching the surface. Starved of nutrients that this water normally supplies, the surface waters are impoverished. Many seabirds starve and the commercial fisheries are seriously affected. It is is to this phenomenon that the failure of the anchovetta catch in 1964-65 is officially attributed. However, there is every reason to suppose that the main culprit was overfishing (*qv*). Indeed, the local fishing industry weathered the failure of El Niño in 1957-58, but when it failed again in 1964-65 the situation was quite different. After the collapse of the Californian sardine fisheries, there was a vast expansion of the anchovetta fishing industry in Peru which, between 1945 and 1962 increased by 200 times, making Peru the top fishing nation of the world.

In 1964-65, stocks collapsed from an estimated 25 million tonnes to 12 million tonnes, with a corresponding fall in catches. Climatic changes occurring in Central and North America and even worldwide have also been attributed to El Niño, but they are probably due to man-made factors such as deforestation and the generation of greenhouse gases.

Electromagnetic pollution

As a result of the "electrical smog" created by today's electro-

Electromagnetic pollution – the peril in power lines

In today's electronic world there are many sources of electromagnetic radiation, including CB radios, radar dishes, microwave ovens and high-voltage power lines. Long-term exposure to the extremely low-frequency (ELF) radiation emanating from power lines has been linked with a variety of illnesses, which in turn have led to court cases. In 1986, Houston Power and Lighting was ordered to pay $25 million in damages to the Klien School District for erecting a 345kV power line within 60m (197ft) of three schools. The company lost its appeal, and has moved the lines.

Low-frequency electromagnetic radiation is often assumed to be harmless because it is non-ionizing, the only danger lying in its ability to heat tissues, as in a microwave oven. But concern over the harm that such radiation can cause varies wildly around the world. The USSR, for example, regards all forms of electromagnetic radiation as potentially harmful. As a result, Soviet microwave exposure standards are 10,000 stricter than those in the West. In addition, the USSR is believed to have a one-kilometre safety corridor for its power lines.

The strength of the field produced by a power line depends upon many interrelated factors: the voltage, local topography, weather conditions, underground water and geological faults. For this reason it is difficult to give an accurate universal measurement of the field strength at varying distances from a power line.

The voltages used in power lines vary from country to country with the USA and USSR using up to 765kV, while the UK uses 400kV.

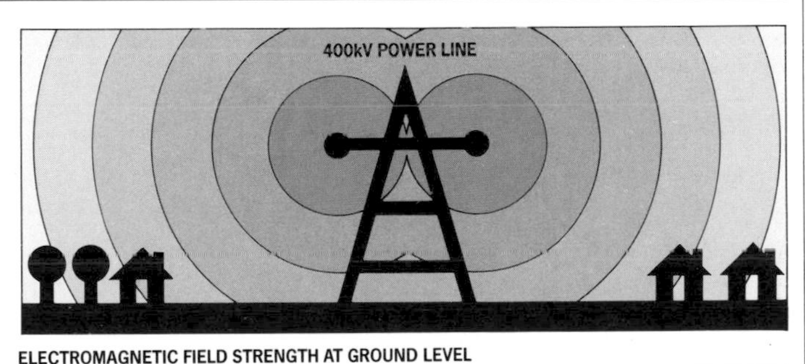

ELECTROMAGNETIC FIELD STRENGTH AT GROUND LEVEL

DISTANCE FROM CENTRE LINE (METRES) 0 100 200 300 400 500

How danger varies with distance

Some of the harmful effects of electromagnetic radiation are shown below. They illustrate the range of health effects that has been reported and give an indication of the distances involved. Included are epidemiological studies under 200kV and 33kV lines that have been appropriately adjusted to merge with the 400kV results. The symbols indicate the sources, which are given at the back of the book.

30m
- ● EPILEPTIC FITS
- ● LOSS OF MUSCLE POWER
- ● BLACKOUTS
- ● ALLERGIES
- ● INCREASED LEUKOCYTE COUNT
- ● EXHAUSTION
- ● PALPITATIONS
- ● LOSS OF CONCENTRATION
- ● HEADACHES

60m
- ● CLUSTERS OF HEART–ATTACK DEATHS
- ● EPILEPSY
- ● CANCER
- ● SEVERE PALPITATIONS
- ● BLACKOUTS
- ● RETINAL DRYING AND BURNS
- ● HEADACHES

105m
- ● RASHES
- ● PALPITATIONS
- ● DIZZINESS
- ● HEADACHES
- ● JOINT PAINS
- ● EYE BURNS
- ■ THYROID DISORDERS

150m
- ▲ INCREASED LEUKEMIA IN CHILDREN AND TUMOURS IN ADULTS

245m
- ● RARE EYE CANCER
- ● CONVULSIONS IN ALLERGIC SUBJECT
- ● EPILEPSY

305m
- ● REACTION IN ALLERGIC SUBJECT

nic gadgetry, the average American now receives a daily dose of electromagnetic radiation 200 times that from natural sources such as the sun. In particular, the extremely low frequency (ELF) radiation produced by high-voltage power lines is becoming a cause for concern (see illustration).

In 1973, a US Navy study reported significant biological effects among volunteers exposed to ELF fields one million times lower than those commonly found under high-voltage power lines. The report was suppressed for two years.

Another study revealed a 40 per cent higher incidence of blood cell disorders amongst employees at the US embassy in Moscow. The study had been commissioned after the Russians were discovered beaming microwaves at the embassy building, as part of their bugging operations.

Most recently, in 1987, a link has been suggested between childhood leukemia and exposure to ELF fields from electricity cables. A US study reveals that cancer rates are five times higher among children living near high-voltage pylons. (See also: *Cancer* Ⓛ.)

Emulsifiers

A large and diverse group of compounds that are added to food products to enable water to form emulsions with oils with which, otherwise, it would not form stable mixtures. They work by breaking up one component so that it can be dispersed in small droplets throughout the other.

Within the UK and EEC, emulsifiers are listed with E-numbers (*qv*) between E322 and E495, along with stabilisers (*qv*) and thickeners. These chemicals are generally controlled by permitted lists, and in most industrialized countries about 60 different compounds are allowed. Some of the most widely used emulsifiers are the lecithins (E322), polysorbate 60 (E435) and mono- and di-glycerides, known as the MDGS, (E471), though emulsifiers are often used in complex combinations.

Emulsifiers are invaluable in the manufacture of products such as margarine, salad dressing, baked products, chocolate and ice cream, but they are also to be found in some meat products, confectionery, and dessert and whipping products. Lecithins are found in egg yolks, but are also extracted commercially from soya bean oil. While these naturally-occurring emulsifiers were generally considered sufficient until the 1960s, subsequent technological developments in the food industry have depended on a complex range of synthetic compounds, such as the MDGS, which have emerged more recently from the laboratory. There is less concern about the safety of emulsifiers than about most other classes of food additives. (See also: *Additives* Ⓛ, *Stabilizers* Ⓛ.)

Encroaching cultivators

Peasant colonists who move into forested areas in order to farm. Also misleadingly referred to "shifting cultivators", they have been unfairly accused of being the prime agents of deforestation (*qv*), particularly in the tropics. In fact, most encroaching cultivators are themselves the victims of the development (*qv*) process. Many have been uprooted from their own lands to make way for plantations, dams, livestock-rearing schemes and other large-scale projects. If they are to survive, they have little option but to colonize the forests. In general, encroaching cultivators clear land using slash and burn (*qv*) techniques, a method of agriculture that has been practised successfully for millenia by "swidden" cultivators. In inexperienced hands and when carried out at an individual, rather than on a communal level, however, slash-and-burn farming can cause widespread destruction, particularly when fallow periods are not observed. (See also: *Swidden agriculture.*)

Endrin

A chlorinated hydrocarbon (*qv*) insectide chemically similar to aldrin and dieldrin (*qv*). It is persistent in the environment and toxic to a wide variety of living organisms. It is highly toxic to humans but has not been used as widely as dieldrin, which it resembles closely in its environmental and toxic effects. (See also: *Biocides* Ⓟ, *Chlorinated Hydrocarbon* Ⓑ Ⓟ, *Dieldrin* Ⓟ, *Insecticides* Ⓟ, *Pesticides* Ⓟ.)

Energy

Literally, the measure of the ability to do work, existing as potential, or stored energy, or as kinetic energy, the energy of motion. Humans derive their bodily energy from the metabolization of carbohydrate foods, converting potential energy to kinetic energy.

For something like 95 per cent of our tenancy of this planet, humans only needed a little extra energy mainly in the form of firewood for camp fires and cooking.

With the Neolithic Revolution and the development of settled agriculture, energy consumption increased dramatically, mainly through the use of draft animals. With the development of industry, the energy of moving wind and water was harnessed and that of fossil fuels, mainly coal (*qv*), burned to convert chemical (potential) energy into heat. The great spurt in economic activity since World War II was based on a massive increase in the use of oil (*qv*) as an energy source. This tapered off in 1973 with the first oil shock. Since then most industrial countries have sought to reduce their dependence on oil by increasing their consumption of other fuels such as natural gas, coal, fissionable uranium (*qv*) and renewable sources of energy, such as wind and water, and by adopting various energy conservation (*qv*) strategies.

In 1973, the total amount of energy used worldwide amounted to 135.2 millions of barrels of oil equivalent. By 1984, this had increased to 163.6. In 1973, oil provided 41 per cent of the energy used, natural gas 16 per cent, coal 25 per cent, renewables 17 per cent, and nuclear one per cent. By 1984, oil provided only 35 per cent of total energy consumption, while natural gas provided 17 per cent, coal 27 per cent, renewables 18 per cent and nuclear three per cent.

Oil, of course, still remains the most important source of energy and, though it can be replaced for certain uses such as electricity generation, is difficult to replace as a liquid fuel for powering motor vehicles. Coal production, which increased between 1973 and 1984 by the equivalent of 10 million barrels of oil a day, is largely being used to fuel new coal-fired power stations. World coal reserves are enormous and the World Coal Study predicts that production will continue to increase at 4.5 per cent a year, which means that consumption of coal would triple between 1972 and the year 2000. Other energy studies have also accentuated the increased role coal is likely to play in world energy consumption in the coming years. Concern over pollution from coal-fired plants, however, may lead to a concerted effort to reduce coal consumption.

In the decade between 1973-84, the consumption of natural gas increased particularly rapidly – by 90 per cent for the world as a whole – except in the USA, where for various technical reasons, it fell steadily during the seventies. World reserves of natural gas are considerable. Estimates have risen 34 per cent in the last ten years. Proven worldwide reserves were 3-4 quadrillion cubic feet in 1985, which is the equivalent of 590 million barrels of oil. They are just 15 per cent lower than oil reserves, but at the present rate of natural gas consumption, provide only half as much energy as oil. According to the World Bank, natural gas can be made available in many parts of the Third World for between $2 and $12 a barrel of oil equivalent, which is far lower than the cost of oil. The potential for increasing natural gas consumption is thus considerable.

Nuclear power was once considered to be the main replacement for oil in many countries. Vast investments were made in nuclear power and indeed nuclear power generation has increased worldwide more than five times since 1973. It is likely to continue increasing because of the many plants ordered but still unfinished. However, because of the high cost of construction, the unsolved problems of disposing of high level waste, the still unknown problems associated with the decommissioning of old reactors and increasing public fears about safety, there is growing disenchantment with nuclear power in many countries. In the US, no new plants have been ordered since 1978 while 108 have been cancelled since 1974. The outlook for the nuclear industry is indeed grim. Christopher Flavin of the Worldwatch Institute in Washington DC does not consider that nuclear power is likely to provide more than six to eight per cent of world energy consumption by the year 2000.

Renewable sources of energy provide about six times as much energy worldwide as do nuclear power plants. Of these, hydropower provides about a third and wood fuel and different waste material provide most of the rest. The contribution of solar collectors (*qv*), wind generators, wave power (*qv*) and other alternative sources of energy is still very small though it is growing rapidly. On the other hand the use of wood fuel is growing slowly, at about 1-2 per cent per annum, largely as a result of the increasing shortage of firewood in Third World countries due to deforestation (*qv*). Many poor people in the Third World are increasingly forced to burn cow dung and other agricultural residues that should be returned to the soil if its fertility is to be maintained.

Perhaps the most important means of reducing our dependence on oil, is energy conservation (*qv*). According to Flavin it has exceeded the contribution of all new sources combined and the potential for further increases remain high. Ecologically this is the most desirable approach to solving the world's energy problems. Indeed, to satisfy ecological requirements, energy use must be reduced rather than increased.

Several studies have argued that energy consumption provides a measure of our impact on ecological systems. They point to the ecological costs not only of producing energy but also of using it. Coal and uranium mining operations, for example, are generally destructive of the environment and detrimental to human health, with miners in both industries suffering high rates of lung disease due to exposure to dust and pollutants. Power generation is also highly polluting: emissions from oil and coal fired power stations are a major cause of acid rain (*qv*) and of the build-up of greenhouse gases; while nuclear power stations give rise to routine radioactive emissions (*qv*) and, when accidents occur, to larger releases of radioactive materials (see essay on *Nuclear Energy After Chernobyl*).

In addition, many of the uses to which energy is put in a modern industrialized society are themselves destructive of the environment: industry, for example, is a major source of pollution, while energy-intensive agriculture is a prime cause of erosion and desertification, particularly in the vulnerable soils of the tropics.

Indeed, if our biosphere (*qv*) is to remain capable of supporting complex forms of life, there is a pressing need to reorganize society so as to reduce energy consumption rather than blindly seeking to increase it in order to accommodate our present destructive and non-sustainable activities. (See also: *Alternative energy*ⒸⒺ, *Energy conservation*ⒸⒺ, *Fuelwood crisis*③.)

Energy conservation

The application of measures designed to increase the efficiency with which fuel is used and so reduce the amount of fuel needed to sustain a particular level of production in an industrial or domestic enterprise.

Energy conservation techniques fall into two categories: insulation (*qv*), and increased mechanical efficiency. Insulation reduces heat losses. Improving the efficiency with which machinery uses energy usually involves modification of the design.

After the dramatic increase in the price of oil in 1973, systematic efforts were made throughout the industrial world to increase energy efficiency, with impressive results. In the US, cars will now travel 29 per cent farther on a gallon of gasoline than they did in 1973. New cars in the US get about

26 miles to the gallon as against 17 mpg in the early 1970s – still a poor performance by European and Japanese standards. A prototype Volvo can now achieve 63 mpg in the city and 80 mpg on the highway, while a Toyota prototype can achieve 89 mpg in the city and 110 mpg on the highway. The engines of such cars make use of plastic and ceramics, which are also used in the bodies. However, new technologies such as direct injection diesel engines are also used.

Similar savings have been made in many other fields, both domestic and industrial. In the EEC, the ratio of primary energy use to gross domestic product (GDP) fell by about eight per cent between 1973 and 1978, while prime energy use increased only by 0.42 per cent. Significantly, about 95 per cent of all EEC economic growth was fuelled by energy conservation and only five per cent by expansion of energy generation. Denmark was able to reduce its total use of direct fuels by 20 per cent between 1979 and 1980.

In the USA since 1973, the amount of energy needed to achieve a given level of gross national product (GNP) has dropped by more than 25 per cent.

More is achievable. In Britain, for example, a study by the Open University Energy Research Group suggests that a programme of energy conservation could reduce peak electricity demand in the year 2,000 by four gigawatts – the equivalent output of four nuclear reactors.

The cost of the programme was estimated at $1,600 million in 1983, but it would save between $4,000 million and $7,000 million in oil and coal costs. (See also: *Alternative technology* Ⓔ, *Energy* Ⓔ, *Oil* Ⓔ.)

E numbers

A system of identification for food additives that was introduced in the EEC during the late 1970s, hence "E" for European. Taste additives (*qv*), the largest group of additives, do not have E numbers. The authority for ascribing an E number to an additive lies with the European Council of Ministers, but it acts on the advice of the European Commission and the Scientific Committee for Food.

Before the introduction of this system of identification, European consumers often received no information on which chemicals were being used in the foods and drinks they were buying. For example, British labelling regulations used to require at most the use of expressions such as "permitted colouring" or "permitted preservative".

The intiative to develop the E numbering system was taken by the food industry in response to proposals to ban all of those food additives that were suspected of provoking symptoms of acute intolerance such as hyperactivity, asthma, eczema or migraine. Industry preferred to reveal their presence on labels, rather than having them banned altogether. The responsibility for protecting consumers has thus devolved from government onto the consumers themselves. (See also: *Additives* Ⓛ.)

Environmental refugees

Term applied to those who have been forced to leave their homes or their lands as a result of environmental degradation, natural disasters or development projects. They are often also referred to as "development refugees".

Accurate figures are hard to obtain but a large proportion of the 2,000 to 3,000 people made homeless every day may be presumed to be environmental refugees. The causes of their plight vary, but the most obvious are large dams (*qv*), plantation projects, livestock rearing schemees, and desertification and soil salinization, often caused by such development schemes.

Industrial accidents that require the evacuation of local inhabitants are also a cause of environmental refugees. More than 220,000 people fled Bhopal (*qv*) in the wake of the 1984 accident at Union Carbide's pesticide (*qv*) plant. In the industrialized world, notable examples of accidents which have led to evacuations include: Love Canal (*qv*), Seveso (*qv*), Three Mile Island (*qv*) and Chernobyl (*qv*).

Erosion

The removal of soil due to the action of the wind and rain. Under natural conditions, erosion is an extremely slow process, but as a result of deforestation (*qv*), overgrazing (*qv*) and inappropriate agricultural practices, the Earth's topsoil is now being eroded at an alarming rate.

Erosion is today a worldwide phenomenon, but is particularly serious in the dry tropics where the soils are very valuable.

- In India, where 800,000sq km (300,000sq miles) of land are affected, 6,000 million tonnes of topsoil are lost annually.
- On the steeper slopes of the Highlands of Ethiopia, annual topsoil losses amount to some 269 tonnes per hectare – 1,600 million tonnes a year.
- In the USSR, the annual loss is 2,500 million tonnes, with 2,320,000sq km (895,000sq miles) – ten per cent of the agricultural land – being affected.
- In Australia, six tonnes of topsoil are eroded for every tonne of produce grown – and in some regions, the figure is 300 tonnes.
- In Britain, which has long denied having soil erosion problems, a 1987 survey reveals that nearly 37 per cent of agricultural land in England and Wales – nearly 20,000sq km (7,700sq miles) – is subject to a "higher than acceptable degree" of erosion. In the European Community as a whole, some 140,000sq km (54,000sq miles) are at risk.
- In the USA, one-third of cropland is seriously affected by erosion. Despite expenditure of more than $15,000 million on soil conservation since the mid-1930s, topsoil losses are 25 per cent greater than in the dustbowl years of the 1930s. By 1981, the US was losing 4,000 million tonnes a year, enough soil to fill a train of freight cars long enough to circle the Earth 24 times.

According to the US Department of Agriculture, (USDA) 430,000sq km (166,000sq miles) of land now suffer from unacceptable levels of erosion. Much of that land may soon have to be retired from agricultural use. Production in the corn belt could decline by 15-30 per cent over the next 50 years due to erosion. In Illinois alone, average grain yields fell by two per cent between 1979-84.

In fact, the state of soil erosion in the USA may be even more severe than the official figures suggest. Many soil specialists argue that the USDA has overestimated the rate at which soil is replenished, leading to over-optimistic "tolerance levels" being set for erosion in many areas.

Belatedly, a few governments worldwide are now beginning to implement soil conservation measures, albeit on a small scale. Remedial measures include: reforestation (*qv*), terracing, reduced tillage, and the reintroduction of rotation with grass leys.

It seems clear, however, that no long-term solution to soil erosion is possible under an economic system that continues to encourage over-production, deforestation, and overgrazing regardless of the long-term consequences. (See also: *Deforestation* Ⓒ ③, *Global 2000* Ⓒ, *Overgrazing* Ⓐ ③, *Sedimentation* Ⓒ ③.)

Eucalyptus

Australia is the least forested of all continents, with the exception of Antarctica. At the time of European settlement in 1788 only nine per cent of the land area was forested. Less than half this area is left in 1987.

Eucalyptus species dominate the Australian forests and to a large extent determine the distinctive character and feel of the continent. In ecological terms eucalypt forests share many characteristics, that in combination, and often singly, make them unique. For example eucalypt forests prosper on extremely poor soils, are outstandingly adapted to frequent fire, are genetically flexible, tolerate an extraordinary range of ecological conditions and support a unique array of arboreal fauna.

Eucalypt forests are being clearfelled and converted into tree farms to be cut every 40-80 years by forestry operations. A radically impoverished habitat results, with soil nutrient levels falling and major ecological imbalances occurring. The re-growth is not suitable for many birds and arboreal marsupials for at least 150 years, which means that many of these animals could be driven to extinction.

The major catalyst for the onslaught against the eucalypt forests is an industry known as "woodchipping". Ancient eucalypt forests are clearfelled and much of the timber is converted to small wooden chips, which are then used to make paper pulp. Most of the woodchips are exported to Japan, with the industry having grown from 770,000 tonnes a year in 1971 to levels in 1987 of five million tonnes a year. More than a kilometre of eucalypt forest is clearfelled every day to feed the woodchip industry.

Environmentalists in Australia are vigorously opposing the current woodchip schemes, as well as plans to expand the industry, and in July 1987 turned woodchipping into a major Federal election issue. In a nationwide "Vote for the Forests" campaign the environmental movement was critical in returning the Australian Labor Party to power. Outside of

Eucalyptus – thirsty trees in arid landscapes

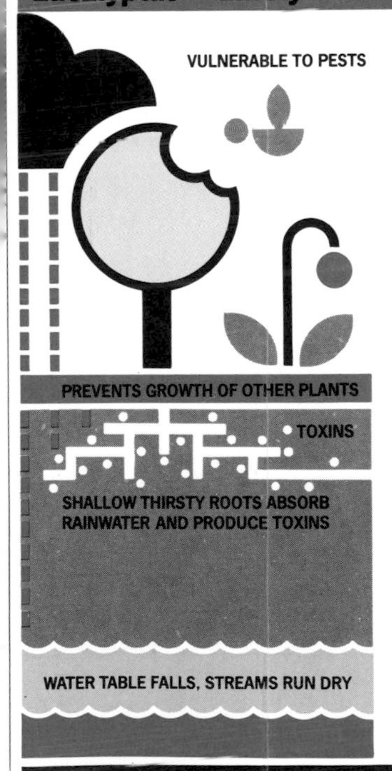

Today there are some 40,000sq km under eucalyptus around the world. This fast growing species was originally chosen to reafforest areas threatened by desertification, but environmentalists now see it as an ecological disaster. In its native Australia it occurs in areas of good rainfall, but planted in many arid Third World countries another side of its nature has been revealed. Its thirst for water inhibits the growth of neighbouring plants, depriving farmers of valuable fodder, and prevents the replenishment of the groundwater. The low rainfall means that the toxins that it produces are not flushed from the soil, remaining for a long time after the removal of the tree. Such is the depth of feeling that villagers in Karnataka, India, have resorted to uprooting the eucalyptus saplings, which are often no faster growing than many indigenous species; India alone has 20 native species with faster growth rates.

Eucalyptus depletes the soil of nutrients, returning far less in leaf litter than it absorbs through its roots. Over a year 100sq m of a hybrid eucalyptus plantation requires:

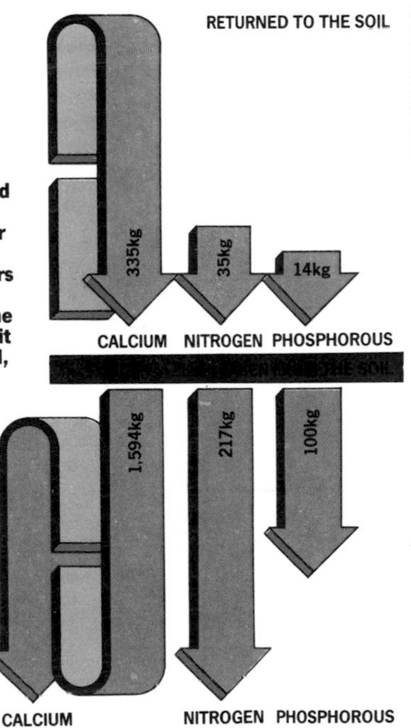

Australia, there is increasing concern over the use of eucalyptus as a timber crop and in reforestation projects (see illustration). (See also: *Deforestation* Ⓢ, *Reforestation* Ⓒ.)

Eutrophication

The process by which a lake or stream becomes richer and richer in plant nutrients until, eventually, plants overgrow and their decomposing remains cause the deoxygenation of the water, which becomes foul-smelling and virtually lifeless.

Nitrate fertilizers draining from fields cause serious eutrophication in some areas, as do nutrients from animal wastes. These, like human sewage, also have a high biochemical oxygen demand (see BOD) which exacerbates their effect on the waterway. In some badly affected areas – such as Britain's Norfolk Broads – the key nutrient is phosphate (*qv*), which is used in detergents as a water softener and enters the waterway via sewage. The North Sea (*qv*) is now also subject to algae blooms, which many scientists blame on the dumping of sewage at sea.

Eutrophication can be prevented by treating sewage to remove nitrate and, in some cases, phosphate. Banning the use of phosphate in detergents would help, but there is strong resistance to this from manufacturers. Restrictions on the use of nitrate fertilizers may be necessary and it is important that slurry and other animal wastes are not allowed to enter streams or lakes. (See also: *Artificial fertilizers* Ⓐ Ⓟ, *Nitrate* Ⓐ Ⓟ. *Biochemical oxygen demand* Ⓕ Ⓟ.)

Export of pollution

As pollution controls are tightened in the industrialized world, many companies are moving their most polluting industries to the Third World, where pollution controls are either less strict or non-existent. Examples include the asbestos (*qv*) industry, the steel industry, and the vinyl chloride (*qv*), pesticide (*qv*), chromate and chloralkali industries. Many have been temped to the Third World by "freedom to pollute" agreements.

The relocated factories operate at far lower standards than would be permitted in their home countries – both in terms of discharges to the environment and worker safety. Dr. Barry Castleman, a US environmental consultant specializing in the export of hazardous industries, reports cases of workers not being informed of the dangers of their work, of defective machinery, of safety codes which are mandatory in the US not being implemented by US subsidiaries, and of high levels of pollution. The Bhopal (*qv*) diasaster resulted from just such double standards.

In addition to exporting hazardous industries, Western companies have also taken advantage of lax safety standards in the Third World to export banned or hazardous produce and goods. Such products include banned pesticides and unwanted drugs. In the wake of Chernobyl, several European companies exported food contaminated with levels of radiation above the 600 becquerels per kilo permitted in the EEC. Milk, lamb and game contaminated with 1,000 becquerels per kilo were exported by Britain but were returned as unfit by the health authorities in Iran, Kuwait, Egypt, Israel, the Phillipines, Hong Kong and Singapore. Britain's Ministry of Agriculture insisted the food was safe – though it would never have been permitted to enter Britain. In West Germany, the Government banned similar exports.

Recently, there has been an increase in the export of hazardous wastes to the Third World, with wastes being shipped from the US to Sierra Leone, Haiti, the Dominican Republic and Mexico. (See also: *Circle of poison* Ⓟ.)

Extinction

Since life first began on Earth, numerous species of animals and plants have fallen prey to extinction, either because they were unable to survive or because they evolved into other forms of life. Today, species are being lost at such a rate that scientists warn that we are heading for the kind of mass extinction that occurred at the end of the Cretaceous Period, some 65 million years ago. But whereas previous mass extinctions have been largely the result of climatic and geophysical factors, today's extinctions are almost exclusively man-made (see illustration).

As wildlife habitats are destroyed, so their indigenous species disappear with them. The loss of the world's tropical forests (*qv*) alone could account for the extinction of a third of the world's species: other critical habitats under threat include coral reefs (*qv*) and wetlands (*qv*). In addition, wildlife must contend with pesticide (*qv*) pollution, hazardous wastes (*qv*), acid rain (*qv*), the degradation of farmland, coastal development and hunting.

Species are now being lost at an unprecedented rate – some 400 times faster than at any other period during recent geological time – and the range of species affected is far wider than ever before. As the US biologist Edward O. Wilson, Professor of Science at Harvard University, points out: "In at least one respect, this human-made hecatomb is worse than any time in the geological past. In earlier mass extinctions . . . most of the plant diversity survived: now, for the first time, it is being mostly destroyed."

The long-term outcome of today's extinctions cannot be predicted with precision: but the loss of species cannot continue with impunity forever. Sooner or later, the Earth's life-support systems will simply be overwhelmed. To avoid such a disaster will take more than the setting up of national parks or better land-use: it will require a fundamental reappraisal of our whole way of life, with priority being given to ecological imperatives rather than the pursuit of economic gain. (See also: *Economics and ecology* Ⓕ, *Genetic diversity* Ⓒ, *Genetic resources* Ⓒ.)

Extinction – pinpointing the vanishing species

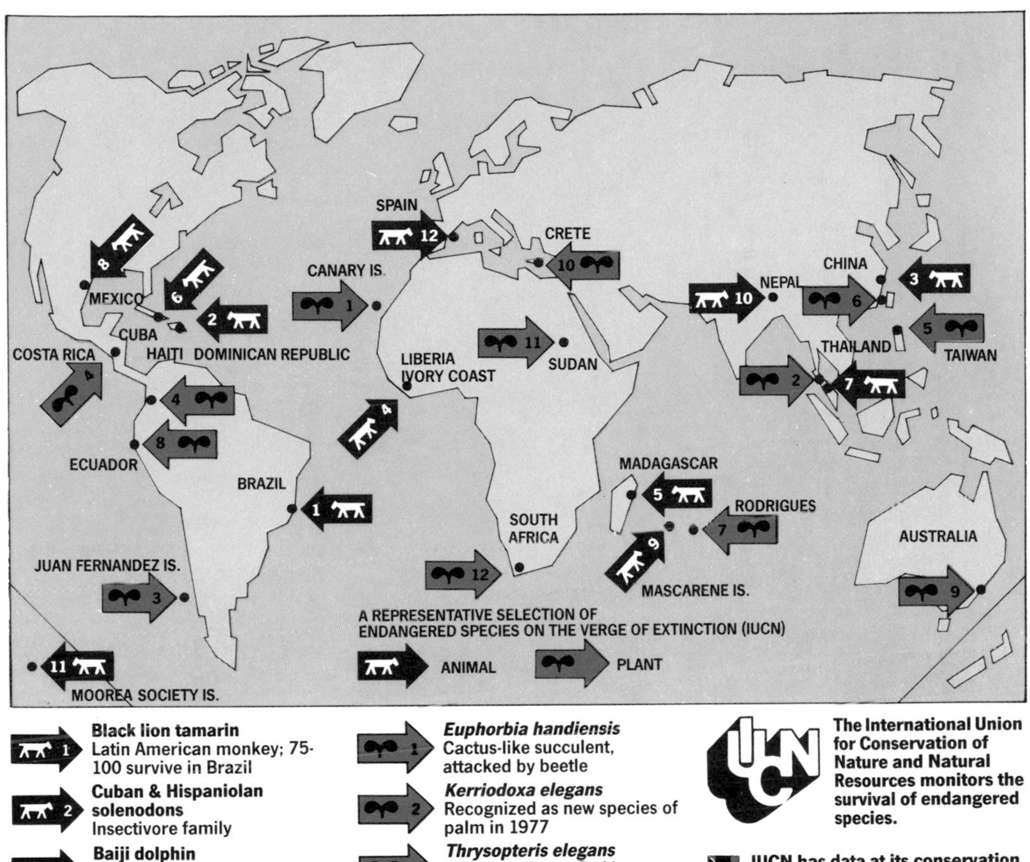

1 **Black lion tamarin**
Latin American monkey; 75-100 survive in Brazil

2 **Cuban & Hispaniolan solenodons**
Insectivore family

3 **Baiji dolphin**
Near-blind dolphin confined to Changjiang River, China

4 **Jentink's duiker**
African forest antelope; more information needed

5 **Madagascar serpent eagle**
Not seen for more than 50 years, but may survive

6 **Ivory-billed woodpecker**
Presumed extinct in USA, sighted in Cuba

7 **Gurney's pitta**
Nesting pair found in Thailand in 1986

8 **Kemp's ridley turtle**
Nests on one Mexican beach, protected by armed marines

9 **Round Island boa snake**
Has divided upper jaw; about 75 survive

10 **Relict Himalayan dragonfly**
Co-existed with dinosaurs, but now endangered

11 **Moorean viviparous tree snail** Threatened by carnivorous snail

12 **Valencia toothcarp**
Victim of mosquito fish and tourism

1 ***Euphorbia handiensis***
Cactus-like succulent, attacked by beetle

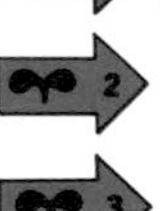

2 ***Kerriodoxa elegans***
Recognized as new species of palm in 1977

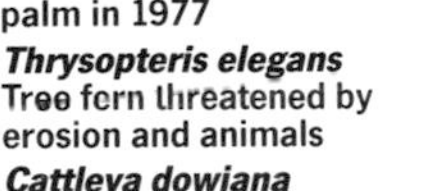

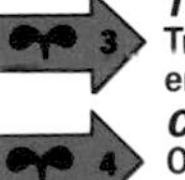

3 ***Thrysopteris elegans***
Tree fern threatened by erosion and animals

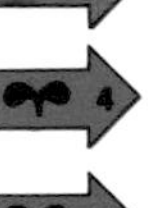

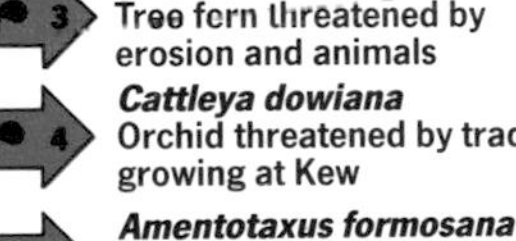

4 ***Cattleya dowiana***
Orchid threatened by trade; growing at Kew

5 ***Amentotaxus formosana***
Rare conifer; habitat needs protection

6 ***Abies beshanzuensis***
Fir in protected reserve; but now threatened

7 ***Ramosmania heterophylla***
Wild coffee tree; one remains. Potted at Kew

8 ***Dicliptera dodsonii***
One vine survives in a biological centre

9 ***Swainsona recta***
Purple pea reduced by heavy grazing

10 ***Cephalanthera cucullata***
Orchid: declined by 90% from 1972 to 1979

11 ***Medemia argun***
Ancient palm; seen 1960, survival uncertain

12 ***Erica pilulifera***
Heather threatened by other plants and trampling

The International Union for Conservation of Nature and Natural Resources monitors the survival of endangered species.

IUCN has data at its conservation monitoring centre on more than 30,000 endangered species. Most of these are threatened by the loss of their habitats through human activities.

Species of fish, birds, mammals, amphibians, reptiles and invertebrates threatened

Species of flowering plants threatened

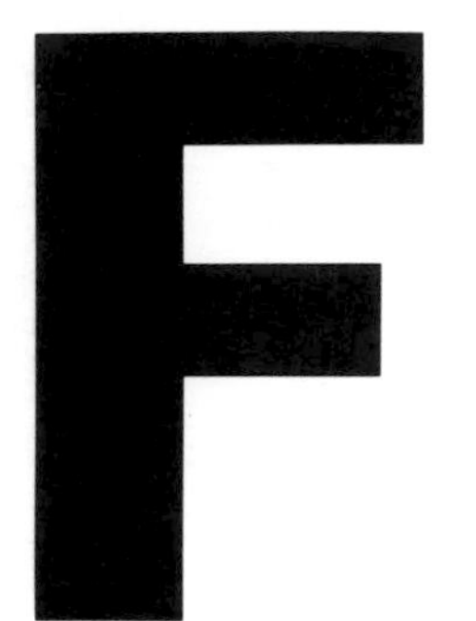

Family planning
The official term for birth control, which has unacceptably authoritarian connotations. Its use implies that it is for a family to decide whether to limit its size, and that the role of government is restricted to providing families with the means of so doing.

In practice, many governments today actively seek to persuade families to limit their numbers, either through advertising campaigns extolling the virtues of contraception, or, as in India during the 1970s, by more sinister methods, including compulsory sterilization.

The emphasis placed on family planning programmes as a means of reducing population growth has been criticized on many grounds. One is that modern methods of contraception can have serious health effects.

Another is that family planning programmes are exorbitantly expensive. Indeed, the World Bank estimates that to achieve "a rapid fertility decline" in Sub-Saharan Africa would mean increasing the funds spent on family planning 20-fold by the end of the century.

But the most serious criticism of all is that family planning programmes fail to tackle the root cause of the population explosion (*qv*) in the Third World. They stand or fall on their success in persuading people to use contraceptives. But ignorance of contraception is not a major problem in the Third World, where traditional contraceptives (including cultural taboos on sex at certain periods) are widely available.

Rather the problem lies in an unwillingness to limit the size of families, with people in many parts of the Third World actively seeking as many children as possible, in part to compensate for the social alienation caused by the development (*qv*) process. Under such circumstances, the distribution of free contraceptives and advice on contraception has little impact.

Family-planning programmes, however, have a role to play as part of an integrated population control campaign. (See also: *Population* ③, *Demographic transition* ③.)

Famine
For the first time in human history we are witnessing famine on a continental scale, with two-thirds of African countries affected. Within a few years, famine is likely to spread to other major areas of the globe – and will eventually become both a global and a chronic phenomenon.

Four factors are officially seen as causes: unfavourable climatic conditions, in particular the terrible drought (*qv*) that has affected many Third World countries over the last decades; the population exlosion (*qv*) which is such that there is no longer enough food to go round; archaic, low-yielding traditional agriculture (*qv*); and poverty, as a result of which people do not have the money to buy food.

The first is seen as "an act of God" against which we can do nothing. The solution to the other three problems, however, is seen to lie in economic development (*qv*). This, it is assumed, will assure that Third World people can afford to modernize their agriculture, buy their food, and acquire the financial security, that, in the industrial world, has made it unnecessary for people to have a large number of children to look after them in their old age.

Each of these arguments can be seriously contested. To begin with, many climatologists now dispute that the drought is "an act of God", arguing instead that it is the result of the terrible deforestation (*qv*) of the last 40 years. This has reduced rainfall in certain areas and has led to the degradation of the soil, which severely reduces its water-retaining capacity, so much that drought occurs even in areas where rainfall is constant.

Secondly, the population explosion is itself the inevitable result of development, which has destroyed culturally determined population control strategies, while creating the insecurity that leads people to have more children.

The evidence also suggests that while traditional agriculture (*qv*) is superbly adapted to local social, ecological and climatic conditions, modern mechanized farming has contributed to malnutrition and famine. In particular, mechanization has made vast numbers of small farmers redundant, depriving them of their land and turning them into landless labourers or slum dwellers, who are vulnerable to famine. In addition, modern agriculture has had a very damaging impact on fragile tropical soils.

A feature of traditional agriculture is that it is geared to subsistence rather than to production for a distant market. According to Karl Polanyi, the economic historian and author of *The Great Transformation*, the development of the market system has been the main cause in the increase in the incidence and seriousness of famines throughout history.

Polanyi showed that the great famines of India during the British Raj were mainly due to the operation of market forces. There had always been crop failures, but these did not necessarily lead to famine, as farmers would conserve sufficient stocks, either in their homes or at a village level. With the development of the centralized market economy, however, peasants, as in most of the Third World today, had to sell their grain to city merchants, usually at a price fixed by the government well below that obtainable on the open market. The result was far greater vulnerability to crop failures. When these occurred, peasants, forced to sell their grain rather than eat it, now had to buy it back at an inflated price which, needless to say, they were not able to do. Hence the extraordinary spectacle of famine in the countryside but not in the cities.

At the same time, the market system, by its very nature, caused radical social changes. The communitarian society that provided people with great security, slowly disintegrated, as vast plantations took over from small farming communities, as peasants moved to the cities in search of jobs, and as the state took over an increasing number of functions that were previously fulfilled at a communal level. This meant that when there was a crop failure, peasants now had to fend for themselves.

Polanyi's analysis applies only too well to the current African famine. In the Sudan, for instance, the recent famine was largely the result of a massive increase in the production of cash crops, in particular cotton, which rose from 518,000 million bales in 1982-83 to 1,200,000 bales in 1983-84. Inevitably, production of sorghum, the staple food for local consumption fell from 3,350,000 tonnes to 1,980,000 tonnes. Moreover, the price of sorghum increased dramatically as a result of food speculation – the inevitable concomitant of famine in a market economy. Stocks, instead of being distributed to the starving, were hoarded by merchants and resold at a vast profit, so much so that price of a bag of sorghum rose 28-fold. How, in such conditions, could famine be avoided?

Famine is then best seen as the result of a constellation of closely associated changes brought about by economic development – changes that adversely affect the land's capacity to feed people, the number of people it must feed, and the way the food is distributed. (See also: *Cash crops* Ⓐ ③, *Demographic transition* ③, *Desertification* Ⓒ ③, *Drought-flood cycle* ③, *Erosion* Ⓒ, *Green Revolution* Ⓐ ③.)

Farakka Barrage – salt takes the place of silt

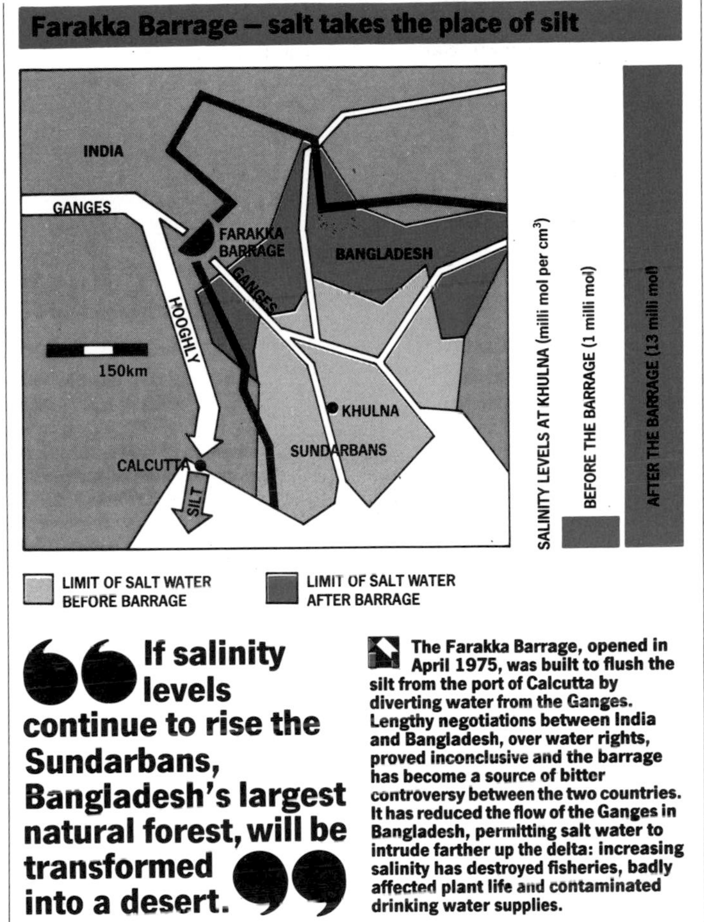

"If salinity levels continue to rise the Sundarbans, Bangladesh's largest natural forest, will be transformed into a desert."

The Farakka Barrage, opened in April 1975, was built to flush the silt from the port of Calcutta by diverting water from the Ganges. Lengthy negotiations between India and Bangladesh, over water rights, proved inconclusive and the barrage has become a source of bitter controversy between the two countries. It has reduced the flow of the Ganges in Bangladesh, permitting salt water to intrude farther up the delta: increasing salinity has destroyed fisheries, badly affected plant life and contaminated drinking water supplies.

FAO

See: *Food and Agriculture Organisation.*

Farakka Barrage

See accompanying illustration. (See also: *Dams* Ⓐ ③, *Salinization* Ⓐ ③, *Water transfer projects* Ⓐ ③.)

Fast breeder reactor

Breeder reactors are often termed fast breeders because they use fast neutrons to bring about the chain reaction. A prime purpose of the fast breeder is to generate more plutonium than is actually consumed in the chain reaction. In time, there should be sufficient surplus plutonium to start up a new fast reactor and give rise to a whole succession of second generation plants.

The period taken to obtain enough fuel for a new fast reactor is known as the doubling time. If the reactor is to make any impact on power requirements, it is essential that the doubling time be as short as possible, a couple of decades at most. To achieve a reasonable doubling time, the spent fuel from fast reactors must be reprocessed – and the sooner after removal from the core the better. However, spent fuel from fast reactors is much more radioactive than that from conventional reactors and reprocessing can present both technical and safety problems. The chief dangers lie in the build-up of plutonium deposits in the reprocessing plant and in the risk of a sudden criticality accident irradiating any operators in the vicinity.

The operation of fast reactors is intrinsically risky. Liquid sodium (or a mixture of sodium and potassium) is the usual choice of coolant, and sodium voids – caused when the sodium overheats, leading to a sharp rise in reactivity – present a major problem, particularly in the larger fast reactors. In that respect, they suffer from the same basic defect as the RBMK reactor at Chernobyl (*qv*). Should the reactor's fuel become distorted through overheating, then its reactivity will also increase, thus compounding the problem of sodium voiding. In theory, fast reactors can explode with the force of a small nuclear explosion, particularly should one lump of molten fuel be forced violently against another. Another hazard is presented by the spent fuel, which must be transported to the reprocessing plant in special flasks cooled by liquid sodium.

A number of serious accidents have already occurred in fast reactors, one a partial core meltdown in the Enrico Fermi plant close to Detroit in 1966 that threatened the entire city. In 1955 the first-ever fast reactor – the EBR-I reactor – suffered a core meltdown and came less than a second away from exploding. To date most fast reactors have had problems in their steam-generating systems, Russia's 350MW Shevtchenko fast reactor by the Caspian Sea developing a serious hydrogen fire in the secondary circuit. Fast reactors are today fitted with hydrogen-flaring systems.

Many countries are now developing experimental fast reactors, but France has now moved ahead with the operation of the 1,240 Super-Phénix plant at Creys Malville (*qv*). Super-Phénix contains some five tonnes of plutonium fuel at its core, enough for several thousand atomic explosive devices, as well as 5,000 tonnes of sodium coolant.

Fast reactors cost between one and a half and two times more to operate than conventional thermal reactors. (See also *Dounreay*Ⓝ, *Nuclear accidents*Ⓝ, *Reprocessing*Ⓝ.)

Feedlots

A system of livestock husbandry, used almost exclusively in the USA for beef production, in which animals are confined in large pens and their food is brought to them. The system is profitable in that it produces large amounts of beef at economic prices, but it is open to a number of objections.

Where cattle graze in the open, the animals are fed on, and help fertilize with their manure, land that is often unsuitable for any other agricultural use. In feedlots the feed brought to them must be specially grown. It often consists of grain, augmented with high-protein foods such as soybean or fishmeal – foods that could be eaten directly by humans.

In effect, the cattle are competing with people for food. A third of the world's fish catch, for example, is used for feeding Western livestock. Since much of the feed used in feedlots is derived from developing countries, those who ultimately suffer are the already undernourished peoples of the Third World.

Cattle are ruminants, and a high protein diet may be harmful to them. In addition, it may induce biochemical changes that alter the composition of the meat.

The concentration in a small area of large numbers of cattle also creates an ever-present risk of epidemic disease, and yields as a by-product vast quantities of manure, whose disposal often creates serious problems and may cause pollution, especially of watercourses. (See also: *Hormones in meat*ⒶⓁ.)

Fenitrothion

An organophosphate (*qv*) used as a general insecticide in agriculture and to control malaria mosquitoes. It is also used to control parasites on animals.

It is not the most toxic chemical in the organophosphorus group, but it is regarded by the World Health Organisation as being at "the limit of acceptable toxicity for conventional indoor application" when used for malaria control. (See also: *Pesticides*ⒶⓅ.)

Fertilizers

See: *Artificial fertilizers*, *Organic farming*.

Fish stocks

Fisheries supply 23 per cent of the world's protein and a livelihood for millions of people. However, world fish stocks are now threatened by a number of forces, notably overfishing, pollution, wetland (*qv*) reclamation schemes, barrages and dams (*qv*).

The modernization of fishing fleets has led to the decimation of fish stocks in many parts of the world. Faced with the need to repay loans, trawler owners have sought to maximize catches, regardless of the long-term consequences. In the North Atlantic, the herring has been fished to the verge of extinction, while stocks of the Atlantic cod, haddock and capelin have been severely depleted. In the South Atlantic, pilchard catches have seriously declined, while in the Pacific, the anchovie, salmon, halibut, King

Crab and Pacific Ocean Perch have all been overfished.

According to a 1983 survey by the UN Food and Agriculture Organisation (FAO) (*qv*), 11 major oceanic fisheries are now overfished to the point of collapse, the annual loss of potential catches amounting to some 11 million tonnes. It is not only old established fisheries which have been affected: as the WorldWatch Institute notes, "Over the last ten years even newly developed fisheries in the more southern latitudes have quickly been fished to the point of collapse."

Pollution too has destroyed many fisheries. In Britain, the Humber estuary used to support a thriving fishing industry: now it is dead, the victim of industrial pollution from the factories, mills and chemical plants that line the Humber.

In the US, fish stocks in the Chesapeake Bay have been all but destroyed by pollution: oyster catches fell by 50 per cent between 1962 and 1984; and catches of striped bass by 90 per cent in the decade fom 1973-1983. The cost of restoring fisheries in the Bay is expected to reach more than $1,000 million. Though a clean-up programme was launched in 1983, little has yet come of it. According to the World Resources Institute, "More than 2,641 facilities hold permits to discharge into the bay or the tidal portion of its major tributaries. The total discharge from Maryland and Virginia amounts to 15 trillion litres of wastewater per year – about one-fifth the amount of water in the bay at any given time."

Unless action is taken soon to ease the pressure on fish stocks, the prospects for world fisheries are not good. Though aquaculture (*qv*) is being promoted as an alternative to ocean fishing, its potential has been overstated and its produce is often not available to the poorer sections of society. (See also: *Aquaculture* Ⓐ ③, *Dams* ③, *Ocean dumping* Ⓟ, *Wetlands* Ⓒ.)

Fission

Process by which atoms are split. A rare isotope of uranium (*qv*), U-235, is the only fissile substance that occurs in usable quantities in nature. It makes up 0.7 per cent of the natural metal, the rest of which is non-fissile U-238. Other fissile isotopes are plutonium (*qv*) 239 and 241, and uranium-233, most of which have been manufactured in nuclear reactors.

On being hit by a neutron, the fissile nucleus splits at random into two nuclei and brings about the release of between two and three neutrons, generating considerable quantities of heat in the process.

The aim in operating a nuclear reactor is to achieve a constant rate of fissioning. One way of achieving this is to employ a moderator in the reactor that will cause the neutrons to slow down from their very fast ejection speeds so that they are more likely to be absorbed by a fissile atom. In Britain's gas-cooled reactors, the moderator consists of blocks of graphite. In light water reactors, water is used. In the Canadian Candu (*qv*) reactor, heavy water is used.

In addition, the nuclear material used to fuel a reactor can be manufactured from uranium that has been "enriched" in uranium-235, thus enhancing the likelihood of fission. Uranium enrichment (*qv*) also means that the reactor can be more compact and, therefore, supposedly cheaper to build. Usually enrichments of between two and four per cent are used in reactors with moderators.

When fuel is enriched above 15 per cent, it is possible to dispense with the moderator. The neutrons maintain high speeds – some 16,000km per second (10,000 miles per second). The advantage of such fast breeder reactors (*qv*) is that some of the fast neutrons are more likely to be absorbed by uranium-238, which then transforms into fissile plutonium. (See also: *Fast breeder reactor* Ⓝ.)

Flavourings

See: *Taste additives.*

Fluoridation

Technique of adding fluorides (*qv*) to public water supplies, with the intention of reducing tooth decay.

Fluoridation transformed a liability of the aluminium industry (bulk sodium fluoride) into a commercial product. Some fluoridation is done with hydrofluosilicic acid, a waste from manufacture of superphosphate fertilizer.

Tooth decay turns out to have declined to about one-third since the 1950s in the wealthy countries that gather dental statistics. This welcome trend is not understood; possible causes may include improved diet and brushing, widespread antibiotics secreted in saliva, and, recently, fluoride in toothpaste, a concentrated direct application that may well be effective. But thorough surveys have corrected the earlier claim that natural fluoride in water is correlated with relative freedom from tooth decay. Trials claimed to demonstrate benefit from fluoridation have been severely criticized for lack of controls, and other major defects.

In some still controversial studies, fluoridation has, on the other hand, been rather closely correlated with cancer. Various other types of harm have been suspected; the one established beyond dispute is dental fluorosis – diffuse white mottling of the teeth, a form of damage commonly observed among children drinking water fluoridated to 1ppm. The margin, if any, is uncomfortably slim between 1ppm and levels known to cause serious damage to bones (skeletal fluorosis). In addition, there is the possibility of adverse synergistic reactions between the fluoride added to water and the thousands of other chemicals to which modern industrial systems expose us.

Fluorides

Simple inorganic compounds of the element fluorine. Soluble fluorides, and in particular the acidic gas hydrogen fluoride, are extremely toxic. Hydrofluoric acid – a solution of hydrogen fluoride in water – will eat its way through most materials, including glass, and causes horrific burns to human tissue. The repeated ingestion of small quantities of fluorides leads to the condition known as fluorosis. This is a crippling disease in which excess calcium is deposited in the bones and ligaments, the bones becoming dense but brittle. Mottled teeth are an early sign of excess fluoride intake.

Fluorides are emitted by two main industrial processes. The smelting of aluminium brickworks also release fluorides into the atmosphere as fluorine compounds naturally present in the clay are driven off as the bricks are baked.

Small quantities of fluorides are needed by the body for the healthy development of teeth. (See also: *Fluoridation*.)

Food additives
See: *Additives.*

Food aid
See: essay on *The Politics of Food Aid.*

Food and Agriculture Organisation (FAO)
The principal international organization for the development of agriculture, fisheries and forests. Its original role was confined to determining food policy and designing appropriate projects, but it has increasingly taken part in the implementation of its policies.

The FAO's activities have also expanded in another direction. Whereas its advisers in Third World countries used to be members of the United Nations Development Programme (UNDP) missions to those countries, and were largely paid by that agency, today the FAO has its own representatives in the field.

FAO policies may be criticized on several grounds. To begin with, it insists on the mechanization of agriculture, which is difficult to justify in Third World countries where farms are small, the soil is delicate, and unemployment is massive. The organization also insists on a maximum use of chemicals. This is equally difficult to justify on many grounds, among them that yields comparable to those achieved with the use of chemicals can be obtained with organic methods, and that chemicals are largely used for export crops, which means that their use is irrelevant to what should be the overriding goal of feeding the starving masses of the Third World.

The FAO also insists on massively increasing the amount of land under perennial irrigation, a policy again difficult to justify in view of the disastrous experience, over the last 40 years, with the large dams (*qv*) and other water development projects needed to provide this type of irrigation.

A further criticizm is that the organization plays an important role in encouraging pesticide-intensive post-harvest technology (see *Post-harvest losses*), rather than encourage methods of cultivation and the use of crop varieties that are less vulnerable to pests during storage. And, finally, it has helped petrochemical companies acquire a large share of the world's distribution of seeds by helping to legalize the patenting of seeds and in general by adopting policies that must lead to the extinction of vast numbers of valuable traditional varieties.

Critics charge that the FAO has lost sight of its original intention to help those countries whose populations are increasingly malnourished, and threatened with starvation, and has instead adopted policies that seem to favour the agro-chemical industry. (See also: *Dams*Ⓢ, *Desertification*Ⓢ, *Erosion*Ⓐ Ⓢ, *Green Revolution*Ⓢ, *Land reform*Ⓐ Ⓢ, *Salinization*Ⓐ Ⓢ, *Waterborne diseases*Ⓢ.

Food dyes
Food dyes are the most conspicuous group of additives (*qv*) and also the most controversial.

There are arguments about both their necessity and their safety. Dyes can, however, be valuable to food manufacturers especially where their use does not have to be fully declared. Consumers tend to associate a bright colour with freshness, wholesomeness and tastiness.

Laboratory experiments have shown that if a range of drinks is presented with identical flavours, most consumers will report that the more darkly coloured they are, the stronger they appear to taste. Moreover, banana-flavoured drinks dyed red will be reported as having a strawberry flavour.

The safety of dyes is highly controversial, especially, but not exclusively, as regards the coal tar dyes (*qv*). Dyes have been linked to symptoms of acute intolerance such as hyperactivity, asthma, eczema and migraine.

In the USA, 28 dyes are positively listed, while in the UK the list consists of 48 distinctive compounds. In 1986 the UK food and drink industry spent about $4,800,000 to buy approximately 700 tonnes of synthetic dyes, and about $3,200,000 buying some 16,000 tonnes of other dyes, mostly caramels. The market for dyes in the USA amounts to about $55 million a year.

One British researcher has estimated that by the age of 12 an average child will have consumed 225g (half a pound) of food dyes. (See also: *Cancer*Ⓛ.)

Food irradiation
The use of ionizing radiation to extend the "shelf life" of food. Proponents of food irradiation claim that it kills insect pests and the bacteria responsible for food poisoning. In addition, food irradiation is held to inhibit the sprouting of vegetables, such as potatoes, and delay the ripening of fruit.

With many chemical fumigants now suspected of causing cancer, food irradiation is now being promoted as a major weapon in the fight against world hunger. In particular, it has been hailed as the solution to the massive "post-harvest" losses (*qv*) of food incurred in the Third World due to rotting during storage. Food that had been irradiated, it is argued, would not even need refrigeration to remain "fresh".

Modern commercial food irradiation plants use cobalt-60, a powerful source of gamma radiation (*qv*) derived from Canadian nuclear reactors. Its gamma-rays are so penetrating that they require two metres of concrete to shield operating staff adequately. Significant leaks of cobalt-60 at food irradition depots have already occurred.

Caesium-137, a major waste product of reprocessed nuclear

fuels, is now being promoted as an alternative source of nuclear material for food irradiation, leading many critics to argue that the nuclear industry is hoping to use food irradiation technology as a "socially aceptable" means of getting rid of its nuclear waste (*qv*). Recently, the US Department of Energy has offered to lease caesium-137 to food irradition companies for ten cents a curie, one-tenth of the price of cobalt.

Food irradiation is carried out in specially-constructed plants. The treated food is typically exposed to between 30,000-1,000,000 rads. By comparison, a chest X-ray delivers less than one-tenth of a rad. A commercial plant may hold the radioactive equivalent of one tonne of radium – more medium-lived radioactivity than was released at Chernobyl.

Tests on rats have shown that irradiated food causes cancer, mutations and perhaps chromosomal disorders. Under some conditions, food irradiation has also been found to cause the production of aflotoxin (*qv*), a highly carcinogenic mould. Given that aflotoxins are a major cause of post-harvest losses in Third World countries, the role that food irradiation might play in cutting such losses is open to doubt.

Where partly rotted food has been irradiated, most of the putrifying bacteria are killed but the toxins remain. Abuse of food irradiation to disguise rotten food has already occurred illegally in Britain and will continue.

At present, the use of food irradiation is limited, though the practice is permitted for most foods in the USA. In the Britain, food irradiation was banned in 1967, but the government is under pressure to lift the ban. In March 1987, the European Parliament refused to clear the sale of irradiated food in EEC countries and called for a ban on the imports of irradiated food to the EEC. The European Commission, however, appears committed to allowing the sale of irradiated imports. (See also: *Industrial Biotest Laboratories*Ⓛ, *Nuclear waste*Ⓝ Ⓟ, *Post harvest losses*Ⓐ ③, *Radioactivity*Ⓝ.)

Food storage
See: *Post harvest losses.*

Formaldehyde
A toxic gas used in many industrial applications, either in gaseous form or in a solution that is sometimes known as formalin. It is an ingredient of resins and glues, such as those used in the manufacture of building board, and is a major component of some foams used to insulate cavity walls. Formalin is also used as a disinfectant and soil sterilant.

The irritant effects of formaldehyde on the eyes, lungs and, in high concentrations, the skin are well known. More recently, concern has been expressed about its carcinogenic potential. Evidence based on experiments with rats in the USA suggests that it may cause lung cancer.

Concern about human health has led the US authorities to ban the use of cavity wall insulation foams based on formaldehyde and to consider restricting its other uses. West Germany is also considering tighter restrictions, but Britain's Health and Safety Executive insists that no human cancers have been formally attributed to formaldehyde. (See also: *Cancer*Ⓛ Ⓟ, *Carcinogen*Ⓕ Ⓛ Ⓟ.)

Fuelwood crisis
Forests have long provided mankind with wood for cooking and heating, and today some 2,000 million people still rely on firewood for their fuel.

In many areas, demand has now outstripped supply. At present, 1,500 million people lack adequate supplies of fuelwood: by the turn of the century, that figure is expected to have reached 2,300 million, with 350 million facing acute shortages.

As fuelwood supplies become exhausted, animal dung is frequently used as a substitute, thus depriving the land of nutrients and, ultimately, the farmer of food. In mountainous regions, such as the Himalayas, the declining fertility of the soil has led to steeper and steeper slopes being cultivated in order to make up for falling yields, thus exacerbating the problems of erosion (*qv*) and deforestation (*qv*).

The roots of the fuelwood crisis are complex. Population (*qv*) pressure is undoubtedly a major factor, but equally important are the degradation of woodlands as a result of commercial forestry and the clearing of forests for plantations, cattle, and other projects.

In terms of volume, firewood accounts for 54 per cent of the wood consumed worldwide every year – some 1,200 million tonnes.

Encouraged by the World Bank (*qv*), several governments have started "social forestry" (*qv*) programmes, ostensibly to provide villagers with firewood trees. Unfortunately, the bulk of the trees

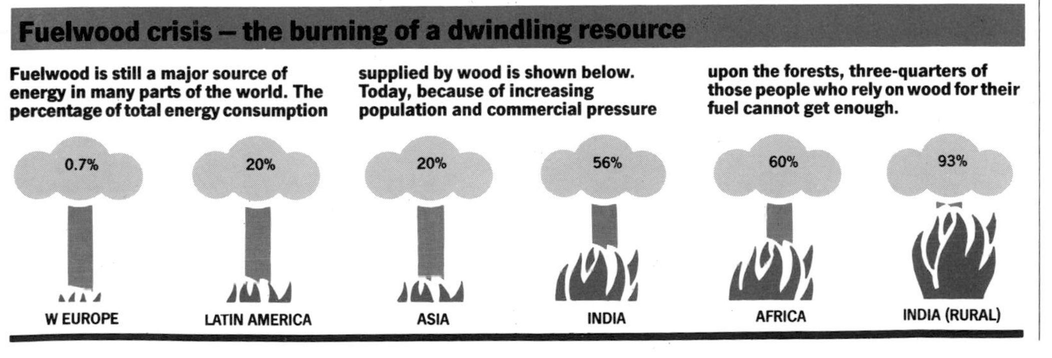

Fuelwood crisis – the burning of a dwindling resource

Fuelwood is still a major source of energy in many parts of the world. The percentage of total energy consumption supplied by wood is shown below. Today, because of increasing population and commercial pressure upon the forests, three-quarters of those people who rely on wood for their fuel cannot get enough.

have almost universally gone for commercial uses, such as rayon production and paper milling.

However, successful reforestation (*qv*) schemes have been implemented by local groups, acting at the village level, and often working without outside funding. Such groups include the Chipko movement (*qv*) in India and KENGO, the Kenya Energy Non-Government Organisation.

Local groups have also helped promote the use of biogas machines (which generate cooking gas by composting dung and organic waste) and better designed stoves, with vastly improved thermal efficiencies. Such technologies have helped reduce the pressure on local woodlots but their overall impact has proved marginal. Indeed, the only real solution to the fuelwood crisis lies in putting an end to further deforestation and in implementing massive village reforestation programmes, using non-commercial species. (See also: *Energy* Ⓔ.)

Fusion

In fusion, light elements such as hydrogen are made to coalesce into heavier elements under extreme conditions of temperature and pressure. Such fusion processes lie behind the enormous releases of energy from stars like the sun. On Earth fusion has been achieved only in thermonuclear bombs in which the initial extreme conditions are brought about through the explosion of an atomic fission bomb.

Since the 1950s, scientists working on internationally funded projects such as JET – the European Torus Project – at Culham in Oxfordshire, have tried, so far in vain, to construct a machine that would produce a net surplus of energy from fusion, in a sustainable fashion.

Currently $1,000 million a year are being spent worldwide on fusion research. Fusion has considerable lure as an energy source. If the deuterium – a heavy hydrogen isotope – found in just one cubic metre of seawater were fused, it would release the energy equivalent to the burning of 2,000 barrels of crude oil, but at the moment the difficulties of building a viable, economic reactor seem insuperable.

A fusion reactor will have simultaneously to operate with temperatures that are both the hottest and coldest achieved on Earth. The fusion reaction will need to take place in temperatures probably higher than 270 million degrees Farenheit (150 million degrees Centigrade), while the giant magnets that keep the plasma away from the reactor walls will have to be superconducting and cooled by liquid helium or some other superconducting substance at temperatures approaching absolute zero.

To compound the problem, most of the energy released during the fusion process will be in the form of fast-moving neutrons that are between 15 and 30 times more energetic than those released in nuclear fission reactions. The fusion neutrons will bombard the walls of the reactor vessel, heating it up – thus offering the opportunity to raise steam and generate electricity – but also knocking each atom of the wall out of its lattice position at least 30 times every year; hardly a recipe for structural longevity.

Meanwhile, the tritium to be used as fuel in the fusion reactor is, in combination with deuterium, at least 100 times more reactive than deuterium alone. The tritium will be bred from a lithium blanket interposed between the reactor wall and the superconducting magnets. However, lithium is highly reactive, burning violently when in contact with either air or water.

Most depressing of all for fusion enthusiasts, the power density of a large fusion reactor will at best probably be one-tenth that obtained in a light water fission reactor, such as the pressurized water reactor (PWR) (*qv*). Therefore the fusion reactor will be considerably larger, and in many respects considerably more expensive to operate, than a fission reactor. Moreover, judging from the extant problems in getting fusion reactors to work at all, the chances of producing a commercial reliable machine appear to be slim. (See also: *Fission* Ⓝ.)

Gaia
Theory put forward by Professor James Lovelock (author of an essay in this book) to explain the ability of the biosphere (*qv*), or world of living things, to create the environment that most favours its own stability, and to maintain that stability in the face of environmental change – such as the 30 per cent increase, since life began, in the heat given out by the sun. Lovelock argues that the biosphere acts as a single living system, which he calls "Gaia", after the Greek Goddess of the Earth. (See also: essay on *Man and Gaia*.)

Gamma-HCH
See: *HCH*

Gamma radiation
Discrete quantities of energy in the range of about 0.01 to 3.3 million electron volts that are without mass or charge and which move at the speed of light in the form of a wave.

Gamma rays are extremely penetrating, being able to pass through steel and, depending on thickness, through concrete. They pass easily through the body, leaving some energy behind in the form of ions that, as with other types of radiation, cause changes and damage to the cellular structure. In general, cells are better able to cope with the damage caused by gamma radiation compared with other types of radiation, such as alpha particles, but there is always the possibility that faulty repair will lead to an aberrant line of cells. (See also: *Alpha radiation*Ⓛ Ⓝ, *Beta radiation*Ⓛ Ⓝ, *Radiation and health*Ⓛ Ⓝ, *Radioactivity*Ⓝ.)

Ganges
One of the holiest rivers of the Hindus. Rising in the Himalayas, the Ganges runs in a broad sweep across north-east India, entering Bangladesh near Farakka and then spilling out through an enormous delta before emptying into the Bay of Bengal.

Deforestation, water diversion schemes and pollution have wrought havoc with the ecology of the Ganges river basin, one of the most fertile and densely populated areas of the world. There has not only been a stupendous loss of topsoil, but also the silting up of the rivers of the Ganges delta and consequent flooding: in 1978 alone, 65,712 villages in Bangladesh were submerged, leaving thousands homeless. 1987 also saw serious flooding.

The Ganges delta is also threatened by rising levels of salinity, the direct consequence of the controversial Farakka Barrage (*qv*) which India has built at the head of the delta.

Pollution constitutes the third major threat to the Ganges. Untreated sewage and industrial effluent – often heavily contaminated with mercury, lead and other heavy metals – are discharged directly into the river, which must also contend with pesticide and fertilizer run-off and other forms of chemical pollution. Some stretches are now so badly polluted that, in the opinion of one expert, "the river is fast losing its regenerative capacity."

Of the 29 largest cities along the Ganges – all with populations of more than 500,000 – only 19 have sewage treatment plants, and these are subject to frequent breakdowns. A recent survey estimates that two-thirds of all illnesses in India stem from drinking contaminated water.

In terms of biochemical oxygen demand (see BOD), sewage discharges account for more than half of the pollution entering the Ganges and its tributaries. The rest comes from industry.

The impact of pollution on fisheries throughout the Ganges basin has been severe. Fish catches in polluted stretches of the Hooghly – the tributary of the Ganges that flows through Calcutta – are reported to be less than a fifth of those in unpolluted stretches. High levels of cadmium (*qv*) in fish have been linked to an increase in the incidence of bone disease in West Bengal, where fish constitutes a major part of the staple diet.

Industrial pollution of the Ganges has also led to contamination of crops and land. Irrigation water drawn from the river has caused the leaves of some crops to turn brown.

In 1984, the Indian Prime Minister, Rajiv Gandhi, announced a $200 million plan to clean up the Ganges. (See also: *Drought-flood cycle*③, *Heavy metals*.)

Genetic diversity
The Earth possesses a staggering diversity of plant, animal and insect species, each equipped with a different genetic make-up. To date, biologists have classified 1,700,000 species, but recent research suggests that the final total could top 40 million, insects alone accounting for 30 million species. Estimates of the number of flowering plants range from 275,000 to 400,000 – a figure which is increased still further if different varieties and strains are taken into account.

Such genetic diversity is vital to the maintenance of ecological stability, enabling the different species to meet different challenges and to fullfil different functions within the biosphere (*qv*). It also provides the redundancy of species that ensures that the loss of a single species is unlikely to threaten the survival of the ecosystem as a whole.

The reduction of diversity greatly increases the vulnerability of an ecosystem. Nowhere is this more evident than in the monocultures (*qv*), favoured by modern farmers, which are vulnerable to pests and disease.

Today, the genetic diversity of the biosphere is being rapidly eroded, 1,000 species becoming extinct every year – more than two a day. By the end of the century, that figure could have risen to 10,000, one species being lost every hour. An estimated 25,000 plants are now threatened with extinction (*qv*) and, at the current rate of destruction, 60,000 plant species may be lost by 2050. Worst hit are small islands: of the original 3,000 plant species in Hawaii, 270 are now extinct.

The implications of such mass extinctions are serious. Though nature has considerable resilience – ensured in part by the sheer number of species on Earth – there is a limit to how far that resilience can be stretched. (See also: *Genetic resources*Ⓒ.)

Genetic engineering

In 1973, Stanley Cohen of Stanford University and Herbert Boyer of the University of California announced that they had performed the world's first successful experiment in "recombinant DNA" – or "genetic engineering" as it is more popularly known.

In 1974, the US biologist Paul Berg, one of the fathers of recombinant DNA research, called for a moratorium on the genetic engineering of cancer-causing viruses or toxin-producing genes that might infect humans. Though the moratorium was supported by many eminent biologists, it did little to halt the onward march of the "biotechnologists". In 1980, the pro-engineering lobby in America won a considerable legal victory when the US Supreme Court ruled that companies had the right to "patent" novel forms of life. Three months after that ruling, Genotech, one of America's first biotechnology corporations, went public: in the first 20 minutes of trading, its shares rose in value by 54 points.

Advocates of biotechnology are starry-eyed about the benefits that genetic engineering will supposedly bring mankind.

In agriculture, it is suggested that recombinant DNA will give non-leguminous plants the capacity to fix nitrogen from the air, thus expanding output.

In the mining industry, there is talk of developing micro-organisms that will mine ores and thus replace miners. One company has engineered a bacterium that produces an enzyme which eats away the salts in copper ores, "leaving behind an almost pure form of copper".

In medicine, it is predicted that "by the year 2000 virtually all the major human diseases will regularly succumb to treatment by disease-specific artificial proteins produced by specialized hybrid micro-organisms".

More ominously, it has been suggested that recombinant DNA be used to screen humans for genes associated with "anti-social behaviour" – the XYY karyotype, for example, which is frequently found in aggressive males.

But what of the dangers? Though the evolutionary success of bacteria is due to their natural abililty to recombine DNA, only a minute number of the millions of possible new strains have survived. Rigorous natural controls have ensured that strains threatening to disrupt the balance of nature have died out. However, in the crude ecosystems developed by man – the monocultures (*qv*) of the prairies, for example – nature's control system has been greatly simplified. The fear is that man has disrupted the natural balance to such an extent that some new opportunistic pathogen – released into the environment either accidentally or on purpose – will play havoc with the biosphere.

Such an accident is inevitable, according to critics of genetic engineering. They point out that absolute containment is impossible; that laboratory leaks frequently occur without being detected until it is too late, and that human fallibility alone will ensure that one day, sooner or later, a genetically engineered organism will escape into the environment. "Once it has escaped, there will be no way to recapture it," argues Professor Robert Sinsheimer, Chairman of the Biology Division at the California Institue of Technology." "And so we have the great potential for a major calamity."

The scale of that calamity will depend on the type of organism released and on its ability to survive outside the laboratory.

Now that biotechnology has come of age, the battle over its applications is intensifying. In April 1986, Jeremy Rifkin, a major US critic of genetic engineering, took the US Department of Agriculture to court to prevent the release of two genetically engineered microbes into the environment. One was a pig vaccine against pseudo-rabies, the other an ice-resistant bacterium that prevents frost damage to crops.

After an extensive legal battle, two strains of ice-resistant bacteria were eventually released in the USA on April 24 and 29 1987. Britain has also licensed the release of genetically engineered bacteria in test trials.

Genetic resources

The genetic variability within wild species has long been exploited by man to create new hybrid strains of domesticated plants and animals and to refine existing ones. As such, the genetic variability of the natural world constitutes a major resource, wild plants and animals providing a huge potential for new crops, medicines, fibres, and foods (see illustration).

The rainforests, for example, have already supplied man with a host of products – from rubber to quinine. Recent surveys reveal that the Indians of Amazonia (*qv*) use more than 1,000 plants of medicinal or nutritional value. Drugs developed from the rosy periwinkle, a rainforest plant, have already saved many children from leukemia. Maize contributes to aspirin, penicillin, plastics, paper manufacture and tyres.

According to the International Union for the Conservation of Nature (IUCN), 40 per cent of the prescription drugs on sale in 1976 contained a drug of natural origin. Wild strains of domesticated plants have also saved some crops from being completely destroyed by pests. In the 1860s, for example, Europe's grape vine was saved from total destruction by phylloxera, an insect pest, only by grafting European vines onto the native American vine, which is phylloxera-tolerant. The practice still continues today.

Unfortunately, the world's genetic resources are rapidly being destroyed. With every acre of rainforest that we cut down, for example, we lose numerous species. "If present levels of deforestation continue," notes E. O. Wilson of Harvard University, "the stage will be set within a century for the inevitable loss of about 12 per cent of the 700 bird species in the Amazon Basin and 15 per cent of the plant species in South and Central America."

Quite apart from their ecological impact, today's mass extinctions of species represent a stupendous waste of resources which, if allowed to survive, could be tapped for the benefit of humanity. (See also: *Extinction*Ⓒ, *Genetic diversity*ⒸⒻ.)

Genetic resources – protecting the work of centuries

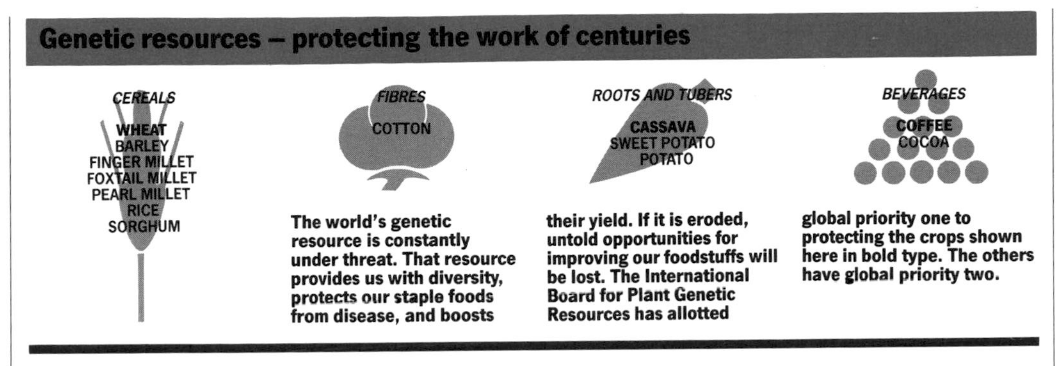

The world's genetic resource is constantly under threat. That resource provides us with diversity, protects our staple foods from disease, and boosts their yield. If it is eroded, untold opportunities for improving our foodstuffs will be lost. The International Board for Plant Genetic Resources has allotted global priority one to protecting the crops shown here in bold type. The others have global priority two.

Geothermal energy

The obtaining of useful heat by exploiting anomalies in the rate at which rocks become warmer with increasing depth in the Earth's crust (the geothermal gradient).

Hot water can be tapped by drilling boreholes directly into it, much in the manner of drilling for petroleum or natural gas. Installations for this purpose exist in many parts of the world where such hot water can be found. Extracting heat from dry rock involves drilling several boreholes into it, fracturing the rock with down-hole explosions, then pumping cold water under pressure down one hole. The water fills the fractures and cavities made in the rock, is heated, and emerges from the other boreholes.

Depending on the temperature, the water can be used for district heating, but more usually it is used to pre-heat water for electrical steam-generation.

Geothermal energy can make only a minor contribution to the satisfaction of our energy requirements as there are very few suitable sites. The leading research sites in hot dry rock geothermal technologies are at Los Alamos in the USA and in Cornwall, England. Test drilling in Cornwall has been linked to an increase in earthquakes. (See also: *Alternative technology* Ⓔ, *Energy* Ⓔ, *Soft energy paths* Ⓔ.)

Ghost acres

Most of the rich nations of the industrialized world depend for their prosperity on importing food from abroad; that is, they depend on exploiting "ghost acres" abroad to grow their agricultural inputs and food. More than a million sq km (386,000sq miles) of arable land in the Third World are exploited in this way.

Georg Borgstrom, the US agronomist who first coined the term "ghost acres", also uses the concepts of "fish acreage" (a measure of the amount of land that would have to be tilled to provide the animal protein provided by imported fish) and "trade acreage", which refers to the net amount of agricultural produce being imported or exported for food and feed.

Borgstrom cites West Germany, the United Kingdom, Japan, the Netherlands and Italy as countries that are all heavily dependent on ghost acres abroad to feed themselves. The United Kingdom, for example, "farms" two ghost acres for every acre of land under the plough in Britain itself.

The USA also exploits vast tracts of ghost acres, importing, for example, 782 million kg (1,725 million lb) of beef in 1977, much of it reared on ranches hacked out of South America's rainforests.

Without access to ghost acres abroad, many industrialized countries would be hard-pressed to feed their population without radical changes to their diet and to their agricultural practices. The extent to which a country relies on ghost acreage is thus a measure of its vulnerability. (See also: *Cash crops* ③, *Malnutrition* Ⓛ ③.)

Glass fibre

Manmade fibre that consists of fine threads spun from molten glass. It can be embedded in resin and used as a structural material where lightness and strength are important or produced as a roll of matting and used for insulation.

Glass fibre resins can be unpleasant to handle before they set because they release toxic fumes. Good ventilation is essential. Once set, they present no hazard unless they are abraded, when an irritant dust may be produced.

The insulating rolls can produce serious rashes and persistent coughs and lung irritation.

In 1987, a study of glass fibre workers in the US found a higher-than-expected rate of lung cancer. The study, undertaken by Philip Enterline, at the University of Pittsbrugh, has been criticized by glass fibre manufacturers, who argue that other carcinogens used during the manufacturing process may be responsible. (See also: *Asbestos* Ⓛ Ⓟ, *Cancer* Ⓛ Ⓟ.)

Global 2000

In 1977, President Jimmy Carter commissioned a study of global environmental trends to determine what would be the main features of life on our planet by the year 2000. This study was directed by an independent consultant, Dr Gerald O. Barney, with the cooperation of relevant US government departments.

Barney's report published under the title of "The Global 2000 Report to the President" was published in 1980. The general conclusion was that "if present trends continue, the world in 2000 will be more crowded, more polluted, less stable ecologically, and more vulnerable to disruption than the world we live in now. Serious stresses involving population, resources, and environment are clearly visible ahead. Despite greater material output, the world's people will be poorer in many ways than they are today.

"For hundreds of millions of the desperately poor, the outlook for food and other necessities of life will be no better. For many it will be worse. Barring revolutionary advances in technology, life for most people on Earth will be more precarious in 2000 than it is now – unless the nations of the world act decisively to alter current trends . . .".

Barney called upon the US government to undertake immediate measures to counteract these devastating trends. A complementary report was written by the Council on Environmental Quality and the Department of Trade, entitled "Global Future, Time to Act." It made various recommendations, mainly for setting up various institutions to monitor the course of environmental destruction.

The report was an immediate success. It sold a million copies in the USA, and spawned such groups as the Global Tomorrow Coalition which, three years later, had some five million members. It also exerted considerable influence on the thinking of some members of the US establishment, such as Senator Edmond S. Muskie, the then Secretary of State, and Mr Robert McNamara, then President of the World Bank.

The Carter administration did not survive long enough to implement the report's recommendations, and the Reagan administration not only ignored the report but set about systematically to dismantle what environmental legislation had been passed during the Carter era.

Various efforts have been made to discredit the Global 2000 Report. These have included the publication in 1984 of a report entitled "The Resourceful Earth", largely inspired by the late Herman Kahn of the Hudson Institute and Professor Julian Simon, a mail-order specialist. Its conclusions are the diametric opposite of the Global 2000 Report, in words that parody those used by Barney. The authors conclude "If present trends continue, the world in 2000 will be less crowded (though more populated) less polluted, more stable ecologically, and less vulnerable to resource supply disruption than the world we live in now. Stresses involving population, resources and environment will be less in the future than now . . . The world's people will be richer in most ways than they are today . . . The outlook for food and other necessities of life will be better . . . life for most people on earth will be less precarious economically than it is now."

Events in the last three years do little to enhance the credibility of Kahn and Simon's conclusions.

Gorleben

See: *Wakersdorf.*

Grande Carajas

Brazil's giant Grande Carajas project is intended to open up the north-east of the country to industry and industrialized agriculture. One-sixth of the whole of Brazilian Amazonia will be affected.

The centrepiece of the project is the Serra dos Carajas open-cast iron ore mine, which is backed by a $600 million loan from the EEC. The railroad and port that will service the mine are to be funded by the World Bank.

Other projects include a bauxite mine capable of producing eight million tonnes of bauxite a year, and an aluminium smelter that will produce 800,000 tonnes of aluminium and 20,000 tons of aluminium oxide a year for sale to Japan.

In addition, some 55,000 sq km (21,000sq miles) will be cleared to make way for export-oriented plantations and biomass fuel farms. A further 30,000 sq km (11,500sq miles) will be given over to ranches. The project as a whole will affect the homelands of 23 tribal groups.

The project is likely to lead to severe water and air pollution problems. Many companies have been enticed into it through the promise of "less stringent" pollution controls. Companies are also being offered a ten-year tax holiday. (See also: *Amazonia*③, *Dams*③ *Hamburger connection*Ⓐ, *Tribal societies*③, *Tropical rainforests*③.)

Gray

The unit of radioactive dose, equivalent to 1 joule per kilogram. The gray is replacing the rad: 1 gray equals 100 rads

Great Barrier Reef

The Great Barrier Reef is a spectacular mosaic of coral, plant and animal life of immense complexity, which was inscribed on the World Heritage list in 1981. It extends for 2,000km (1,240 miles) along the north-east coast of Australia, the largest reef ecosystem in the world. Comprising about 2,600 separate reefs, shoals and coral formations, and covering an area of 348,700sq km (134,600sq miles) it stretches from Fraser Island to the Torres Strait near the coast of Papua New Guinea.

The living reef, built up from tiny coral polyps over millions of years, is the richest ecosystem on earth. About 400 different coral types form the reef structure, home to a variety of plant life, thousands of marine invertebrates, 1,500 fish species and 240 species of sea birds. The Great Barrier Reef also includes the major feeding grounds for the endangered dugong and the nesting grounds for many turtles, including the endangered loggerhead turtle.

Reefs are extremely vulnerable to changes in their environment. They require warm shallow, clear waters to grow. In particular, they cannot tolerate pollution, which can cloud the waters reducing light and changing the water temperature, choke the pores of marine animals and coral, or simply poison reef organisms.

During the late 1960s, public concern for the future of the reef began to grow and focused on proposals for oil drilling and mineral exploitation of the reef. A long battle by a young Australian environmental movement ensued as the beauty and biological significance of the Great Barrier Reef was brought to the attention of the Australian people. In 1975, the Great Barrier Reef Marine Park Act was passed, excluding mining from all sections of the reef, and a joint Federal-State government authority was established to manage the region. By 1987, 95 per cent of the reef lay within the marine park.

Current management involves zoning the reef to allow access by various user groups, for commercial fishing, shell collecting, tourist operations, recreational and scientific uses.

Unfortunately, the authority has not eliminated all threats to the reef environment. Areas of the reef above the low water mark are not adequately protected under the Act and many developments continue to pose a threat to this marine ecosystem. By far the greatest and most immediate threat to the reef and its environment comes from uncontrolled tourism pressures and speculative developments.

Great Lakes

Known as "North America's inland seas", the five Great Lakes – Superior, Michigan, Huron, Erie and Ontario – contain 18 per cent of world's liquid fresh water, supplying drinking water to 21 million people and a habitat for many fish.

Strict pollution controls have helped to rid the lakes of conventional sources of pollution, such as sewage, which causes eutrophication *(qv)* and which threatened to turn the lower Great Lakes – notably Lake Erie – into "dead lakes". However, there is growing concern about levels of toxic chemicals in the lakes, chemicals which have already led to a US ban on the sale of several species of fish. Acid rain *(qv)* damage to the streams that feed the lakes is also an increasing problem.

In 1985, the eight US states and two Canadian provinces that border the five Great Lakes signed the Great Lakes Charter, in which they agreed to consult each other on any plans to divert water from the lakes. The charter was followed a year later by the Toxic Substances Control Agreement, which aims at reducing toxic discharges into the lakes "to the maximum extent possible". Under the agreement, discharges from the Niagara River into Lake Superior, for example, are to be cut by 50 per cent by 1995. A bilateral agreement between Michigan and Ontario binds both states to eliminating toxic compounds from areas bordering their lakes. Oil drilling in all the lakes is banned under the Toxic Substances Control Agreement. (See also: *Love Canal* Ⓟ.)

Greenpeace

See: *Direct action.*

Green politics

The first Green party was founded in New Zealand in 1972. Though the Values Party, as it was called, never gained any seats in national elections, it did work out a detailed programme that became the first statement of Green politics.

The programme explicitly stated the need for a steady-state population and economy, decentralization of government, human-centred technology, soft energy *(qv)* systems, and such values as co-operation, nurturing and peacemaking.

In early 1973, a green party called "People", inspired by *The Ecologist*'s Blueprint for Survival, was set up in the United Kingdom. By 1987, its name changed to the Green Party, it was able to field 133 candidates at the general election of that year and to obtain 1.36 per cent of the vote.

On the Continent, however, the Greens have been more successful. In 1983, the West German Green Party won five per cent of the national vote in the General Election, which, thanks to proportional representation, gave it 27 seats in the Bundestag. Contrary to the predictions of Christian Democrats and Social Democrats, the Greens increased their representation to 42 seats in 1987.

The success of the German Greens lies partly in their being a coalition of groups spanning the entire political spectrum.

Their ideas of ecology, social responsibility, grass roots democracy and non-violence, which hold this diverse coalition together, attract a growing number of young voters born after 1945, and they have also become the political voice of the antinuclear movement.

In Belgium, the Greens gained national representation two years before the Germans, and enjoyed the success of becoming the first Greens to participate in governing when they went into coalition with the Social Democrats after municipal elections in Liege. In Italy, too, Greens have been elected to Parliament, winning 15 seats in the 1987 general election. In the US, a Green party has been formed, and has held its first convention in San Francisco. (See also: *Conservation* Ⓒ, *Deep ecology* Ⓕ, *Ecological movement* Ⓕ).

Green Revolution

The growth of world hunger, and the spectre of still worse hunger to come, led the United Nations Food and Agricultural Organization (FAO) *(qv)* to organize its first World Food Congress in 1963 and to set up its Freedom from Hunger Campaign. On the assumption that world population would grow by 2.6 per cent a year, and that incomes would also grow, FAO argued that if hunger was to be overcome, an annual increase in food supply of 3.9 per cent was required as compared to the 2.7 per cent increase of the years 1956-66. This meant in effect increasing world food production by 140 per cent by 1985.

To achieve this, FAO launched its "Green Revolution", known formally as the Indicative World Plan for Agricultural Development. The Green Revolution involved the introduction of high-yielding cereal varieties or HYVS *(qv)* developed by the International Rice Research Institute (IRRI) in the Philippines and a similar institute in Mexico. The hybrid cereals were short-stemmed to avoid lodging (being blown over by the wind), and quick-maturing, making it possible to grow three crops on the same land every year. Their yields were impressive.

Most people were persuaded that the Green Revolution would provide the solution to the world food problem and, significantly, Norbert Borlaug, generally regarded as the father of the Green Revolution, was awarded a Nobel Prize. The plan, however, was and still is criticized on a number of counts.

The first criticism is that the Green Revolution affects only cereals, which are easily stockable and saleable on the world market, no interest being shown in vegetables, fruit or any of the staple crops grown in different parts of the Third World for local consumption. This alone suggests that the Green Revolution cannot overcome the problem of hunger.

In addition, the food quality of the hybrids is lower than that of traditional varieties, their taste rarely being appreciated by rural people. They are also difficult to stock because of their high moisture content, which not only makes them vulnerable to moulds, such as aflotoxins (*qv*), but reduces their keeping qualities – a serious problem since, according to the FAO, something like 30 per cent of the food produced in the Third World is destroyed by pests during storage.

Furthermore, high yields are obtained only in highly favourable soils and climatic conditions. On the marginal soils, where most Third World people live, the results have been far less impressive.

Worse still, the hybrids provide high yields only when used in conjunction with heavy applications of fertilizer. In India, the Green Revolution has increased wheat yields by 50 per cent and rice yields by 25 per cent. However, this has required a more than twenty-fold increase in fertilizer use (from 212,000 tonnes in 1960-1961 to 4,264,300 tonnes in 1982-83). The abandonment of the traditional crop varieties in favour of the high-yielding hybrids has also led to a terrible loss of genetic diversity (*qv*) and a corresponding increase in vulnerability to pest infestations.

Moreover, the hybrids have also proved vulnerable to pests and diseases, which has meant a massive increase in the use of pesticides (*qv*). A 1981 survey undertaken in the Philippines reveales that expenditure on insecticides increased from two pesos to 90 pesos per hectare between 1966 and 1979 – an increase of 4,500 per cent. The use of pesticides has had serious consequences, including the destruction of fish stocks (*qv*), with severe nutritional implications for Third World peoples who depend on fish as a major source of animal protein.

In addition, high yielding varieties require irrigation water, which has helped justify massive dam-building programmes with terrible ecological consequences such as salinization (*qv*) and an increase in waterborne diseases (*qv*).

Finally, the cost of the chemical inputs and irrigation water have proved to be beyond the means of small farmers, many of whom have fallen into terrible debt, while others – the majority – have simply been put out of business to beome landless labourers and slum dwellers.

The costs of fertilizer, pesticides and irrigation water has also caused an escalation of the Third World debt, making it imperative that the food produced be exported. As a result, despite higher yields, less food is available to feed local people. This in itself makes a nonsense of the claim that the Green Revolution has provided a means of feeding the rural masses of the Third World. (See also: *Cash crops*, *Development* ③, *High-yielding varieties* Ⓐ ③.)

Greenhouse effect

See: *Gaia*, and essay on *Man and Gaia*.

Greenhouse gases

See: essay on *Man and Gaia*.

Groundwater

The technical name for water stored within the pores of soil and rock formations.

Worldwide, groundwater supplies are being depleted at an alarming rate, mainly for agricultural purposes. In addition, many groundwater sources are being polluted by the dumping of hazardous wastes and by the slow seepage of agricultural chemicals

Groundwater — pressure on a renewable resource

There are between 40 and 60 million cu km of water in the ground, but only about 40,000cu km of it are readily accessible and rechargeable. That tappable amount is an important supplement to the main water supply from lakes and reservoirs. But there are problems linked to its use. It takes time to recharge it, so it can be depleted. In regions that are water-deficient, such as Africa and large parts of Asia, not enough is known about where it is or how much there is. And, finally, there is increasing concern in industrial countries about how pure it may be, because it is being polluted by agro-chemicals and hazardous wastes.

Applying the pressure

1 **Pollution of groundwater is a major problem. California's main water wells now have pollution levels above safety limits and in Florida, where 90 per cent of the population relies on groundwater, more than 1,000 wells have been closed due to pollution by a carcinogenic worm killer.**

2 **Agro-chemicals are also a pollutant of groundwater. The most common contaminants are nitrate and nematocides. In California, 57 different pesticides have been detected in groundwater and in Britain more than 100 wells have levels that exceed the EEC safety limit.**

3 **Badly planned and executed irrigation leads to high levels of salt in groundwater. Water from rivers, or recycled groundwater, dissolves natural salts in rocks and soil as it finds its way down to the watertable. Re-used, it eventually renders the land unfit for agriculture.**

4 **Irrigation has led to excessive use of groundwater and consequent loss of agricultural production. In the High Plains of the USA the area under irrigation dropped by 5,920sq km (2,285sq miles) between 1978 and 1982 due to excessive "mining" of the vast Ogalalla aquifer.**

5 **If too much groundwater is withdrawn, the aquifers begin to compact, causing the land above to subside. Shanghai has sunk 2.63m (8.6 ft), and some areas of the San Joaquin Valley in California have sunk by 9m (29.5ft). In London, loss of industry has caused the watertable to rise.**

6 **Excessive deforestation has reduced the rate at which groundwater is recharged. Trees create an environment that allows rain to percolate down through the soil; when they are cut down, the bulk of the rain runs off the denuded topsoil to feed rivers, not the ground beneath.**

through the soil (see illustration).

In the Third World, groundwaters are being depleted at an unsustainable rate. In the southern Indian state of Tamil Nadu, water tables fell by 80-100ft (25-30m) between 1975 and 1985, while production in many areas of the Punjab and Haryana – the breadbasket states of northern India – is at risk due to the mass use of tubewells. The problem is compounded by excessive deforestation (*qv*) which has reduced the rate of groundwater recharge.

One consequence of groundwater depletion is that aquifers begin to compact, causing the land above them to subside. The classic case is the city of Venice, which has sunk by ten inches (25cm), resulting in widespread flooding during spring tides.

The pollution of groundwater is now a major problem in industrialized countries – and, increasingly, in the Third World. One-fifth of California's major drinking water wells have pollution levels above the safety limits set by the state, carcinogenic chlorinated solvents being the most common pollutants.

Hazardous waste dumps are the most common source of groundwater pollution in the US, but agricultural chemicals also pose a major threat. Half of the so-called "Superfund" (*qv*) sites in the US are now known to be leaking.

In Britain, there has been no national survey of groundwater quality near hazardous waste sites. However, a 1985 report from the Department of the Environment reveals that ten per cent of Britain's potable aquifers could be contaminated with concentrations of chlorinated solvents above the World Health Organization's "safe" limits.

Toxic agrochemicals are also a major source of groundwater pollution. Each year, American farmers spray or spread some 260,000 tonnes of pesticides and 42 million tonnes of fertilizers. Twenty-three states now report finding groundwaters contaminated with a wide range of agrochemicals. The most common contaminants are nitrate (*qv*) and the carcinogenic nematocides EDB, DBCP and aldicarb (*qv*).

In Britain, nitrate pollution is already a major problem in such agricultural areas as East Anglia, the East Midlands, Lincolnshire, and Essex.

Experts are agreed that worse is yet to come. Because water moves so slowly through the ground, many of the chemicals sprayed or dumped on the land two or more decades ago have yet to reach underlying groundwaters.

Once polluted, groundwaters are difficult (in some cases impossible) to cleanse. There are no microorganisms to break down the pollutants, and groundwater moves too slowly to permit any significant dilution and dispersal of contaminants. In Britain, an aquifer near Norwich that was polluted with whale oil in 1815 still contained significant toxic residues when wells were dug in 1950. (See also; *Aldicarb*ⒶⓅ, *Deforestation*③, *Hazardous waste*, *Landfill*, *Nitrate*ⒶⓅ, *Ogallala aquifer*Ⓐ, *Superfund*Ⓟ.)

Growth promoters

The main use to which hormones (*qv*) are put in the livestock industry is as growth promoters. For a long time Diethystilbestrol (DES) (*qv*), a synthetic oestrogen, was in wide use. It permitted a weight gain of 15-19 per cent and increased feeding efficiency by ten per cent in steers.

It was banned in 1979 when it was found to cause clear cell adenocarcinoma, a rare form of cancer of the vagina, in some daughters of women exposed to it. However, it continued to be widely used illegally.

The incentive to use these growth promoters is considerable. According to Dr Garry Duhl of Kansas State University, for a cost of $2 they make it possible to increase the weight of an animal by 18kg (40lb), which means a $10 (1983 prices) return for each dollar invested.

Lutalyse has been known to cause abortions and changes in the menstrual cycle. Melengesterol acetate (MGA), a synthetic progesterone manufactured by a division of the Upjohn Corporation, is, like other progesterones, a suspected co-carcinogen; that is to say, it can become carcinogenic when used in conjunction with another substance that alone is not a carcinogen (*qv*).

In 1982, there was found to be a considerably increased incidence of premature thelarche (breast development) in Puerto Rico. Some of the children affected also showed other signs of precocious sexual development. Little girls of no more than three or four, for instance, developed pubic hair and even menstruated. In most of these children increased oestrogen levels were found. This problem was ascribed it to the use of oestradiol in locally produced pork and chicken. Significantly, many of the affected children also developed ovarian cysts. This fits in with the result of a study undertaken by the Commonwealth Health Department, which showed a higher incidence of cancer in precisely those areas in Puerto Rico where most of the children with thelarche and precocious puberty came from.

We know something of the effects of oestrogens from the experience of young women taking the birth control pill, which leads to an increased incidence of thromboembolism and is suspected of causing melanoma, a skin cancer that can often be fatal. High levels of oestradiol have also been held responsible for heart attacks.

In the UK, the use of growth promoters such as Zeranol and trenbolen acetate, or Finaplix, has been encouraged by the Agricultural Development and Advisory Service (ADAS) of the Ministry of Agriculture, Fisheries and Food (MAFF). However, in October 1985 the European Parliament banned the use of zenarol (Ralgro) and Finaplix and the three so-called natural hormones testosterone, progesterone and 17-beta-oestradiol, which are in reality synthetic compounds identical to the sex hormones common to many mammals.

This ban will take effect at the end of 1988, in spite of fierce opposition by the then British Minister for Agriculture. Britain and Denmark are at the moment trying to obtain a reversal of the decision in the European Court at the Hague.

Hamburger connection
Since 1960, thousands of square kilometres of tropical forest in Central America have been cleared to make way for cattle ranches (see illustration). In Latin America as a whole, according to Dr. Norman Myers, author of *The Primary Resource*, "the cattle raiser is accounting for at least 20,000sq km of forest a year." If current rates of clearance continue, most of the remaining forests in Central America – some 200,000sq km – will have been destroyed by the mid-1990s.

Most of the beef raised on the cattle ranches of Latin America is exported, 85-90 per cent of exports going to North America. Because the meat is too lean for American tastes, it is primarily used to make hamburgers for sale by such fast-food chains as Burger-King, Roy Rogers, Bob's Big Boy, Jack-in-the-Box and Hot Shoppes. McDonald's deny using beef from Central America – a claim which it is impossible to disprove since, once it has entered the USA, much of the beef enters the domestic market where it is

Hamburger connection – the cost of fast food

Fast food habits contribute to the greenhouse effect. Demand for cheap hamburger meat has led to the clearing of Central American forests for cattle ranching; an estimated 9sq m are cleared for each hamburger. The burning of the forest and natural digestive fermentation in cattle together produce 39% of the methane rising from the tropics: methane is a greenhouse gas.

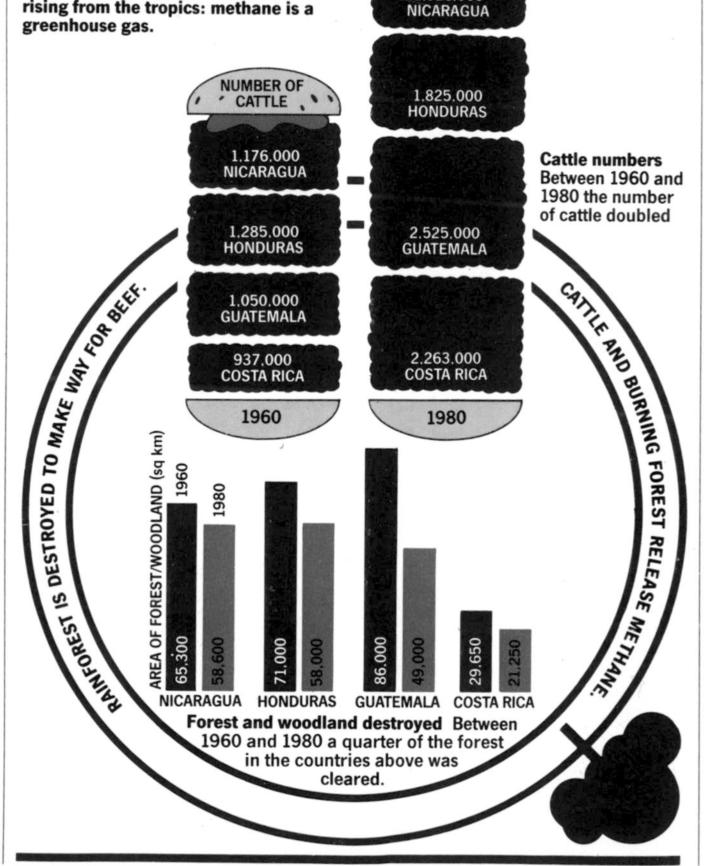

indistinguishable from home-raised beef.

Cheap beef imports from Latin America have reduced the price of hamburgers in North America by five cents a pound. However, the cattle trade has done little to help the poor of the cattle-exporting countries, few of whom can afford to eat beef. In Costa Rica, per-capita beef consumption fell by more than 40 per cent between 1960 and 1979, while production rose three-and-a-half times. "By 1978, the average Costa Rican was eating only 28lbs (13kg) of beef a year," comments Catherine Caufield, in her book *In the Rainforest*, "The average American pet cat ate more."

The ecological destruction caused by ranching programmes is long-term and sometimes irreversible. To date, more than 5,000 million trees have been cut down in Central America alone.

Once deprived of its tree cover, the land is quickly depleted of nutrients and invaded by toxic weeds. In a few years, it is so degraded that it must be abandoned – a fate that has befallen nearly all the cattle ranches established in Amazonia before 1978. The ranchers have not been deterred: they have simply moved.

For billions of birds that migrate each year from North America to over-winter in Central America, the loss of their forest habitats is proving calamitous. More than 150 species are affected, including warblers, peewees, vireos and kingbirds. Scientists at the Smithsonian Institute in Washington warn that the numbers of migrating species are declining by between one and four per cent a year.

Recent research suggests that the adverse climate effects of tropical forest clearance is likely to be exacerbated by ranching projects. As unlikely as it may sound, cattle add appreciable amounts of methane to the atmosphere through belching. Like CO_2, methane is a greenhouse gas (*qv*) and will therefore contribute to the general warming up of the Earth's surface. (See also: *Amazonia*③, *Cash crops*③, *Deforestation*③, *Development*③, *Tropical forests*③, *World Bank*③.)

Hanford

Major nuclear complex in Washington State. There are now nine plutonium reactors, plus reprocessing plants and a large number of storage tanks for high level waste at Hanford, which at present covers some 1,490sq km (575sq miles).

Since the early 1970s, the Hanford site has become notorious for the failures of its high-level nuclear waste storage tanks, some of which are made of ordinary carbon steel, rather than stainless steel. One leak in 1973 released 118,000 US gallons (450,000 litres) of high level waste containing some 55,000 curies of radioactivity, much of which leaked into the Columbia River.

In 1977, a controversial study of the health of workers at the plant found a 25 per cent excess of fatal cancers among those occupationally exposed to radiation. Attempts were made to suppress the study, the results of which have subsequently been confirmed by several independent experts. (See also: *Low level radiation*Ⓝ, *Nuclear accident*Ⓝ, *Nuclear waste*Ⓝ.)

Hazardous waste

The improper disposal of hazardous wastes is proving an increasing problem throughout both the industrialized world and the Third World.

Worldwide, some 325-375 million tonnes of hazardous waste were generated in 1984, 90 per cent arising in the industrialized world. In the USA, some 260 million tonnes are generated each year – the equivalent of one tonne for each US citizen. In 1984, the ten nations which then made up the EEC generated an estimated 160 million tonnes of industrial waste a year, of which 30 million tonnes were classified as "toxic and dangerous". Conservative estimates put the amount of hazardous waste generated in England and Wales at some four million tonnes.

Unless disposed of with great care, hazardous wastes pose a severe threat to public health and the environment. In particular, the use of landfill (*qv*) has led to the pollution of surface and groundwaters, to the contamination of land, and to the mass exposure of whole communities to the dangerous effects of highly toxic chemicals.

Numerous "problem" sites have now been unearthed, the most infamous being Love Canal (*qv*) in New York State. All told, the USA has an estimated 50,000 hazardous waste landfills, 20,000 of which are thought to pose a potential threat to human health. In addition, there are some 170,000 industrial impoundments – used to dump liquid wastes – and 7,000 deep wells into which hazardous wastes have been injected under pressure. According to the EPA, 2,000 sites require immediate clean-up under federal "Superfund" (*qv*) legislation. The Office of Technology Assessment estimates that clean-up costs could total $100,000 million.

In Europe, the problem of old sites is also acute. West Germany has closed 6,000 dump sites as potential hazards, 800 being deemed a threat to public water supplies. The clean-up bill is estimated at $10,000 million.

In the Netherlands, a national survey of old sites was initiated after a housing estate at Lekkerkerk had to be evacuated due to a leaking dump. Within six months, 4,000 illegal dumps had been found. Clean-up costs are put at $1,500,000.

No one knows how many polluting dumps exist in the United Kingdom. The British Government has yet to undertake a survey. Nonetheless, an independent survey puts the figure at 600 sites – including potentially dangerous domestic landfills.

If properly treated, most of today's wastes can be rendered harmless before disposal, primarily through high-temperature incineration (*qv*). However, the long-term solution to the problem of hazardous wastes lies as much in reducing the production of wastes, not least through recycling (*qv*), as in the use of more sophisticated disposal technology. (See also; *Groundwater*Ⓟ, *Incineration*Ⓟ, *Ocean dumping*Ⓟ, *Ocean incineration*Ⓟ, *Recycling*Ⓟ, *Waste reduction*Ⓟ.)

Heart disease
See: *Coronary heart disease.*

Heat pump
A device that abstracts heat from a cold area, cooling it further, and releasing it in a warmer area. (See also: *Energy conservation* Ⓔ, *Soft energy paths* Ⓔ.)

Heavy metals
A group of elements with a wide range of chemical properties and biological effects. Some of them, such as manganese, copper and zinc, are essential trace elements in the diet and their absence can lead to serious illness. Others, such as mercury, lead and cadmium, have no biological function and their presence in all but very small quantities can cause poisoning.

Human activities such as mining, smelting, waste dumping, burning rubbish and the addition of lead to gasoline have greatly increased the amounts of heavy metals circulating in the environment, and much harm has been caused as a result.

Because they are elements, heavy metals cannot be destroyed: they can only be converted from one chemical compound to another. They tend to build up in soils, watercourses and living organisms. Some plants can accumulate high levels of heavy metals from the soil in which they grow, while animals can build up dangerous levels through eating contaminated food. Marine animals – such as shellfish – are particularly efficient at concentrating heavy metals from the water and sediments in which they live.

Heavy metal wastes are difficult to dispose of safely. If burned, they cause air pollution and if dumped at sea – a process banned (except in trace quantities) in the North Sea (*qv*) and North Atlantic as far as the more toxic heavy metals are concerned – they can accumulate in the environment. Dumping on land can lead to pollution as heavy metals in leachate (*qv*) may contaminate groundwater or rivers. Some heavy metals can be recycled, but as they often appear as complex mixtures this is often difficult. Some countries – notably Sweden – are restricting the use of certain heavy metals to prevent their appearing as wastes and subsequently causing pollution. (See also: *Cadmium, Hazardous wastes* Ⓟ, *Leachate* Ⓟ, *Lead, Mercury, North Sea* Ⓟ).

Hedgerows
Since 1945, thousands of kilometres of hedgerows have been lost in Britain alone (see illustration). The process is encouraged by government grants and justified by civil servants and farmers alike as a move towards more "efficient" farming. Hedges, it is argued, take up potentially productive land – 3,000 square feet for each hundred yards of hedge (280sq m for each 90m) – and make plowing more difficult: their removal thus saves both time and energy.

Numerous studies have shown the value of hedges in providing shelter to both livestock and crops. If a hedge is semi-permeable, then wind speeds can be reduced by as much as 80 per cent for distances up to five times its height. Where such protection is present, the gain in production can be between two and four times greater than the loss incurred by keeping the land beneath the hedge out of production.

Hedgerows also act as vital shields against erosion, either by wind or rain, because runoff is significantly reduced, and storms appear to be less severe in hedgerow country.

But perhaps the most overlooked benefit of hedgerows lies in the habitat they provide for wildlife. Though they harbour some insects and fungi that can harm crops (the bean aphid and wheat rust are two examples), the greater part of their population are the natural allies of the farmer – from the bumble bee, which plays such a vital role in pollinating flowering crops, to birds of prey and insect predators, which keep farm pests in check. Once the habitat for these natural enemies of farm pests have been destroyed the only course open is chemical treatment (see *Insecticides* and *Herbicides*).

If domesticated livestock is included as part of the hedgerow ecosystem, then the case in favour of hedgerows is still more overwhelming. The removal of hedges has been associated with a rise in bovine tuberculosis, grassland tetany, and bruscellosis. Hedgerows not only help break the cycle of parasite diseases but, in hot countries, they reduce cattle mortality from sunstroke. (See also: *Conservation* Ⓒ, *Erosion* Ⓐ Ⓒ.)

Herbicides
Chemicals used to kill weeds and other undesirable plants. There are many different types, varying in their mode of action, the plants against which they are used and their persistence in soil. Modern

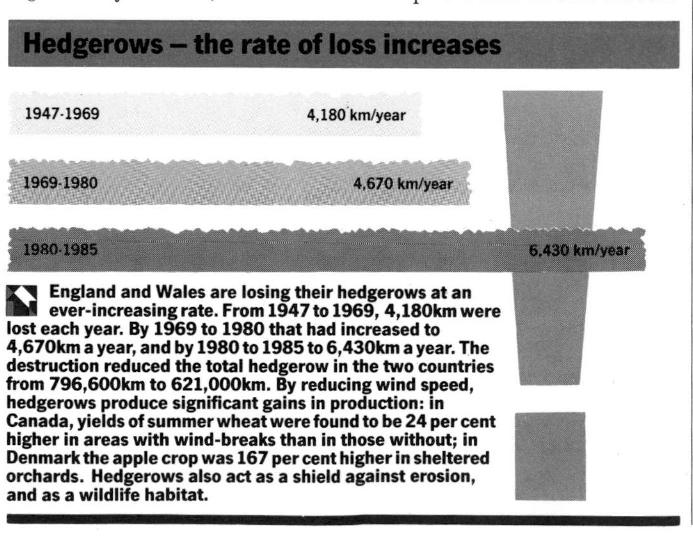

England and Wales are losing their hedgerows at an ever-increasing rate. From 1947 to 1969, 4,180km were lost each year. By 1969 to 1980 that had increased to 4,670km a year, and by 1980 to 1985 to 6,430km a year. The destruction reduced the total hedgerow in the two countries from 796,600km to 621,000km. By reducing wind speed, hedgerows produce significant gains in production: in Canada, yields of summer wheat were found to be 24 per cent higher in areas with wind-breaks than in those without; in Denmark the apple crop was 167 per cent higher in sheltered orchards. Hedgerows also act as a shield against erosion, and as a wildlife habitat.

Western agriculture, with its large fields and intensive applications of fertilizer, requires a substantial input of herbicides to keep down weeds, which would otherwise thrive, and few fields remain unsprayed.

Some herbicides – sulphuric acid, for example – destroy only the plant tissue they actually touch. These are known as contact herbicides and contrast with the translocated herbicides, such as glyphosate, which are absorbed by the leaves and carried within the plant to other parts, thereby killing the whole weed. Another class of weedkiller is the hormone type of which the phenoxy herbicides (*qv*) are the most important. These mimic the effect of plant hormones and cause the weed to overgrow and die. They are translocated.

Most of the herbicides in general use are not persistent in the environment, though some are deliberately used because they do persist. An example is simazine, used to kill weeds on paths and tennis courts, which leaves residues in soil for two years or more. The conversion product of one fairly persistent herbicide, aminotriazole, is thought to be carcinogenic.

Herbicides are not usually acutely toxic to humans or other mammals, the main exceptions being paraquat (*qv*) dinoseb and DNOC, all of which are poisonous.

Some species are more susceptible to herbicide poisoning than others. Fish are very sensitive. The widespread use of herbicides has also led to reductions in the numbers of insects in fields in many parts of the world, thereby restricting the food supplies of insectivorous birds.

Modern herbicides are extremely powerful and some are volatile. Serious damage has been caused when warm weather has evaporated sprayed herbicide from one field and allowed it to drift onto crops growing on adjacent farms.

Agriculture is currently highly dependent on the use of herbicides, but weeds in some areas are becoming more resistant to their effects. More research into alternative methods of weed control is becoming an urgent priority. (See also: *Biocides*Ⓛ Ⓟ, *Paraquat*Ⓛ Ⓟ, *Phenoxy herbicides*Ⓛ Ⓟ.)

Hexachlorocyclohexane (HCH)

An organochlorine insecticide used for a wide variety of agricultural and public health purposes as well as for preserving wood. It exists in several chemical forms, known as isomers, and is used as a mixture of those isomers. Products containing more than 99 per cent of the gamma isomer are known as lindane. HCH was formerly known as benzene hexachloride (BHC), but this name has been dropped as it is chemically inaccurate.

Like other chlorinated hydrocarbons (*qv*) HCH persists in the environment and can accumulate in food webs. It is especially toxic to fish. Birds of prey have also been seriously harmed by lindane and HCH, as have bats when rafters have been sprayed to protect them against insect attack.

The toxicity of HCH depends to some extent on the isomers present. The mixture has been shown to cause cancer in laboratory animals and has been linked to anaemia in humans. Liver and kidney damage have been reported in people exposed to HCH over a long period.

Lindane is generally more dangerous than mixtures of HCH. It has also been linked with leukaemia and is embryotoxic, causing spontaneous abortion. It also damages to nervous system.

Because of their environmental and health effects, the use of lindane and HCH is banned or severely restricted in Scandinavia, the USA, New Zealand, the USSR, France, Germany and many other countries – though they are aggressively marketed for use in the Third World. In the UK, however, despite EEC restrictions, products containing lindane or HCH can still be bought freely by gardeners and farmers. (See also: *Chlorinated hydrocarbons*Ⓟ.)

High level waste

See: *Nuclear waste.*

High yielding varieties (HYVs)

Modern hybrid strains of cereals that can produce higher-than-normal yields. They were mainly developed by the International Rice Research Institute (IRRI) in the Phillipines, and provide the basis of the Green Revolution (*qv*).

The first successful HYV was a rice variety, IR8, developed in 1966. It was a cross between the Indian tall rice plant and a Taiwanese dwarf variety. It produced a high yield, but needed a great deal of fertilizer. This has been a feature of all HYVs. In addition, it could support such heavy fertilization without lodging (flattening), which traditional varieties could not do.

HYVs also generally require irrigation (*qv*), which goes a long way toward explaining the proliferation of large-scale irrigation schemes throughout the Third World. A further feature of HYVs is that they are more susceptible to disease. This is fully admitted by Dr M.A. Swaminathan, one of the fathers of the Green Revolution, and a director of the IRRI. He notes how the IR8 rice proved particularly vulnerable to bacterial blight (BB) both in 1968 and to rice tungro virus (RTV) in 1970 and 1971 in the Philippines. It was largely because of its great vulnerability to such diseases that it was replaced by IR20 in 1971. Though IR20 was resistant to BB and RTV, it proved vulnerable to the brown plant hopper (BPH) and the grassy stunt virus (GSV), which in 1973 devasted vast areas planted with that variety. In 1973, the IRRI was thus forced to develop another variety, IR26 which had built-in resistance to BPH. For two years it was used throughout the Philippines, but in 1966 it proved vulnerable to a new type of BPH, so it too had to be replaced too with a new variety, called IR36. This has done well since. However, it is now being threatened by yet another biotype of BPH as well as by two totally new diseases.

The experience in India has proved similar. A taskforce of eminent rice breeders notes how the introduction of HYVs has brought about "a marked change in the status of insect pests like gall midge, brown plant hopper, leaf folder, whore maggot etc." Most of the HYVs released so far are susceptible to major pests with a crop loss of 30-100 per cent.

There is now a growing trend to breed for resistance to possible pests rather than for increased yields only. However, breeders are unlikely to develop a variety that displays resistance to more than a few specific pests. The only strategy for achieving real security is by planting a large number of different varieties that between them ensure against pest losses. Traditional agriculturists had already achieved this. Unfortunately, the vast stock of valuable plant genetic material that they had developed over hundreds, if not thousands, of years has been largely lost with the introduction of the Green Revolution as farmers switched to HYVs. Traditional varieties have even on occasion been wilfully destroyed in order to increase the market for patented seeds (*qv*).

Holism

Holism is best seen as the opposite to reductionism, or the "analytic method" that underlies modern science.

The holistic approach consists in studying complete living systems, rather than studying their component parts in isolation from each other "in controlled laboratory conditions", as modern science does.

This is justified on the grounds that natural systems are more than the sum of their constituent parts. By studying the parts separately, reductionist science does not take into account the way these parts are organized.

To insist in trying to understand biological and cultural phenomena in terms of physics or molecular biology, which is still the fashion in scientific circles, is thus considered by holists to be a vain pursuit that does not enable scientists to understand the effects of the changes they are bringing about to our health, our society and to the biosphere (*qv*), whose true nature can only be understood in the light of a holistic view of the world.

Hormones in meat

Hormones are chemical substances secreted by the glands. Their role is to regulate such essential functions as growth, metabolism and reproduction. The endocrine system that handles hormones is highly sophisticated, delicately balanced and still not fully understood. Yet we are interfering with it by treating animals with hormonal drugs.

Such veterinary drugs are used for a number of purposes. One is to ensure that all the cows in a herd come on heat at the same time. This enables the cattleman to rationalize his breeding programme so that all the calves are born at the optimum time from the point of view of weather and pasture conditions. In the USA, a product called Lutalyse, manufactured by the Upjohn Corporation, is highly regarded for this purpose. It is a prostaglandin, a member of a group of endocrine compounds similar to hormones. Prostaglandins are only now being seriously examined. There appears to be a large number of them and they affect all sorts of biological functions, including respiration, blood clotting, digestion, blood circulation, nerve responses and reproduction.

Another purpose for which hormones are used is as abortifacients – to induce miscarriage in feedlot heifers which do not put on enough meat to satisfy commercial requirements when in calf. Lutalyse is also used for that purpose.

Hormones are also used for caponizing cockerels. An example is diethylstilbesterol (DES) (*qv*), an established carcinogen which has been blamed for a wide range of unpleasant side effects. Farmers who have come into close contact with it have developed symptoms of impotence, infertility or gynecomastia (tender breasts), and also changes in voice register. DES is even supposed to have affected the menstrual cycles of farmers' wives. Men who have eaten the necks of chicken treated with DES, and male dogs which had been fed on chicken wastes, have also shown signs of developing feminine characteristics. Such side effects led the USA to ban the use of DES implants for caponizing cockerels and to spend $10 million to remove contaminated chickens from the market.

In 1983, a new synthetic hormone that acts by blocking the production of testosterone was developed as a means of castrating young piglets without surgery.

The main use of hormones in livestock rearing, however, is as a growth promoter (*qv*).

Hydro-electricity

In the last 25 years, installed hydro-electric capacity has increased by 262 per cent – from 149,571 megawatts in 1960 to 541,976 megawatts in 1984.

Even so, much of the hydro-electric potential of the world's rivers remains untapped.If all the rivers of the world were to be dammed, then an estimated 73,000 terawatt-hours of electricity – an output equal to that of 12,000 nuclear reactors – could be produced every year. (One terawatt is equivalent to 10^{12} watt-hours). Technical difficulties preclude all of that energy being exploited. Nonetheless, according to the World Energy Conference, 19,000 terawatt-hours could be tapped, as against the 1,300 terawatt-hours produced today.

For those countries that lack conventional sources of energy – and which must therefore import their fuel – the lure of hydropower is evident. In 1984,for example. Brazil had to find $25 million a day to pay for the one million barrels of oil imported daily. Yet, as the science writer Catherine Caufield points out, the energy potential of the rivers of the Amazon basin, excluding the Amazon itself, is 100,000 megawatts. By the turn of the century, Brazil hopes to generate 22,000 megawatts from the Amazon Basin. The Itaipu Dam on the Parana River will alone generate 12,600 megawatts. By comparison, most nuclear reactors generate a mere 1,500 megawatts.

Nonetheless, large dams (*qv*) carry heavy environmental and social costs, which, especially in the tropics, invariably far outweigh the benefits of hydro-electricity. Moreover, the industries attracted to Third World countries by the promise of cheap electricity are often highly polluting. (See also: *Export of pollution* Ⓟ ③.)

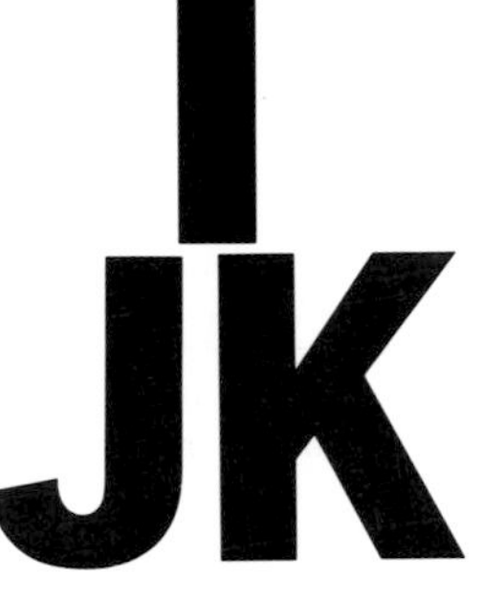

Incineration
Potentially the safest method of destroying many of the more toxic of today's hazardous wastes – particularly such substances as organic solvents, chlorinated hydrocarbons and oily wastes.

There are four major categories of incinerators: high-temperature incinerators; industrial boilers and furnaces; industrial kilns; and municipal incinerators.

Of these, only high-temperature incinerators are purpose-built for the incineration of toxic wastes. By contrast, industrial boilers burn certain toxic wastes – flammable solvents and oily wastes, for example – because they constitute a cheap and efficient fuel. In Norway and Sweden, PCBS (*qv*) and other "difficult" wastes have been successfully incinerated in kilns used for the manufacture of cement.

Effective incineration requires sophisticated equipment, including such anti-pollution devices as scrubbers (*qv*), and depends critically on good management. Properly operated, the most modern high-temperature incinerators have "destruction efficiencies" of 99.99999 per cent.

To ensure the complete and even combustion of a waste, the incineration chamber must be kept at the right temperature; the waste must be kept in the chamber for the right length of time; and there must be adequate turbulence within the chamber to ensure that the waste is well mixed with oxygen.

Failure to observe the "three Ts" of incineration carries two major dangers. First, inefficient incineration means that not all of the original waste will be destroyed; and, second, it can lead to the formation of new products, known as PICS or Products of Incomplete Combustion.

According to the US Science Advisory Board, some 24 organic compounds have been discovered in emissions from high-temperature incinerators. Most of those compounds have yet to be identified chemically, let alone assessed for their health of human health and the environment.

Partially incinerated PCBS have long been known as a source of highly toxic dioxins (*qv*) and polychlorinated dibenzofurans. Both chemicals are also released in significant quantities from municipal incinerators, which operate at temperatures typically 500°C (932°F) too low to destroy the plastics in domestic rubbish. Dioxins – thought to result from the incomplete combustion of plastic bone replacements – have also been found in emissions from crematoria.

Industrial Biotest Laboratories
Private Illinois-based laboratory found to have falsified hundreds of safety tests on a wide range of products – from pesticides (*qv*) to irradiated foods (*qv*).

Investigations by the US Food and Drugs Administration revealed that safety tests on 200 pesticides had also been faked or misread. As a result, according to the US Environmental Protection Agency (EPA), 15 per cent of all the pesticides now on the market are suspect.

Some of the suspect pesticides have still to be retested and most are still in worldwide use, though Sweden and Canada banned a number because of their reliance on IBT data. In Britain, no pesticides approved for use on the basis of IBT tests was withdrawn. Though the EPA has published a list of the suspect pesticides, Britain has refused to do so. (See also: *Cancer*Ⓛ Ⓟ, *Cancer research*Ⓛ Ⓟ.)

Infectious diseases
In spite of the successes of medical science, the incidence of infectious diseases is increasing, and many are spreading to parts of the world where they were previously unknown.

Three different factors are involved in the incidence of infectious disease. The first is the pathogen itself, the second is the host and the third is the environment in which infection occurs. Unfortunately, present research is primarily concentrated on the first of these factors to the near exclusion of the others.

Man harbours in his body a vast number of different microbes. In fact, he is made up of more microbial cells than animal cells and without these microbes his basic physiological functions would be impaired. In the course of evolution, man has learned to live with these microorganisms and it is only in aberrant conditions that they harm him. Thus the herpes simplex virus is often acquired early in life and persists in latent form within the body. It only becomes "activated", and thereby harmful, when certain physiological disturbancees occur which upset its equilibrium with the host.

Lower resistance to disease can occur for a host of reasons, such as the use of certain drugs like cortisone or imuno suppresants, exposure to various chemical pollutants and to low-level radiation, and also because of emotional disturbance.

Ironically, excessive hygiene can also be regarded as a cause of infectious disease, since it reduces resistance to microorganisms that are inevitably present and that in normal conditions are relatively harmless – indeed often essential – to the normal functioning of the human organism.

Environmental destruction is also a basic cause of infectious disease. Among other things it interferes with established relationships between parasites and their non-human hosts, which can cause them to shift their attention to human hosts. Thus bubonic plague developed as a disease of rodents, yellow fever and malaria as diseases of monkeys, rabies of bats. Once man destroyed the habitat of the host animals and modified his own to create a niche for the microorganisms involved, the diseases were quickly transferred to him.

The creation of vast urban conglomerations has also provided a perfect niche for the vectors of infectious disease, including the rat that transmits bubonic plague. It has also put us in close contact with parasites that had previously established stable relationships with the animals we had domesticated. An example is smallpox, a variant of cowpox, which is a disease of cattle.

With the present trend towards urbanization (*qv*) and the development of ever-increasing mobility, ideal conditions are being created for the rapid spread of infectious disease, throughout the world. (See also: *Waterbourne disease* Ⓛ③.)

Insecticides

Chemicals used to kill harmful or irritating insects, though often used against other animals, such as mites, as well.

Before World War II, the insecticides in common use were mainly inorganic poisons, such as arsenic, or natural organic products such as nicotine and derris. The development of two groups of new insecticides, the chlorinated hydrocarbons and the organophosphates (*qv*), revolutionized pest control but brought with them serious environmental problems.

Insecticides are rarely selective: they can thus harm species other than the target insect. If the predators of an insect pest are killed, then the pest problem may be made worse as natural controls on their numbers are removed. Animals at higher levels in ecological food webs are particularly at risk from the more persistent insecticides, such as chlorinated hydrocarbons, since residues pass from one species to another during predation. Examples of such damage include the decimation of birds of prey in the US and Europe, the decline of the English otter, and the deaths of seals in the Dutch Waddensea, all of which can be attributed in whole or in part to insecticide residues.

Human beings are also at risk from insecticide poisoning, since many of these substances are highly toxic to people.

Insecticides in common use can be divided roughly into four groups: the chlorinated hydrocarbons; the organophosphates; the carbamates; and those based on or adapted from natural products. The synthetic compounds in the first three groups have caused the most environmental and health problems, but the fourth group is not without risk. Some, such as nicotine, are highly poisonous and few are selective, though modern insecticides developed from natural compounds called pyrethrins can be used against greenfly without killing their ladybird predators. Most natural insecticides, however, are not persistent in the environment.

A major problem with the use of insecticides is the development of resistance by the pests. Natural selection can rapidly produce a pest population that is unaffected by an insecticide. The amount used is then increased – which causes even more ecological damage – or an alternative found. There is only a limited number of alternatives and the development of a new insecticide takes many years; thus farmers and public health officials in some areas are facing a crisis as pests and disease-carrying insects become impossible, or very expensive, to control. The development of more subtle strategies including biological controls (*qv*) is thus seen as essential if pests are to be kept to manageable levels in the long term. (See also: *Biocides* Ⓟ, *Pesticides* Ⓟ.)

Insulation

The use of materials that are poor conductors of heat to reduce heat loss, especially from buildings. In an average uninsulated home, about 25 per cent of all the energy used for space heating is lost through the walls, 17 per cent through the roof, five per cent through the windows, and about six per cent by too frequent air changes.

Insulation has a key role to play in energy conservation (*qv*) programmes in temperate areas. (See also: *Energy conservation* Ⓔ, *Soft energy paths* Ⓔ.)

Integrated development

See: *Sustainable development.*

Integrated pest control

See: *Biological control.*

Intermediate level waste

See: *Nuclear waste*

International Commission on Radiological Protection (ICRP)

Founded in 1928, the ICRP is responsible for advising national governments on acceptable exposures to ionizing radiation. It is composed of leading radiation specialists from many different countries, and most countries derive their radiological standards from recommendations laid down in a 1977 report known as ICRP 26. The commission's views on maximum acceptable radiation doses are governed by the assumption that a one in 80,000 chance of death per rem of exposure represents an acceptable risk for the general public, with a one in 10,000 chance being acceptable for radiation workers. The resultant dose limits are five milli-Sievert (0.5 rem) a year for the public, and 50 mSv (five rem) per year for industry workers. Allowable exposures to various body organs are also defined. The ICRP recommends that members of the public should not be exposed to more than 0.1 milli-Sievert a year from the operations of the nuclear industry.

The UK's policy is to keep all doses "as low as reasonably practicable" (ALARP), the maxima being one-tenth the ICRP five-rem limit. (See also: *Nuclear accidents* Ⓝ, *Low level radiation* Ⓝ, *Routine discharges* Ⓝ.)

International Atomic Energy Agency (IAEA)

The agency, founded in 1956, has headquarters in Vienna. Its two main roles are to act as an international information exchange on civil nuclear applications (a role that embraces the Comecon countries), and to administer the Non-Proliferation Treaty and that treaty's safeguards.

International Monetary Fund (IMF)

Institution set up, together with the World Bank, its sister body, at the Bretton Woods Conference in the US after World War II. Its purpose was to promote international monetary cooperation and the expansion of international trade, to facilitate monetary convertibility and to ensure financial stability.

The IMF is an international body with 140 members. Power resides in theory with the 21 executive directors but, is, in fact, mainly exercised by the five members with the largest quota, the United Kingdom, West Germany, France, Japan and, in particular, the US which has the largest quota of all. The IMF is primarily the lender of the last resort to Third World governments that are experiencing balance of payment difficulties and are in debt to private international banks. A debtor government can borrow up to the limit of its special drawing rights (SDRS) in exchange for its own currency. If this is not sufficient, it can borrow over and above its quota, but then it must accept IMF "conditionalitics". The more it borrows above its quota, the stricter the conditionalities and the more vigorously they are enforced. If a country borrows the equivalent of double its quota, and still needs more money, then the IMF will effectively take control over economic and, indeed social, policies. Since IMF approval is critical to obtaining loans from other sources, its powers are considerable.

The "conditionalities" imposed on debtor countries by the IMF take several forms. To begin with, the debtor government is made to reduce expenditure – in particular "non-productive" expenditure on welfare, education, health and food subsidies. This has caused a great deal of hardship, particularly in the slums of the Third World where thousands of people (many of them environmental refugees (*qv*) from development schemes) depend on food subsidies for their very survival. It is not surprising that such policies have given rise to food riots among the urban poor in Peru, Turkey, Egypt and elsewhere.

The debtor country is also made to reduce real wages, which clearly creates further hardships. Thus when the Jamaican government under Prime Minister Michael Manley applied to the IMF for funds in 1977, it was made to reduce real wages by 25 per cent. In 1980, after Jamaica had adopted further IMF imposed restrictions, wages fell by another 45 per cent. This goes some way towards explaining why Manley lost the subsequent election to Edward Seaga.

A third feature of IMF imposed policies is an insistence on the expansion of cash crops (*qv*) for export, at the expense of food crops grown for local consumption. This has caused still greater human misery, particularly in those countries which already suffer from serious malnutrition and famine. Yet the IMF has been uncompromising in its imposition of this conditionality. When, for example, the Sudan cut back on cotton production in order to feed its starving population, the IMF insisted on the policy being reversed before it would grant any further loans.

In order to further boost exports and, by the same token, limit imports, a debtor country is frequently required to devalue its currency. Significantly, this is as far as the IMF permits debtor countries to go in reducing expenditure on imports in order to solve their balance of payment problems. The IMF strictly prohibits the setting up of import quotas or other constraints on the import of manufactured goods from the industrialized world – even though runaway expenditure on superfluous imports is, more often than not, the main cause of the debtor country's financial plight.

Not surprisingly, critics of the IMF have accused it of placing the short-term economic interests of the industrialized world above those of debtor nations in the Third World – to the detriment of the health of millions and the ruin of the environment. (See also: *Cash crops* Ⓐ③, *Development* ③, *World Bank* ③.)

International Whaling Commission

See: *Whales.*

Irradiated food

See: *Food irradiation.*

Irrigation

One of the most productive forms of agriculture known to man. In the USA, for example, a Nebraskan corn farmer can produce 40 bushels of corn a year on unirrigated land. By introducing sprinkler irrigation, that yield can be increased to an average of 115 bushels an acre.

Not surprisingly, the UN Food and Agriculture Organisation (*qv*) sees the expansion of irrigated agriculture as being vital to the battle against world hunger. In 1982, an estimated 2,200,000sq km (850,000sq miles) were irrigated worldwide. By the end of the century, FAO hopes to have increased that area by a million sq km (386,000sq miles).

There are good reasons to suppose, however, that, far from relieving world hunger, FAO's strategy will exacerbate the problem. It is not irrigation itself which is at fault but FAO's attempts to practise it on an intensive scale.

In arid areas, good management is essential to successful irrigation. Traditional methods of irrigation can draw on centuries of experience of the best ways to ensure the equitable distribution of water or to regulate such mundane tasks as clearing drainage ditches.

By contrast, the modern irrigation schemes being promoted by FAO are large-scale and perennial, with two to three crops being grown on the same land year after year without respite. Drainage is rarely installed – due largely to cost – and salinization (*qv*) is near universal. Hundreds of thousands of hectares of agricultural land are thus lost to farmers every year. Indeed, as much land is now being taken out of production due to salinization as is being brought into production through new irrigation projects.

In addition, by creating a permanent niche for the vectors of such waterborne diseases as schistosomiasis (*qv*) and malaria (*qv*), perennial irrigation projects pose a major threat to public health. Agricultural pests also thrive in perennial irrigation systems, benefiting not only from the permanently moist atmosphere but also from the presence of a permanent food supply.

Modern irrigation schemes are notorious for being badly managed. Problems with machinery are common and disputes over water rights are rife: bribery is frequently the only means of obtaining water in the right quantity and at the right time.

In India, only 65 of the 246 large-scale surface irrigation projects initiated since 1951 have been completed. In fact, the last people to have benefited from irrigation projects – whether in India or elsewhere in the Third World – have been the rural poor.

Setting up large-scale irrigation projects is exorbitantly expensive – in some areas, it costs as much as $10,000 to irrigate a single hectare of land – and to earn the foreign exchange to pay the bills, irrigated land is invariably used to grow cash crops (*qv*) for export. Even where the project is specifically intended for peasant farmers, plantation agriculture frequently takes over. The water from Iran's Dez Dam, for example, was supposed to irrigate land belonging to small farmers in Khuzestan. In the event, however, the land went almost exclusively to foreign-owned companies, which cultivated crops for export.

Irrigation

Irrigated land now produces about 30 per cent of the world's food. This highly productive form of agriculture is practised around the world; the figures below show the percentage increase in irrigated land between 1950 and 1980, by continent. But the gain in food production is often at a high cost, with large areas of irrigated land now being lost through salinization, waterlogging and alkanization. Irrigation also places a massive demand upon the world's freshwater reserves.

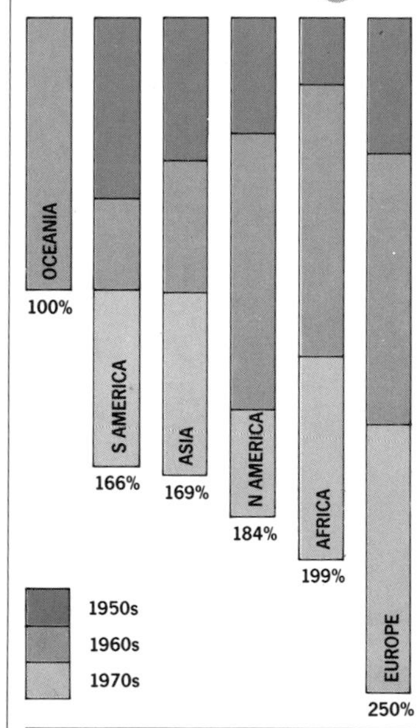

The problems associated with large-scale, intensive irrigated agriculture are not new: history records the disastrous fate of the monumental irrigation schemes set up in ancient Mesopotamia. There are signs, however, that some irrigation engineers are learning the lessons of the past and rediscovering the merits of traditional techniques. Such experts are beginning to realize that successful irrigation is only possible where fallows are observed and where management is based firmly within the local community. The pressure, however, is still to build massive perennial irrigation schemes that by their very nature must be run by large bureaucracies. (See also: *Dams* ③, *Development* ③, *Green Revolution* Ⓐ ③.)

Isotope

See: *Radioactivity*.

Junk food

Modern convenience foods not only lack nutritional quality, but their consumption contributes to a wide range of the so-called "diseases of civilisation" (*qv*) – from tooth caries to cancer, diabetes and diseases of the bowel.

Most processed foods consist of just four basic ingredients – corn, wheat, soybean and sugar beet – which are then dressed up with additives (*qv*) to create artificial foods that look, taste and have the same "mouth feel" as the real thing.

The technology often owes more to heavy industry than to the kitchen. Thus Professor Ross Hume Hall of Canada's McMaster's University describes how the "salami" filling of a typical frozen pizza is manufactured from soyabean by dissolving the milled soy flour in lye and precipitating out the protein in an acid bath "in the form of fine threads that are wound on a bobin, cemented together, dyed, flavoured and cut into hamburger-sized chunks." Additives then give the texture and taste of salami.

Since Roman times it has been known that processing foods reduces their nutritional quality – white bread being nick-named "castratus". Modern processing techniques ensure that most convenience foods have little nutritional value. After milling, for example, modern refined flour has lost up to 80 per cent of its known vitamins and minerals.

Freezing also reduces nutritional quality: when scientists at Rutgers University in the USA tested samples of frozen chicken pies, they found that the pies contained no vitamin C at all – this despite their being packed with vegetables. Even when the scientists added vitamin C, they found that 25 per cent of it had been lost after the pies had been thawed and reheated.

Though vitamins are added to "enrich" a whole range of processed foods, studies have cast doubt on the effectiveness of the practice. Thus a Canadian survey found widespread thiamine and iron deficiencies among the Canadian population despite both thiamine and iron being added to all bread sold in Canada.

Most damaging of all, food processing reduces what Ross Hume Hall has called the "biological sophistication" of food. Processed foods contain only a minute fraction of the chemicals present in natural foods: synthetic orange juice, for example, consists of just ten ingredients, while a natural orange has many hundred.

Just as important, the molecules in natural foods have their own chemical architecture that ensures, for example, that the right nutrients are released in the gut at the right time and in the right quantities. Even the smallest adjustments to that chemical architecture can convert essential nutrients into valueless products.

Kakadu – tribal culture versus uranium

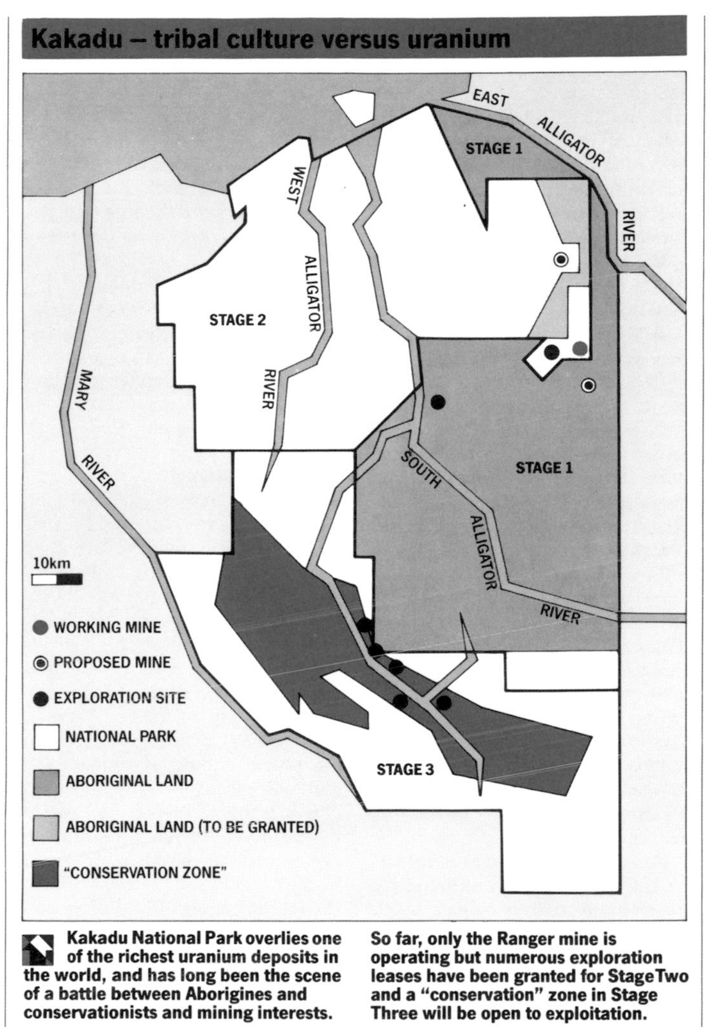

Kakadu National Park overlies one of the richest uranium deposits in the world, and has long been the scene of a battle between Aborigines and conservationists and mining interests. So far, only the Ranger mine is operating but numerous exploration leases have been granted for Stage Two and a "conservation" zone in Stage Three will be open to exploitation.

Kakadu

Located east of Darwin in the Alligator Rivers Region of Arnhem Land, in the Northern Territory of Australia, Kadadu is a beautiful wild area of spectacular and varied landforms, vegetation and wildlife. A land of contrast, it experiences the extremes of monsoon floods and scorching droughts. It is a region steeped in Aboriginal history and legend, and has rightly gained a place on the World Heritage List for its natural and cultural significance.

Kakadu abounds with rich and varied flora and fauna; from the mangrove tidal flats, idyllic billabongs and watercourses, through open grassland, to pockets of eucalypts and rainforest. The wetlands are frequented by large numbers of waterbirds and the total area provides habitats for approximately one-third of Australia's bird species.

Kadadu's cultural significance lies in the rich and ancient Aboriginal sites and paintings to be found there. They trace at least 20,000 years of continuous human culture, record extinct animals, and show environmental and social changes in northern Australia during the last ice age through to the coming of Europeans.

There are many sacred sites that integrate a living spiritual tradition within the landscape for the descendants of the original Australians who continue to live in the region.

Kakadu National Park has been established in three stages (see illustration). Stage One encompassing the headwaters of the river systems, was declared in 1979, and inscribed on the World Heritage List in 1981.

Stage Two declared in 1984, covers most of the flood plains of the South Alligator and other rivers, running north to the sea.

Mining interests attempted to block nomination of Stage Two for the World Heritage List in 1986, forcing the Australian Government to withdraw the nomination. A largely reduced Stage Three, which provides a vital buffer to the south west of Kakadu, was declared a national park in June 1987.

Kakadu has long been the scene where Aboriginal traditional owners and conservationists have battled against mining interests, for beneath its natural wonders is one of the richest uranium (*qv*) deposits in the world.

There is currently one uranium mine operating in the national park. Two more are planned, but currently not operating, because the Australian Government has refused to grant export licenses.

In addition, 14 granted exploration leases and 300 exploration applications exist for Stage Two. The mining industry also has interests farther south, in Stage Three of Kakadu. Concerted lobbying led to only 65 per cent of this section being declared national park, with the remaining area rather ineptly called a "conservation zone", to be opened up for exploitation.

Apart from the obvious impact on the landscape, uranium mining poses many serious threats to the whole of Kakadu region. Many tribal people, who have been put under incredible pressure to allow mining, are deeply concerned at the potential for intrusion and destruction of sacred sites. Accidents, spillages, the release of toxic waste or contaminated water could all have disastrous long-term impacts on this incredibly rich and valuable national park.

With the inclusion of the "conservation zone" in Stage Three, Kakadu would total an area of about 19,000sq km (7,300sq miles) embracing almost the whole catchment area of the South Alligator River.

Conservation of such whole systems is rare in the world. Australia has the choice – to allow further uranium mining and exploitation to proceed, at the risk of damaging Kakadu, or to take this rare opportunity to preserve a unique catchment.

Kalkar

Prototype fast breeder reactor (*qv*) and site of major anti-nuclear demonstrations. In 1972 West Germany, Belgium and Holland began construction of SNR-300 at Kalkar, on the German side of the border with Holland. Comparable in size to the British prototype fast breeder reactor at Dounreay (*qv*), Kalkar has so far failed to obtain a licence to operate because of concern over its safety. Its cost have risen from an initial estimate of $800 million to more than $3,500 million – the plant barely escaped cancellation in 1982 and 1983.

Kesterson reservoir

Californian reservoir, built in 1971, and now thoroughly polluted by contaminated drainage water from the irrigated farmlands of the Central Valley Project. In particular, the reservoir's 12 ponds now contain high levels of selenium – a natural element which, though essential for animals in trace quantities, is toxic in large doses. Other pollutants in the water include pesticides, salt and toxic wastes.

Kesterson lies on a major migration route for birds and, in 1971, was declared a national wildlife refuge. Studies have shown high rates of malformation among birds at the reservoir.

The contaminated water entered Kesterson via the San Luis Drain. Alarmed at the extent of selenium pollution, the US Department of the Interior (DOI) ordered Kesterson reservoir to be closed in March 1985. In addition, farmers were ordered to find alternative means of draining 42,000 acres served by the San Luis Drain. The drain itself was plugged in 1986.

Plans to clean up the reservoir are still under discussion. The only safe long-term solution, say environmentalists, is to remove the bottom mud altogether. The cost could be as high as $146 million.

According to the DOI, nine other areas in six other states suffer from potentially dangerous levels of selenium.

Kesterson is not the only wildlife site in the USA to be affected by pollution. Nationwide, ten other reserves have been identified as having contaminant problems – and a further 74 are under investigation. (See also: *Hazardous wastes* Ⓟ, *Salinization* Ⓐ.)

Krill

A shrimp-like crustacean (*Euphausia Superba)*, which grows to a maximum size of about two inches (6cm) and is found in most of the Southern Ocean south of the Antarctic Convergence. It plays an essential role in Antarctica's marine ecosystem, being a major source of food to many species of fish, animals and birds. These include fin fish, six species of seals, several species of great whales and more than 50 bird species.

Estimates of the total biomass of krill, and the annual rate of production, are difficult to make due to the complex variables involved. Exploitation of krill, however, for which the USSR is 90 per cent responsible, has risen from an exploratory four tonnes in 1961-2 to more than 520,000 tonnes in 1982.

Such heavy exploitation must be reduced until more is known about the rate of regeneration. Catches for 1983 and 1984 have fallen to 233,800 tonnes and 128,329 tonnes respectively. This already suggests that krill stocks are heading for collapse, as has already happened to Antarctic fin fish. (See also: *Antarctica* Ⓒ, *Extinction* Ⓒ, *Overfishing* Ⓒ, *Whales* Ⓒ.)

Kyshtym

Site of a major nuclear accident in the Urals in 1957, involving an explosion at a nuclear waste repository. Though rumours of the accident were quick to circulate in the West, the facts did not become known until 1976, when the exiled Soviet physicist Dr. Zhores Medvedev revealed the full extent of the explosion.

Medvedev reported that the accident had led to the permanent evacuation of 30 communities and to a great many radiation victims, although he was unable to give precise figures. According to Medvedev, more than 400sq km (150sq miles) were heavily contaminated by fall-out.

Medvedev's claim that the accident involved a nuclear waste dump was greeted with scepticism by the nuclear industry in the West. Sir John Hill, the then Chairman of the United Kingdom Atomic Energy Authority, dismissed Medmedev's account as "rubbish" and "a figment of the imagination". In his view, "the burial of nuclear waste could not lead to the type of accident described."

Subsequent studies have left Sir John in a minority. The most detailed report yet – issued by the US Oak Ridge National Laboratory in 1980 – concludes that an explosion involving high-level nuclear wastes was the most likely cause of the accident. (See also: *Nuclear accidents*Ⓝ, *Nuclear waste*Ⓝ.)

Land disposal

According to the Organisation of Economic Co-operation and Development (OECD), more than 70 per cent of the hazardous waste (*qv*) generated by its 24 member states (which include the USA, Great Britain, West Germany, the Scandanavian countries, and Australia) is disposed of on land. In most countries, the figure for municipal wastes is even higher – though there are notable exceptions, Switzerland, for example, incinerating 90 per cent of its municipal wastes.

The principal methods of land disposal are landfill (*qv*), surface impoundment, deep-well injection (*qv*) and storage in underground mines. All three methods pose a major threat to groundwaters and surface water supplies.

In the US, the greatest volume of hazardous wastes (58.7 per cent, according to a 1981 survey) were until recently disposed of via underground injection, with 34.9 per cent dumped in surface impoundments and 5.5 per cent in engineered landfills. Amendments to the Resource Recovery and Conservation Act, however, have virtually banned landfills and surface impoundments, and severely limited the use of deep-well injection. In Britain and many other OECD countries, however, landfill and impoundment remain the major form of land disposal.

Landfill

Method of waste disposal, used to dispose of most of the world's domestic and hazardous wastes.

At its most primitive, landfilling consists of no more than tipping waste into a hole in the ground. Modern "sanitary" landfills, however, are specially engineered, the waste being tipped into prepared cells and covered daily.

Numerous pollution incidents – most notoriously Love Canal (*qv*) – have made the landfilling of hazardous waste the subject of bitter controversy.

Several countries have already outlawed the use of "dilute and disperse" sites (see illustration) for the disposal of hazardous wastes. Since such sites make no

Landfill – dumped waste that poisons water

"DILUTE AND DISPERSE" LANDFILL

CONTAINMENT LANDFILL

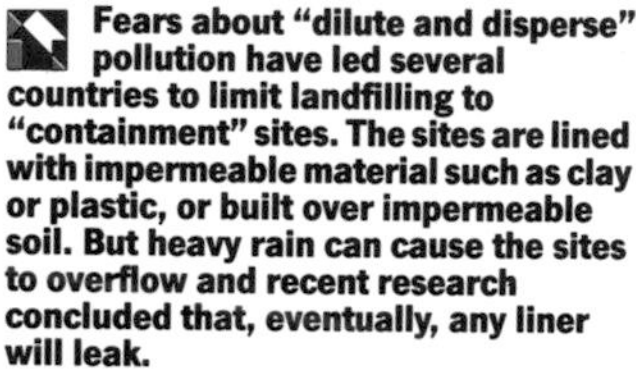

More than 90 per cent of the world's domestic and hazardous wastes are disposed of in landfills. Unlined "dilute and disperse" sites often allow toxic chemicals to pollute groundwater, even though the theory behind them holds that natural chemical and biological processes will render the wastes harmless as they seep through underlying soils.

Fears about "dilute and disperse" pollution have led several countries to limit landfilling to "containment" sites. The sites are lined with impermeable material such as clay or plastic, or built over impermeable soil. But heavy rain can cause the sites to overflow and recent research concluded that, eventually, any liner will leak.

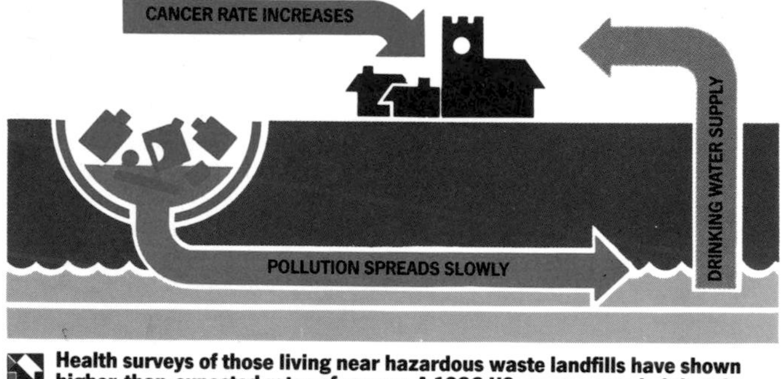

Health surveys of those living near hazardous waste landfills have shown higher-than-expected rates of cancer. A 1986 US survey revealed that the cancer rate in rural Louisiana, which has a disproportionate number of hazardous waste sites, was 19 per cent higher than expected. Decades may pass before toxins that have escaped from a leaking landfill are detected in the water supply.

effort to prevent wastes from seeping out of the landfill, all too often the wastes simply migrate undiluted into local groundwaters (*qv*). One US survey found that of 50 sites studied, 40 had contaminated groundwaters. Nonetheless, dilute and disperse sites remain the most common form of landfill in the world.

Fears over groundwater pollution have led several countries to limit the landfilling of hazardous wastes to "containment" sites, the aim being to prevent the migration of leachate (*qv*) from the landfill (see illustration). In France, only sites with a base of impermeable material 16 feet (5m) thick are permitted to take "special" wastes.

But containment sites have been plagued by problems. Certain chemicals – particularly industrial solvents – can transform even clay liners into sieves. Synthetic materials are no more reliable. A 1980 study of four "secure" landfills in New Jersey discovered that the liners at all four leaked, one within months of being installed.

The landfilling of hazardous wastes has been severely restricted in the Netherlands following a spate of pollution incidents. In France, it is seen as "a necessary and technical acceptable solution" – but only "in the short term" and only for wastes of low toxicity. In the USA, the landfilling of untreated hazardous wastes is to be phased out altogether by 1990. A ban on the land disposal of liquid wastes is already in force.

Britain, however, insists that, if properly sited and managed, landfill is a safe means of hazardous waste disposal. Confidence in that position has been somewhat eroded by two reports from the Hazardous Waste Inspectorate (HWI), an official watchdog agency, both severely critical of the standard of management at British landfills. The Chemical Industries Association has also expressed concern about standards, though, like the HWI, it remains committed to landfill.

Leachate

The liquid that drains from a waste dump when water percolates through the waste, picking up polluting materials as it does so. The water may be from rain or from streams flowing over or through the surface. Poor landfill (*qv*) management can give rise to large volumes of leachate.

The composition of leachate depends on the age of a dump, with fresh refuse producing the strongest leachate. Very high biochemical oxygen demands (see BOD) are characteristic of fresh refuse leachate which, coupled with high levels of ammonia and sometimes heavy metals, result in massive pollution if the liquid enters streams in any quantity. Older leachates have lower BODs, but still have large amounts of dissolved ammonia, a species especially toxic to fish.

Leachates from industrial waste sites may contain a wide range of pollutants such as heavy metals, solvents, ammonia, phenols and cyanides, though they may have lower BODs than those from domestic waste sites.

The migration of leachate from both domestic and industrial landfills has led to serious contamination of groundwaters (*qv*) in many parts of the world.

In Britain, most of the research done into the movement of leachates has dealt with industrial waste sites. Limited studies have been used to justify the widespread landfilling of domestic and industrial wastes in sites where leachate is intentionally allowed to migrate from the site, relying on natural processes of "attenuation" to render it harmless. Considerable disquiet has been expressed at this view and in West Germany, much of the USA, Italy and the Netherlands, sites where leachate is collected and treated are the rule. Britain, too, is moving in this direction, helped by the success of several leachate treatment plants. (See also: *Hazardous waste*Ⓟ, *Landfill*Ⓟ.)

Lead

A poisonous heavy metal that has been in widespread use for thousands of years. A powerful neurotoxin, even in trace quantities, it is currently used in batteries, alloys, shot, paints and gasoline, and as a building material.

Lead plumbing, particularly in soft-water areas, causes raised lead levels in drinking water and, in some instances, poses a serious threat to human health. A British survey, carried out in 1975-76, revealed that between seven and eight per cent of water samples taken from taps in England exceeded the EEC limit for lead in drinking water. In Scotland, the corresponding figure was 34.4 per cent. Though steps were subsequently taken to reduce lead levels in the worst affected areas, hundreds of thousands of homes in Britain still have water with lead levels that exceed the EEC safety standard. To ensure that all drinking water sources in Britain complied with that standard would cost between $1,500 million and $5,000 million.

Lead in paint, particularly the older high-lead varieties, has caused numerous fatalities. In the USA, the law restricts the amount of lead in paint to 600 parts per million. In Britain, there is no law restricting the amount of lead in paint, though paints containing large quantities must be labelled. A 1984 survey found that 80 per cent of primers, undercoats and gloss paints sampled had lead levels above the US limit. The paint industry is now taking voluntary steps to phase out lead additives.

The addition of tetra-ethyl lead to gasoline in order to prevent engine knocking has resulted in the universal distribution of lead in the environment, with particularly high concentrations occurring in urban air and dust. A recent study in Turin showed that at least 30 per cent of the lead in the blood of people living there came from gasoline.

Levels of lead in food are also raised by airborne lead. Indeed, such is the pervasiveness of lead that a 1975-76 study found few places left in the British Isles where leafy vegetables could be grown with lead levels below those considered fit for consumption by babies. The problem of lead in food is aggravated by the use of lead solders in food cans – a practice now being phased out in some countries.

Acute lead poisoning causes stomach pains, headaches, tremor, irritability and, in severe cases, coma and death. Much controversy surrounds the effects of low levels of lead on the developing brains of children. Lead is known to affect the nerves and the brain at very low concentrations and many authorities now accept that levels of lead commonly found in the blood of urban children are having a small but significant effect on their mental functioning.

In the USA, research at Harvard Medical School in Boston has found that the mental development of babies is affected by blood lead levels as low as ten micrograms per decilitre. More than a quarter of babies born in urban areas in the USA are exposed to such levels.

Lead-free gasoline has been available in the USA since 1975 and its use has been encouraged by strict car-emission standards. In West Germany, the lead content of gasoline has been severely limited since 1976 and the government now intends to phase out leaded petrol altogether. In Britain, it took a major campaign by environmentalists and doctors before the Government announced in 1981 that the lead limit for gasoline would be reduced from 0.4 grams per litre to 0.15 grams per litre by 1985. Britain is opposed to an outright ban on leaded gasoline. Lead-free gasoline is to be available in all EEC countries by 1989. (See also: *Car emissions*ⓁⓅ, *Heavy metals*ⓁⓅ, *Water quality*ⓁⓅ.)

Leukemia clusters

Significantly higher-than-expected rates of leukemia in local areas. The most notorious cluster in Britain is near the Sellafield (*qv*) nuclear complex in Cumbria.

In November 1983, Yorkshire TV broadcast "Windscale – the Nuclear Laundry", having unearthed evidence that the cancer and leukemia rate among young people living near Windscale (now renamed Sellafield) was ten times higher than expected. Bootle and Waberwaite, villages that should have had no more than one case of cancer in children under 18 over a 30-year period, had instead four cases in 20 years: whereas Seascale, just one and half miles from Sellafield, had had 11 cases of childhood cancer since 1950. The chances of there being 15 childhood cancers in the three villages rather than the expected three cases was no more than one in 50,000 – a very low probability indeed.

The British government responded by setting up an independent advisory panel, headed by Sir Douglas Black, to investigate. Black concluded that a link with Sellafield discharges was possible but unproven. The report gave a "qualified reassurance" to local people.

The report has been criticized, first, because seven cancer cases were left out of the analysis. Secondly, many children may have spent much longer in the sea near Sellafield than Black's assumption of 25 hours per year. Thirdly, Black's conclusions as to the induction of cancer were based wholly on the International Commission on Radiological Protection's (*qv*) model of the relationship between radiation dose and cancers – a model that has been criticized for underestimating the carcinogenic potential of radiation by a factor of ten to 20. Fourthly, as was later revealed,

Black was himself misled over the atmospheric discharges of uranium from the Sellafield plant. British Nuclear Fuels, the operators of the plant, had told him that between 1952 and 1955 only 14oz (400g) of uranium had been released. In fact, 44lb (20kg) had been discharged.

The Sellafield cluster was the first to come to public attention in the UK. A five-fold increase in childhood cancer near Dounreay (*qv*) has since been identified, as well as clusters around the nuclear establishments at Aldermaston and Burghfield. It is difficult to prove the relevance of radioactive discharges (or any other factor) to observed leukemia clusters, partly in view of the large number of other carcinogens to which people are exposed, but the case against Sellafield is generally accepted to be convincing.

A high incidence of cancer among people living near the Rocky Flats nuclear weapons plant in Colorado is also considered to be linked to amounts of plutonium and other radionuclides in the local topsoil. (See also: *Radiation health effects*Ⓝ Ⓛ.)

Lindane
See: *HCH*

London Dumping Convention
The Convention on the Prevention of Marine Pollution by Dumping of Waste and Other Matter – better known as the London Dumping Convention or LDC – came into force in 1975. An international treaty, covering all the world's oceans, its main aim is to control the dumping of wastes from ships and aircraft. It was originally signed by 33 countries. Others (notably Belgium) have ratified the treaty in recent years.

The convention specifically prohibits the dumping of certain "black list" substances into the sea, except in trace quantities. The black list includes organohalogen compounds; mercury and mercury compounds; cadmium and cadmium compounds; persistent plastics and other persistent synthetic compounds; crude oil, fuel oil, heavy diesel oil, hydraulic fluids and lubricating oils, if taken on board for the purpose of dumping; high-level radioactive waste (*qv*) or other radioactive material, where deemed unsuitable for sea disposal by the International Atomic Energy Agency (*qv*); and materials produced for biological or chemical warfare.

In addition, there is a "gray" list of substances that may be dumped only with a specific permit. Grey listed substances include arsenic, lead, copper and zinc; organosilicon compounds; cyanides, fluorides; and pesticides (*qv*) and their by-products.

In recent years, signatory nations to the LDC have clashed over several issues, notably ocean incineration (*qv*) and the dumping of nuclear waste. Thus, in 1985, 25 countries, led by Australia, Spain and New Zealand, voted for an indefinite ban on the dumping of nuclear wastes at sea. The proposal was opposed by six countries, including the USA, Britain and France. When the vote was passed, Britain insisted that the resolution was not legally binding and announced that it would continue to view ocean dumping as a possible option for nuclear waste disposal. (See also: *Heavy metals*Ⓟ, *Ocean dumping*Ⓟ, *Oslo Treaty*Ⓟ.)

Love Canal
Dump site in Niagara City, New York, that gained notoriety as America's first major dumping disaster.

First excavated in the late 19th century, as part of a hydroelectricity and industrial project planned by an eccentric entrepreneur called William T. Love, Love Canal was used by the Hooker Chemicals and Plastics Corporation as a dump for its chemical wastes from the early 1940s to 1952. More than 43,000 tonnes of waste, much of it carcinogenic, was dumped in the canal.

In 1953, Hooker Chemicals sold the site to the local Board of Education for one dollar, on condition that the company was absolved of future liability for any injury that might result from the chemicals in the dump. A school was built on the site, followed by a housing estate.

In 1977, the dump was found to be leaking. Tests revealed the air, soil and water around the dump site to be heavily contaminated with a wide range of toxic and carcinogenic chemicals. At first, the local authorities refused to act, but in April 1978 the New York State Commissioner for Health ordered the evacuation of some 240 families. The dump was cordoned off and the site declared a Federal Disaster Area – the first time in the history of the USA that a national emergency has been declared as a result of chemical pollution.

Later, in 1980, after further studies had revealed the pollution to be more extensive than previously thought, still more families were evacuated.

Three other Hooker dumps in the Niagara Falls area are now known to be leaking. Wastes from one, at Hyde Park, are oozing directly into the Niagara River, the source of drinking water for some six million people. The site contains the largest deposit of dioxin (*qv*) in the world. A second dump, at Bloody Run Creek, is also leaking. It is sited just across the road from a water-treatment plant serving 100,000 people.

Quite apart from Hooker's dumps, there are 212 other hazardous wastes sites in the Niagara Falls area, containing an estimated total of eight million tonnes of waste, including one tonne of dioxin (*qv*).

Some $14,000 million-worth of law suits have been filed against Hooker. (See also: *Hazardous waste*Ⓟ.)

Low level radiation
See: *Radiation and health.*

Magnox reactor

Nuclear reactor that uses natural uranium metal fuel, encased in a magnesium alloy cladding (hence the name "magnox"), using graphite bricks as a moderator (*qv*) and carbon dioxide to take the heat out of the reactor core.

There are now 26 Magnox reactors operating in Britain. In addition, Britain has exported two Magnox stations, each consisting of a pair of reactors; one went to Japan in 1965 and one to Italy in 1962. France also developed a similar design for its first programme of ten reactors, but switched to pressurized water-reactors in 1969.

Early Magnox reactors in Britain had steel pressure vessels, but the last two to be built – at Oldbury in Gloucestershire and Wilfa in Anglesey – were concrete. After the Chernobyl (*qv*) accident, some Magnox reactors were criticized for having inadequate, or no, secondary containment (*qv*).

Magnox reactors produce more plutonium per unit of electricity generated than any other commercial reactor type, and the plutonium has a higher concentration of fissile Pu-239. As a result, Britain's Central Electricity Generating Board (CEGB) now owns the largest stockpile of separated plutonium in the West.

With an average thermal efficiency of 25.8 per cent, Magnox reactors have never been able to compete with large modern coal fired stations, whose efficiency is 36.1 per cent. Their electrical output is modest because they use natural, rather than enriched, uranium as fuel, and they were downrated in 1969 by an average of 18 per cent because of corrosion problems. They now generate about 12 per cent of the CEGBS electricity.

The Magnox reactors were originally built for 25 years operation but most of them have now been running for nearly 30 years. It has been suggested that, coupled with their age, serious corrosion problems make their continued use particularly hazardous. (See also: *Advanced gas-cooled reactor* Ⓝ, *Pressurized water reactor* Ⓝ.)

Malaria

One of the most widespread and lethal diseases in the world. It kills at least one million people a year and at any given moment 160 million people suffer from it.

Swamps and stagnant pools are ideal breeding grounds for malarial mosquitoes. Irrigation ditches, water storage ponds, and the marshy shores of reservoirs are also favoured breeding grounds. Not surprisingly, by creating permanent and vastly expanded habitats for the anopheles mosquito, irrigation schemes have greatly contributed to the spread and incidence of malaria.

Once established, malaria is extremely difficult to control. Draining mosquito breeding grounds is one solution – but this can have severe ecological effects and is clearly not an option in areas under irrigation. A more common method of eradication is to spray breeding grounds with insecticides. Again, the ecological costs are high: moreover, the success of spraying programmes generally proves temporary. One problem lies in the remarkable ability of mosquitos to develop genetic resistence to insecticides. In 1981, the World Health Organizatiron reported that 51 species of mosquito were resistant to one or more insecticide.

Resistance is also growing to anti-malarial drugs. Resistance to chloroquine is now recorded in ten South American countries, ten African countries, and nine South-East Asian countries. Resistance to Fansidar, a drug used in chloroquine-resistant areas, is widespread throughout Vietnam, Kampuchea and Burma. Fansidar-chloroquine resistance has also been reported in West Papua and along the coasts of Kenya and Tanzania. (See also: *Dams* Ⓛ③, *Insecticides* Ⓟ, *Pesticides* Ⓟ, *Wetlands* Ⓒ.)

Malathion

An organophosphate (*qv*) used to kill a variety of insect pests in agriculture, public health and the garden. It is not generally considered to be very toxic to humans. Malathion is unselective and has caused damage to wildlife. (See also: *Pesticides* Ⓟ.)

Malnutrition — the countries where hunger kills

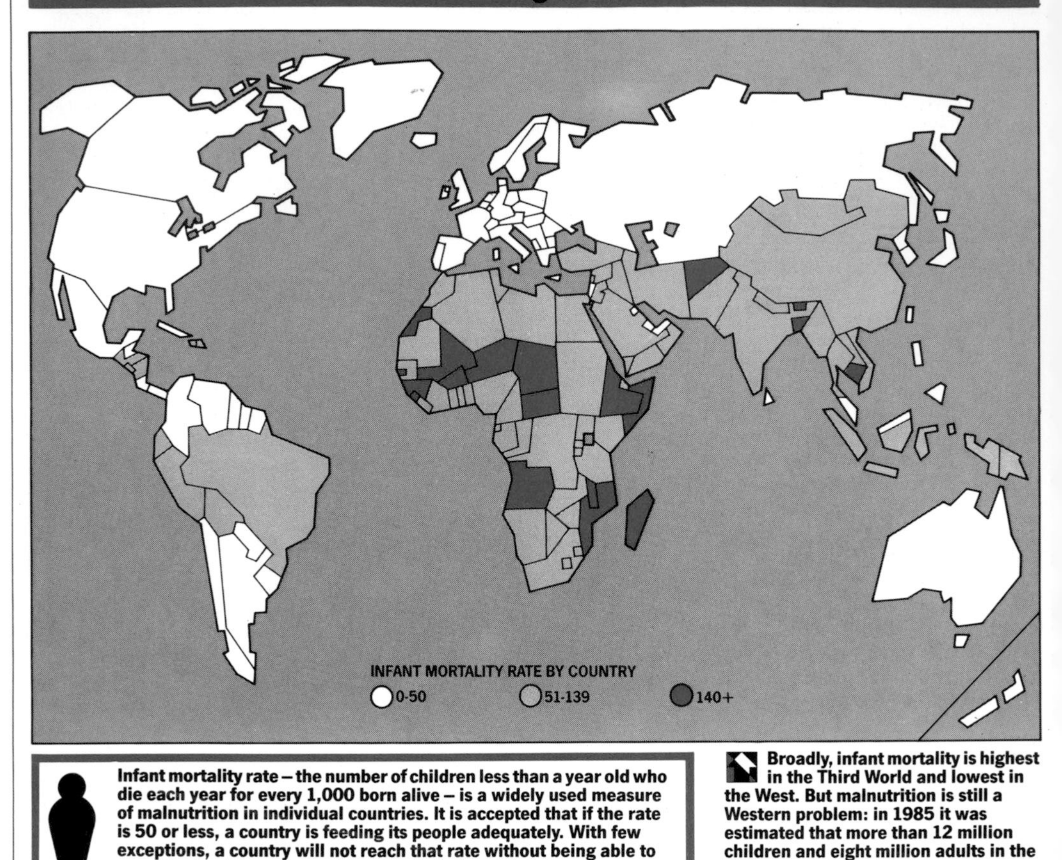

Infant mortality rate – the number of children less than a year old who die each year for every 1,000 born alive – is a widely used measure of malnutrition in individual countries. It is accepted that if the rate is 50 or less, a country is feeding its people adequately. With few exceptions, a country will not reach that rate without being able to grow the food it needs or, alternatively, being able to buy it.

Broadly, infant mortality is highest in the Third World and lowest in the West. But malnutrition is still a Western problem: in 1985 it was estimated that more than 12 million children and eight million adults in the USA were malnourished through hunger.

Malnutrition
A word that means "bad feeding" and thus both overeating and undereating. Where malnutrition is caused by a chronic lack of nutrients, it leads to listlessness, damage to the immune system, impairment of mental functions, and, if severe enough, to death. At the other end of the scale, excess eating causes obesity, heart disease, and other "diseases of civilisation" (*qv*). An imbalanced diet also causes malnutrition and can result in such deficiency diseases as beri-beri (lack of thiamine), rickets (lack of vitamin D) and pellagra (lack of niacin).

Between 1970 and 1980, the number of people in the Third World not getting enough to eat rose by 14 per cent to a total of 340 million people. Since then, according to the World Bank, their numbers have risen inexorably and continue to do so. Today, estimates of the number of people "chronically deprived of the food necessary to enjoy an active life" vary between 512 million and 730 million. Though malnutrition in Africa has received the lion's share of publicity, Africa is not the worst affected continent. A 1987 UNICEF report on world hunger notes: "In the last two years, more children have died in India and Pakistan than in all the 46 nations of Africa together. In 1986, more children died in Bangladesh than in Ethiopia, more in Mexico than in the Sudan, and more in Indonesia than in all eight drought-stricken countries of the Sahel."

On the face of it, low incomes are the most immediate cause of hunger-related malnutrition in both the US and the Third World. But corrosive as the effects of poverty undoubtedly are, the problem is more deep-rooted than simply a lack of cash.

In the slums of the USA and other Western countries, malnutrition is as much a symptom of social alienation as it is of outright poverty. Surveys confirm that, all too often, what money is available is spent on junk foods of dubious nutritional value, on alcohol and narcotics, and on such trappings of affluence as television sets, videos and tape recorders.

For the poor of the Third

World, the lack of cash is a more immediate cause of malnutrition than in the industrialized world, where welfare payments provide a cushion between poverty and outright starvation. Nonetheless, their poverty can also be seen as a symptom of a deeper problem; in this case, the social and ecological disruption caused by the introduction of the market economy.

Previously, though they had no cash, traditional peoples were well capable of feeding themselves. Indeed, anthropological studies repeatedly show a decline in nutritional standards when traditional peoples begin to work for wages and to grow cash crops. Where such peoples are still independent of the market, nutritional standards are high, with diets generally being varied and healthy.

In effect, the very development programmes being touted as the solution to Third World malnutrition often lie at the roots of its spread. Examples include large-scale dams (*qv*) and agricultural projects aimed at promoting the Green Revolution (*qv*). (See also: *Development* ③.)

Maralinga

Nuclear test site in the South Australia desert used by Britain from 1952 until 1957. Other tests were carried out in the Monte Bello Islands and at Emu Fields, also in the South Australian desert. In all, 12 atomic bombs were exploded with a total explosive force of some 180 kilotonnes, the equivalent of 180,000 tonnes of conventional TNT.

The last major atomic test in Australia took place at Maralinga in October 1957. But this was not the end of Britain's experiments with nuclear weapons in Australia. "Minor trials" continued at Marlinga until 1963. One particular experiment consisted of setting fire to a quantity of plutonium (*qv*) to see how it would react.

In 1967, it was decided to evacuate Maralinga and Operation Brumby was launched to clean up the range. The procedures followed were totally inadequate. Any plutonium that could not be recovered was ploughed into the soil. Particularly contaminated areas were covered in topsoil. This ignored the fact that duststorms and high winds frequently occur at Maralinga. Furthermore, some of the contaminated areas were left unfenced.

More than 24kg (53lb) of plutonium were used in the Maralinga "minor trials" but only 900 grammes (32oz) were returned to Britain. The remaining 23 or so kilogrammes remain at Maralinga, either on the range or in burial pits. A recent study, using the latest equipment, found up to 100,000 small metal fragments contaminated with plutonium scattered over the range.

The whole period of the British nuclear tests is marked as one in which scientific integrity, and the safe health of servicemen and the Australian population in general, were sacrificed to the political urgency of developing a credible nuclear force.

Though safety standards on radiation exposure existed at the time of the Australian atomic tests, they were inadequate and frequently ignored. As a result, servicemen were subject to high levels of radiation. Many have subsequently developed cancer.

The measures taken to safeguard the aborigines who inhabited the area surrounding Maralinga and Emu fields were totally inadequate. For the first four years, searching for aborigines in the 800,000sq km (309,000sq miles) range (an area five times the size of the United Kingdom) was the job of just one man. Air searches were executed, but the chances of finding aborigines, who always extinguished their fires when they heard aircraft, were extremely remote. Moreover, it was not until 1956 that it was realized that aborigines, who wore no clothes and slept in the open, were more vulnerable to radiation than Westerners. Only then was a lower maximum permissible dose set for them.

It is known that aborigines strayed well into the range. In 1957, one particular family was found to have camped near a crater caused by a bomb that had been exploded a few months earlier. The wife, Edie Milpuddie, was pregnant at the time and later had a miscarriage. Her next child, born in 1961, died two years later of a brain tumour. Her husband died in 1974 of pneumonia and heart failure. It is difficult, however, to establish for certain how many aborigines were affected by the atomic tests. At the time, very little was known of their numbers and their movements, though, according to one estimate, as many as 1,500 aborigines were living in and around the Maralinga test zone. Also aborigines, for cultural reasons, are reticent about acknowledging deaths and vague about precise dates.

In 1984, an Australian Royal Commission was appointed to investigate the entire history of Britain's atomic tests in Australia. When the commission ended its inquiry in December 1985, it found that the tests had constituted a threat to the health of the Australian population, especially to those directly involved. Some of the tests should never have been held due to the weather conditions at the time. One particular test was in breach of the Atomic Weapons Tests Safety Committee rule that no explosion should occur if there was rain within 1,300km (800 miles). The "minor trials" at Maralinga should never have been held, according to the commission, which ruled that the British must accept the cost of cleaning up the area, as well as the islands of Monte Bello. The commission also found that the Australian government should accept responsibility for compensating the servicemen and aborigines whose health was affected by the tests.

Whether these recommendations are acted upon remains to be seen. It is estimated that the cost of cleaning up the land at Maralinga, which has now been legally returned to the aborigines, could be as high as £120 million ($A300 million). Britain claims to have been discharged from further responsibilities for the site in two agreements, signed in 1968 and 1979. Australia claims that Britain did not fulfil her obligations in those agreements since the sites are still contaminated.

In the meantime, Australia has

set up a Maralinga Commission to supervise the whole process and a technical assessment group, which has recommended a series of studies on how best to go about cleaning up the site. Meanwhile, aborigines continue in their attempts to force the authorities to act. It looks, however, as if a lot more time and a lot more paperwork needs to be got through before they will be able to return to their rightful lands. (See also: *Atomic tests*Ⓝ, *Radiation and health*ⓁⓃ.)

Marshall Islands
Group of 30 atolls in the Pacific used by the USA as a nuclear test site from 1946 to 1958. Seventy atmospheric tests took place, the bombs being exploded on the atolls of Bikini and Enewetek.

Throughout the test programme, scant regard was paid to the health of the Marshalese or their way of life. The inhabitants of Bikini and Enewetek were forcibly removed from their homelands and resettled on two barren atolls which proved unable to support them. Malnutrition was rife.

Other islanders were evacuated when their atolls became too radioactive for human habitation. Many are now dying of cancer.

Few warnings were given on the dangers of radiation. The inhabitants of Rongelap Atoll were exposed to fall-out from the "Bravo" hydrogen bomb test for three days before they were evacuated. Many suffered symptoms of acute radiation sickness.

Three years later, the islanders were allowed to return to Rongelap – even though radiation levels were still ten times higher than those on uncontaminated islands. An official memorandum records: "The habitation of these people on the island will afford most valuable ecological radiation data on human beings."

Other islanders fared no better. In 1978 – less than ten years after they had been allowed to return home – the inhabitants of Bikini were again evacuated. The islanders were found to be taking in dangerous levels of plutonium-239, caesium-137 and strontium-90 through their food. The US Department of Energy now estimates that it will be another 50 years before Bikini will again be safe for human habitation. (See also: *Atomic tests*Ⓝ, *Radiation and health*ⓁⓃ.)

Medfly epidemic
The epidemic of Mediterranean fruit flies – or "medflies" – that hit California in the early 1980s, causing millions of dollars worth of damage to fruit crops, provides a classic example of how biological control can go disastrously wrong.

The control programme involved the release of sterile medflies to disrupt the breeding cycle, a techniques that had worked remarkably well with other insect pests in the past. However, one batch of supposedly sterile males from a laboratory in Peru had been improperly treated: as a result more than 100,000 fertile medfly were released onto crops.

Eventually, the then Governor, Jerry Brown, agreed to permit the aerial spraying of malathion (*qv*) – despite protests from local residents. The governor's action was in response to a Federal threat to quarantine all California fruit produce. (See also: *Biological control*ⒶⓅ.)

Mediterranean
See: *Regional seas programme.*

Mercury
Mercury is a highly toxic heavy metal (*qv*) which has caused considerable damage to human health and the environment as a result of its misuse.

Mercury attacks the nervous system. Small doses cause irritability, nervousness and headaches, while large doses cause convulsions, coma and death. It is toxic when swallowed or if particles of mercury compounds are inhaled. Mercury metal is easily absorbed through the skin and its vapour is highly toxic.

Most inorganic mercury compounds are poisonous, but organic mercury compounds, formed when mercury compounds combine with certain organic molecules, are even more so. These organic mercury are soluble in fat, which means they accumulate easily in the body and so penetrate readily into the nerves and brain. Methyl mercury, one such organic compound, is also teratogenic (*qv*). Under certain conditions, methyl mercury can be formed in the environment from inorganic mercury and organic matter, a potentially very dangerous phenomenon.

Like all heavy metals, mercury can accumulate up food chains and species such as fish and birds of prey at the top of the chain may contain large quantities.

A major use of mercury compounds, which still continues, is as a seed dressing to prevent the growth of moulds. Hundreds of cases of mercury poisoning were caused in Iraq when dressed seed was cooked and eaten by mistake and the use of such dressings has also caused many wildlife casualties.

The most famous mercury pollution incident was at Minamata Bay in Japan when mercury discharged by a factory was passed up the marine food chain to fish, which formed the staple diet of local people. Brain damage, delirium and death were caused while exposed women produced deformed babies. The form of mercury involved was methyl mercury.

Mercury metal is still in use in thermometers, barometers and even children's toys. Spillage of even small quantities of this metal in a warmish room can cause mercury poisoning as the material vaporises. (See also: *Heavy metals*Ⓟ, *Minamata*Ⓟ.)

Michigan disaster
Mass poisoning of cattle in Michigan resulting from a fire-retardant being accidentally packaged as a cattle feed. The fire-retardant contained polybrominated biphenyl (PBB), a highly toxic chemical which is closely related to polychlorinated biphenyls (PCB) (*qv*).

The contaminated feed was fed to tens of thousands of farm animals, causing stillbirths, abortions, tremors, anorexia, liver damage, abcesses and other symptoms of poisoning. The affected animals were buried in mass pits.

It was nine months before the cause of the poisoning was discovered. By then, nine million people in Michigan had eaten contaminated meat and dairy products. In 1982, a study by New York's Mount Sinai School of Medicine revealed that 97 per cent of the population were contaminated with PBBS – the highest levels being found in children and men. (See also: *Cancer* Ⓛ Ⓟ.)

Microchip pollution

The use of microchips in electronic devices is bringing a number of environmental benefits, but, unfortunately, the industry itself has caused health and pollution problems and may continue to do so if adequate controls are not established. Toxic chemicals are used, the effects of which are not always fully understood, and even well-understood chemicals, handled carelessly, have caused pollution that could have been prevented.

In California's Silicon Valley, massive solvent pollution of groundwater has occurred because storage tanks built to hold these materials were badly designed and leaked. Water supplies have had to be taken out of use because of toxic and carcinogenic contamination, while adverse health effects have been claimed by people living in the vicinity.

In a semiconductor plant in Massachusetts, production workers have suffered from increased headaches, sore throats and nausea, which have been blamed on the chemicals used. More serious was the finding that women working in the production area were more likely to miscarry than workers working in non-manufacturing areas of the plant such as offices. The exact cause of this effect has not been discovered, but the company concerned has encouraged women of childbearing age to work elsewhere in the plant.

Silicon, the basis of the microchip, is not harmful in the form in which it manufactured for use in microelectronics, but some other materials which are used in smaller amounts are poisonous. Arsenic and selenium are potentially harmful, but their properties are reasonably well understood. Some other substances in increasing use, such as gallium and germanium, have not been distributed or studied so widely and much less is known about their toxicological and environmental properties. (See also: *Chlorinated solvents* Ⓟ, *Groundwater* Ⓟ.)

Micropollutants

Some 1,200 unidentified chemicals have been detected in public drinking water supplies. Little is known about the effects on health of these "micropollutants". In 1975, the World Health Organization (WHO) reviewed the safety data on 289 compounds commonly present in drinking water and found that no toxicological data of any sort existed for more than half of them.

Six years later, a British study came to the same conclusion: of the 343 compounds it identified "most" had never been evaluated for safety.

Among the micropollutants that scientists have identified are chloroform, benzene, trichloroethylene and carbon tetrachloride – all suspect or proven carcinogens. Ironically, many micropollutants result from the chlorination (*qv*) of water supplies to protect public health. Certain organic pollutants – some of natural origin, others industrial – react with chlorine to form trihalomethanes (THMS), several of which – chloroform, for example – are supected of causing cancer (*qv*) in humans. Though micropollutants are present only in low concentrations, repeated small doses may well cause chronic toxic effects.

Some countries have taken steps to minimize exposure to THMS. Municipal water treatment plants in the USA are required to use granulated active carbon to remove THMS from drinking water. In Germany, the government has limited the concentreation of THMS in drinking water to 25 micrograms per litre – and many German treatment plants are now using ozone instead of chlorine as a disinfectant. (See also: essay on *Drinking water*, *Water quality* Ⓛ Ⓟ.)

Minamata

Fishing village in Japan whose inhabitants suffered an epidemic of nervous diseases from eating fish contaminated by industrial discharges from a nearby chemical plant owned by the Chisso Corporation. Victims suffered convulsions, blindness, brain damage and, in some cases, death. The Minamata tragedy was one of the first incidents to alert people to the dangers of chemical pollution.

The first cases of Minamata disease were diagnosed in 1953. By 1966, 43 people in Minamata had died and a further 68 were permanently disabled – mainly from celebral palsy. By 1983, more than 300 people had died and nearly 1,500 were officially recognized to be suffering from the disease – though 6,000 claimed to have been affected.

The cause of the disease was methyl mercury poisoning. Initially, it was thought that the mercury (*qv*) had been discharged in inorganic form and had been transformed by marine organisms into the more toxic methyl mercury, an organic form of mercury. Later, it emerged that Chisso had in fact been discharging methyl mercury directly into Minamata Bay – though biotransformation undoubtedly compounded the problem.

Though Chisso paid token damages to Minamata victims (£100/$160 to adults and £30/$48 to children), it consistently denied responsibility for the disaster. Those who received compensation did so only after signing a document that precluded them from taking further legal action. (See also: *Bioconcentration* Ⓟ, *Mercury* Ⓟ.)

Mineral resources

Industrial man has an energy- and resource-intensive way of life. It has been estimated, for instance, that an average American will use, directly and indirectly, during the course of his or her life, some 1,600 tonnes of various materials, all of which have to be extracted from the earth, processed and delivered – that is, about 50 times more resources than are used by an average Indian. The

Mineral resources

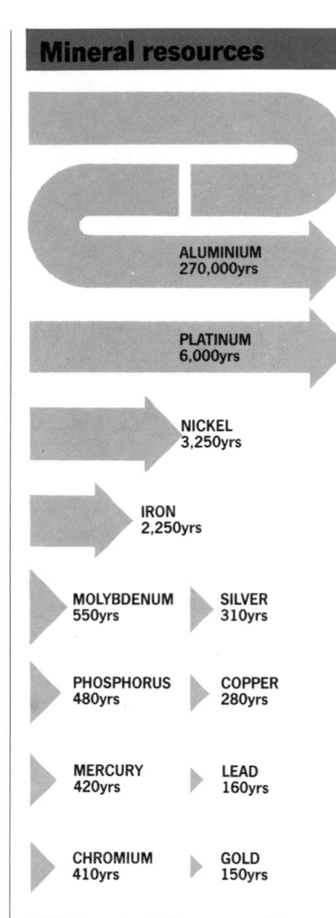

The potentially recoverable resources for a number of the major minerals are shown above. They are expressed as the number of years that they could last at current rates of consumption.

The "resource" of a mineral is an estimate of its geological abundance within the Earth's crust. It is far greater than the "reserve" – the amount which has already been discovered and is recoverable at an economic cost. Much of the world's mineral resources lie in geological formations under the sea bed and other inaccessible sites. It has been argued that new technologies will permit the mining of these inaccessible resources along with the exploitation of lower grade ores, but both of these possibilities would require the use of considerable amounts of energy.

Once the relatively small proportion of mineral resources that are concentrated in accessible and suitably rich deposits has been used up, the costs of recovering new deposits will increase substantially. At current rates of exploitation, it has been estimated that such easily accessible deposits will be exhausted for all but a few redundant minerals within the next 50 to 100 years.

Mineral resources – consumption

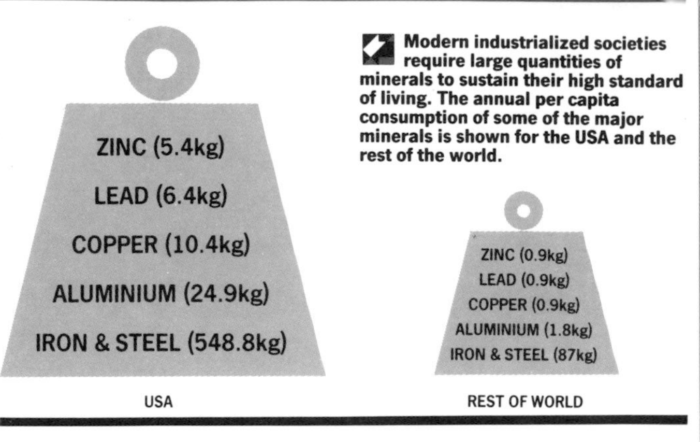

Modern industrialized societies require large quantities of minerals to sustain their high standard of living. The annual per capita consumption of some of the major minerals is shown for the USA and the rest of the world.

contrast between American consumption and the rest of the world is shown in the illustration.

Are there enough resources in the world to support the resource-intensive way of life of the First World and, at the same time, enable people in the Third World to increase their consumption of materials, which they would have to do if they are to develop? How long, in fact, are the Earth's mineral resources likely to last (see illustration left)?

It can be argued that if we are willing to pay a higher price for mineral resources, more can be made available through new explorations. This is only true up to a point. Thus the massive increase in the price of oil (*qv*) in 1973 has not made very much more oil available. On the contrary, it has encouraged us to use less of it.

Indeed, only a few countries, such as Japan and West Germany, could have afforded to pay for oil at $36 a barrel for very long, since at that price, in many countries, it was absorbing a significant part of foreign earnings. The same must be true for many minerals; once their price has reached a certain level, many of their uses will cease to be economic.

It can also be argued that, with the development of the "service economy" and with the increased use of plastics and of new technologies such as microelectronics, the demand for minerals will fall. This is probably true. It would also fall, of course, if there were a major economic slump. However, if consumption continues to increase at the current rate, shortages should begin to occur early next century; by then, most Third World countries will also be suffering serious shortages of land, water, food and firewood, making the pursuit of an industrialized way of life difficult to continue.

Moderator
See: *Fission*

Molluscicides
Chemicals used to kill snails, mussels and other animals with shells. There are two applications of these chemicals that have environmental significance: disease control and anti-fouling treatments for yachts.

In slow-moving tropical freshwater, snails frequently carry the parasite that causes the crippling disease schistosomiasis (*qv*). One of the measures by which this disease is controlled is the destruction of the snail host by the application of molluscicides to the water. Copper sulphate and a pentachloropheol derivative were originally used but these tended to kill other forms of life. More selective compounds are now available.

Molluscs of a different kind have proved to be a nuisance to yacht owners. To prevent their clinging to the bottoms of boats anti-fouling paints, toxic to the animals concerned, have been developed. The principal compound used is tributyl tin (*qv*) and its use has caused serious problems to other organisms, notably oysters. (See also: *Biocides* Ⓟ, *Waterborne diseases* Ⓛ③.)

Monoculture

The cultivation of large stretches of a single crop, which is harvested all at once. It is a necessary feature of modern mechanized farming and clearly the best way to reduce labour costs, make the maximum use of expensive machinery, and maximize marketing efficiency.

Ecologically, however, monoculture is the least desirable agricultural strategy, with few of the advantages of traditional polyculture.

Polyculture minimizes vulnerability to pests by providing the smallest niche for pests affecting specific crops. What is more, if rotation is practised, then this niche is reduced both in time and space. When monoculture is introduced, then the niche is maximized and the pest population expands to occupy it as fully as possible. If, as is usually also so with modern agriculture, the same crop is grown year after year, then the massive niche provided for the pest becomes a permanent one, encouraging the pests' proliferation.

Polyculture, especially when it involves the cultivation of many different strains of the same crops, also provides insurance against droughts, heavy rains, frosts, and pest infestations. The reason is that while some crops, or strains of a particular crop, are affected, many others are likely not to be. Monoculture, by contrast, is highly vulnerable to discontinuities of all sorts because it consists of only one crop.

Polyculture also often involves growing crops that are complementary, such as cereals interspersed with legumes, the latter providing the nitrogen which is needed to grow the former. Crops of different height are also grown, the higher ones providing shade for the lower ones and also helping to keep down weeds.

In addition, the different crops grown provide effective cover to the soil and reduce loss of soil moisture. Hedgerows (*qv*) and windbreaks also help provide protection from the winds, reduce moisture loss, and help prevent erosion.

Not surprisingly, though yields of specific crops may not be high, total yields of all the different crops grown are often higher than those obtained when monoculture is practised. This is not surprising, since under monoculture all the subtle devices for maintaining soil fertility, soil structure and soil moisture, and for reducing vulnerability to pests and other discontinuities, are sacrificed to short-term economic efficiency.

In partial compensation, more nutrients in the form of artificial fertilizer and trace elements are added, as well as pesticides (*qv*). However, these have adverse ecological consequences, and make no contribution to the maintenance of soil structure, which can only deteriorate.

Moreover, the tendency, under a regime of monoculture, to leave the soil bare between crops, especially during the rainy winter months as now occurs in the North American cornbelt, must lead to erosion (*qv*).

Monosodium glutamate (MSG)

An additive (*qv*) used as a flavour enhancer. It has only a slight flavour on its own, but when added to food it stimulates the taste buds and so deceives consumers into thinking that the foods have more flavour than they really do.

Worldwide, MSG is probably used in more than 15,000 different processed food items. It is also added to some animal feeds to stimulate appetites. By the mid-1960s, the Japanese were producing 63 million kg (140 million lb) of MSG a year, and the Americans were consuming some 20 million kg (45 million lb) annually, equal to more than a kilogramme per person per year. Since then consumption has increased rapidly.

For much of the last 25 years, MSG has been among the more controversial additives. During the early 1960s, evidence emerged that MSG could not only have profound effects on the chemistry of the brain, but also that high doses could provoke undesirable changes in the brains of mice. There was further evidence that glutamic acid (to which MSG is closely related) may concentrate in the placental fluid of pregnant women. Furthermore, in the late 1960s, studies revealed that in some (but not all) tests, MSG and glutamic acid damaged the reproductive processes of rabbits and chickens, though in previous tests on rats no problems had emerged. In the late 1960s, some direct human evidence indicated that, on an empty stomach, glutamates can provoke a range of sub-lethal effects – from headaches to respiratory difficulties and muscular tightness.

Several official bodies have recognised that MSG can provoke some symptoms of discomfort and intolerance. MSG has been particularly popular with some Chinese restaurants, and customers and doctors came to recognize what has been called Chinese Restaurant Syndrome, which is attributed to MSG.

Official bodies have generally argued that since glutamilc acid is a normal component of proteins, it, along with MSG, can be adequately metabolized as long as the digestive system is not overloaded. They conclude a "safe level" for daily intake can be established. MSG is thus permitted and used in the UK, the US and much of the EEC, but it does not have an E-number (*qv*). In the UK, it is recommended, however, that MSG should not be included in commercial baby foods. (See also: *Acceptable daily intake* Ⓕ Ⓛ, *Flavourings* Ⓛ.)

Mururoa

Nuclear test site in French Polynesia in the south Pacific. It was established in the face of bitter opposition from the local Territorial Assembly after Algerian independence prompted France to move its previous nuclear test site from the Sahara.

Testing began in 1966, with 41 above-ground tests being carried out in the first nine years, (France did not sign the Partial Test Ban Treaty). The testing programmme was condemned by the UN conference on the human environment at its 1972 Stockholm Conference (*qv*) and, with France undergoing diplomatic pressure from Pacific states and from environmental groups such as Greenpeace (whose yacht *Vega* was illegally attacked and detained on the high seas by the French navy), and with legal action being taken at the International Court of Justice, the testing programme was moved underground in 1975. Subsequently, 84 underground tests have been carried out – the latest being on May 6 1987.

President De Gaulle, visiting the island in September 1966, refused to delay a test for a change in wind direction towards the uninhabited Southern Ocean. Consequently, all the islands as far away as Western Samoa (3,000 km/1,800 miles) received high levels of radiation.

Underground testing has proved hazardous also. The island has sunk more than 1.5m (5ft) since testing began and a crack 35-50cm (14-20in) wide and 1km (1,100yd) long has developed in the atoll beneath sea level. Radioactivity has been leaking into the ocean for several years. Particularly damaging was a bomb which was detonated, after it had stuck half-way down its test shaft, in the coral rather than in the underlying basalt bedrock. A large slab of the internal wall of the atoll was dislodged and resulting tidal wave injured seven people on the beach. The north beach of the atoll was contaminated in 1975 by several kilogrammes of plutonium, after a series of tests. The authorities attempted to "bury" the plutonium under a layer of asphalt. In March 1981, however, a tropical storm tore apart the asphalt covering, washing radiactive debris into the sea.

Radioactivity from the tests (either leaking into the sea or from fallout) is reconcentrated in the seafood that forms the staple diet of the people of the Pacific. No one knows how many have developed cancer, however, since the French refuse to disclose the health statistics for French Polynesia since the tests began. Attempts to set up an independent inquiry have come to nothing. In the 1981 French elections, Mururoa became a major political issue and Francois Mitterand promised the anti-nuclear and environmental groups that the testing programme would be reviewed. The decision to continue testing came just four days after the review began.

The widely scattered population, subject to colonial rule, and hindered by press and radio censorship, have only recently begun to mount organised opposition to the testing. Despite growing worldwide condemnation, however, France looks set to continue its testing programme. (See also: *Atomic tests*Ⓝ, *Underground nuclear tests*Ⓝ.)

Murray River

The Murray River flows through New South Wales and the State of Victoria before crossing into South Australia. The Murray River Basin extends over 2,600,000sq km (one million sq miles) and the river is a major source of water for agriculture, industry and domestic consumption.

Rising salinity levels within the Murray have been causing increasing concern in South Australia, which depends on the Murray for 60 per cent of its water supplies – a figure that can rise to more than 80 per cent in dry years. An estimated 1,300,000 tonnes of salt enter the river each year. Though New South Wales and Victoria contribute at least 60 per cent of the Murray's salt load, salinity levels do not reach worrying concentrations until the lower reaches of the river. Where the Murray crosses into South Australia, the average salt concentration is just under 200 parts per million (PPM). By the time the Murray reaches Adelaide, salt levels have risen to 400ppm – only 100ppm short of the World Health Organization's maximum recommended limit for drinking water.

Some of the salt entering the Murray is natural. The groundwaters in the region are typically highly saline (in some areas they are more saline than seawater) and the seepage of groundwater into the river inevitably adds to its salt load. However, in recent years, the natural flow of groundwaters has been greatly increased through the setting up of numerous irrigation schemes, which, by raising the water table, increase seepage. Drainage water from the irrigation schemes also carries a large volume of salt into the river.

In an attempt to cut down on the salt entering the Murray, tiled drains have been introduced below many irrigated areas, the water being pumped into "evaporation basins" set away from the river. Those basins, however, have not proved watertight and highly saline water has been discovered seeping out of them into the Murray.

With ever-greater demands being made on the waters of the Murray for domestic, agricultural and industrial purposes, the river's flow has inevitably been reduced. Since the amount of salt carried by the river tends to remain constant – about 3,000 tonnes a day – salinity levels can only increase unless action is taken soon.

Time alone will tell whether the 30-odd anti-salinity programmes which have been launched will prove successful. If they do not, the prospects for South Australia are grim. As Michael Butler, a geographer at Adelaide College, warned in 1980, "Irrigated lands will eventually be abandoned and farmers will lose a way of life". (See also: *Groundwater*Ⓟ, *Irrigation*③, *Salinization*Ⓐ③.)

Mutagens

Chemicals that cause mutations in the offspring of living organisms. A mutation is a change in the chemical structure of the genetic material of an organism and though a few mutations may be advantageous, most are harmful or neutral in effect. Chemicals capable of combining with DNA – the chemical that carries genetic information – are often mutagenic; a common example is the insecticide dichlorvos (*qv*). Mutagens are often carcinogens (*qv*) as well, a fact used as the basis of the Ames test, which uses bacteria to screen chemicals for possible carcinogenicity. Ionizing radiation is one of the best documented mutagens.

Screening for mutagenic properties is still not mandatory in most industrial countries, despite 1,000 new chemicals entering the market each year. (See also: *Radiation and health*Ⓛ Ⓝ.)

Mutual defence agreement

Signed by the UK and and US Governments in 1958, the Agreement for Co-operation on the Uses of Atomic Energy for Mutual Defence Purposes lays down the terms governing the export of plutonium from one country to the other and its use for military purposes.

Much controversy surrounds the origin of the British plutonium exported to the USA. Until 1986, the government denied that any plutonium from civil nuclear reactors had been used for mutual defence purposes. However, investigation by independent physicists, aided by information released under the US Freedom of Information Act, revealed that up to seven tonnes of plutonium may have been exported and used in US weapons manufacture, a significant proportion of that plutonium coming from civil Magnox reactors. This use of civil plutonium contravenes the Non-Proliferation Treaty (*qv*) to which both Britain and the USA are signatories.

The agreement was renewed in 1984 for ten years. (See also: *Magnox reactors*Ⓝ, *Nuclear proliferation*Ⓝ.)

Mutualism

Theory of the interdependence of all living things.

The idea of the world as a "vast cooperative enterprise" is an ancient concept. The principle was embodied in the "Oeconomy of Nature", a term first used in 1658 by Sir Kenelm Digby, before becoming, a century later, the title of Linnaeus's famous essay.

"By the Oeconomy of Nature", Linnaeus wrote, "we understand the all-wise disposition of the creator in relation to natural things, by which they are fitted to produce general ends and reciprocal uses". He saw living beings as "connected, indeed so chained together that they all aim at the same end, and to this end a vast number of intermediate ends are subservient".

Eighteenth century naturalists also adopted this view of our planet, but with the industrial revolution the accent shifted from cooperation or mutualism to competition and aggression.

Thus Adam Smith, Thomas Malthus, Herbert Spenser, Charles Darwin and T.H. Huxley, Darwin's most distinguished disciple, each saw competition and the "struggle for survival" as the natural tendency among living things. Huxley regarded cooperation as something so totally foreign to our planet that it had to be introduced by man – via the agency of scientific, technological and industrial progress.

Ecology developed very much as a reaction against this view of the world as atomized, random, and in perpetual flux. But after World War II, which accentuated human aggression and triggered off an unprecedented euphoria for economic development, the accent, within the ecological community, as among theoretical biologists, shifted once more towards competition and aggression.

To quote Professor Boucher, author of *The Biology of Mutualism*, "twentieth century ecology . . . has continued the tradition of seeing antagonistic interactions as the basis of community organization."

Professor Rickless, author of a famous textbook on ecology, makes the same point. The notion of competition, he writes "as a major organizing principle in ecology, is so widely accepted that it has achieved the status of a paradigm."

It is only in the last few years that this assumption has even been questioned. It has been found that the evidence for competition as the driving force of nature is more than slim.

That competition serves a function is clear, but it may by no means be the ubiquitous force it is supposed to be, still less the only tool for assuring the evolution of species and hence the structure of the biosphere.

In the 1970s there was a sudden resurgence of interest in mutualism. Boucher, in an article written jointly with the ecologists Drs. James and Keller, for instance noted "that gut flora are involved in breaking down cellulose and related substances in mutualism with many vertebrates as well as with termites and other arthropods. Urea is broken down and its nitrogen recycled by rumen bacteria and by the fungal components of some lichens".

Interest in mycorrhizae – the fungi that live in mutualistic association with the roots of plants – was also revived. Mycorrhizae increase the ability of plants to extract minerals from the soil, and in return are provided with photosynthate. Many trees will not grow without mycorrhizae.

Interestingly, an increasing number of parasitic or predatory relationships have turned out, on closer examination, to be mutualistic. Thus the ecologist Professor McNaughton has pointed out that the normal view of the relationship between grazers and grass is false. He identifies nine different ways in which this relationship can be regarded as mutualistic. They include evidence that grazing increases the water-use efficiency of grass, and that saliva may increase grass growth by as much as 50 per cent above control levels.

Boucher sees present theories of mutualism as "still basically mechanistic, mathematical, fitness-maximizing, and individualistic". This seems inevitable. A realistic theory of mutualism belongs to a holistic or really ecological paradigm and is irreconcilable with today's reductionistic and mechanistic science.

Boucher refers to the difficulties ecologists have experienced in this respect: "While arguing that nature is an integrated whole and that everything is connected to everything else, we continued researching with theories that said that communities are no more than sets of individual organisms. The problem in other words, is one of cognitive dissonance – the difficulty of working with two sets of ecological ideas, based on different fundamental assumptions and ultimately in conflict."

Narmada Valley Project

India's Narmada Valley Project is one of the largest and most controversial development programmes ever undertaken in the world. Over the next 50 years, 30 major dams, 135 medium-sized dams and 3,000 smaller dams are to be built on the Narmada and its tributaries. The project will cost tens of billions of dollars. The bulk of the funding is to come from the World Bank (*qv*). The dams have been vigorously opposed by environmentalists throughout the world.

Twenty million people inhabit the Narmada River basin, which stretches more than 1,300km (800 miles) from the Arabian Sea to the heart of India, passing through the north-eastern states of Gujarat, Maharastra and Madhya Pradesh. Hindus consider the Narmada to be the holiest of all rivers: even the Ganges is said to travel to the Narmada every year, disguised as a cow, in order to purify itself in the river's waters.

Of the planned major dams, five are hydroelectric, six multipurpose and 19 are for irrigation. According to the Narmada Valley Development Agency (NVDA), the project, when completed, will irrigate 50,000sq km (19,000sq miles) of land, and supply 2,700MW of electricity. The NVDA claims that 11,500,000 local villagers will benefit, in addition to millions of others in the cities.

But the ecological and social costs of the project are devastating. The project will submerge 3,500sq km (1,300sq miles) of forest – 11 per cent of the forest in the Valley – 200sq km (77sq miles) of cultivated land and some 400sq km (150sq miles) of grazing land. As much as 40 per cent of the land to be irrigated by the project is "black cotton soil", and is thus particularly susceptible to waterlogging and salinization (*qv*).

According to the Institute of Urban Affairs, New Delhi, one million people will be displaced by the dams, though some put the figure at nearer 1,500,000. Many of those who will be moved are tribal peoples, who have cultivated the forests for centuries. Because they have no title to land, they will not receive compensation. This is not only in contravention of the International Labour Convention but of the World Bank's own guidelines on tribal peoples. There are few who doubt that the resettlement programme spells cultural death for these tribal people.

In the face of growing international concern over the adverse environmental and social impact of Narmada Project, the World Bank temporarily suspended its $450 million funding for the first two dams to be built – the Sardar Sarovar dam in Gujarat and the Narmada Sagar dam in Madhya Pradesh – pending a full review of "resettlement and rehabilitation" plans. As a result, the Indian Government commissioned an environmental impact statement from the Department of the Environment in New Delhi. The review was strongly critical, pointing out, for example, that the authorities have yet to identify sites for the rehabilitation of those to be resettled at Narmada Sagar.

Nonetheless, on April 14 1987, Prime Minister Rajiv Gandhi announced that the Narmada Valley Project would proceed. The decision was purely political: after a series of electoral defeats throughout India, Gandhi needed to secure Gujarat and Madhya Pradesh for his Congress party.

Unless the World Bank can be persuaded to withdraw its funding, more than one million people and some of the best forest in India thus look set to be sacrificed on the altar of short-term political expediency. (See also: *Dams* ③, *Development* ③, *Resettlement* ③.)

Nitrate

A naturally occurring substance that is essential for plant growth. Nitrate itself is not particularly poisonous to humans, but if converted into nitrite (*qv*) or nitrosamines (*qv*) it can present a health risk.

Large quantities of nitrate, in animal manures and more particularly in artificial fertilizers, are added to growing crops to improve yields, but often lead to pollution.

Nitrate is highly soluble and easily washes off fields into rivers, where eutrophication (*qv*) may result. It can also be carried down through the soil by percolating water and may contaminate groundwater to dangerous levels.

In Britain, an estimated 1,500,000 people are now exposed to nitrate levels in drinking water that exceed the limit of 50 milligrams per litre (mgl) set by the EEC. Loopholes in the EEC directive have allowed 52 water sources in the Anglian, Thames, Yorkshire and Severn-Trent areas to be exempted by the British Government from having to comply with the new limits. The exemptions were granted on the grounds that the affected areas had high nitrate levels due to natural conditions. (See also: *Artificial fertilizers*Ⓐ Ⓟ, *Groundwater*Ⓟ.)

Nitrite

A substance found in the soil and formed in the human gut by the action of bacteria on nitrate (*qv*). In nature it forms part of the nitrogen cycle and is rapidly converted to nitrate, which is absorbed by plants. However, in humans it can cause problems.

Nitrite combines with the red blood pigment haemoglobin. This prevents the blood from carrying oxygen around the body efficiently, and if this happens to a significant degree the body may suffocate from within, a condition called methaemoglobinaemia. This is rare in adults, but occurs in babies and small children who absorb excessive amounts of nitrate in drinking water and made-up feeds. Babies are particularly at risk because conditions in their digestive system favour the rapid formation of nitrite from nitrate. They also have a type of haemoglobin in their blood which combines especially readily with nitrite. Methaemoglobinaemia can be fatal and doctors in areas where nitrate levels in water are high are warned to be on the lookout for it. Nitrite may also present a cancer risk. (See also: *Cancer*Ⓛ Ⓟ, *Carcinogen*Ⓕ Ⓛ Ⓟ, *Nitrosamines*Ⓛ Ⓟ.)

Nitrosamines

Chemical compounds formed when nitrites react with certain organic nitrogen compounds in an acid solution. More correctly called "N-nitrosamines", they have been shown to cause stomach cancer in laboratory animals and there is growing concern that they may also do so in humans who absorb nitrite (*qv*) or nitrate (*qv*) in their diet. (See also: *Cancer*Ⓛ Ⓟ, *Carcinogen*Ⓕ Ⓟ, *Nitrate*Ⓐ Ⓛ Ⓟ, *Nitrite*Ⓛ Ⓟ.)

Nomadic pastoralism

For many years, nomadic pastoralism has been denigrated as both unproductive and environmentally destructive, the nomads themselves commonly being described as ignorant, irrational and unproductive.

To improve productivity, vast sums have been spent on livestock management programmes, intended to transform pastoral nomads into settled cattle ranchers. Almost without exception, the projects have proved unmitigated failures.

Well-digging programmes have led to overgrazing (*qv*), while bans on the burning of grassland prior to grazing have increased the habitat for tetse fly and reduced the nutritional quality of dry-season grasses.

Attempts to prevent pastoralists moving their herds from one area to another have also failed. During periods of drought, mobility enables the pastoralists to exploit those few areas where the rains fall. As Michael Horowitz, Professor of Anthropology at the State University of New York, points out, research in the Ferlo region of Senegal reveals that such is the variation in local rainfall that in 1972 (a year of catastrophic drought) "a herder locked into the Ferlo without the right to move would have lost 100 per cent of the herd." Nonetheless, the fencing of rangelands continues: under the World Bank's (*qv*) Senegal Project, for example, herders will be forbidden to stray from their allotted grazing units.

In the light of the abject failure of modern ranching programmes, traditional practices appear increasingly sound by comparison. (See also: *Desertification*Ⓐ Ⓒ ③, *Development*③.)

Non-Proliferation Treaty

The Treaty for the Non-Proliferation of Nuclear Weapons was signed by the USA, USSR and UK (the depository signatories) in 1968, and came into force in March 1970. One hundred and twenty-one other states have also signed to date.

Among the states with nuclear weapons, France and China have not signed. Notable non-signatories among the states without such weapons are India, Israel, South Africa, Pakistan, Argentina, Spain and Brazil, all countries with known nuclear competence. (See also: *Nuclear proliferation*Ⓝ, *Nuclear safeguards*Ⓝ.)

North Sea

Opinion is deeply divided over the health of the North Sea. An official Dutch Government report, prepared for a 1987 conference of European environment ministers on the protection of the North Sea, calls it "one of the most polluted seas on Earth" and warns that if its state does not allowed to deteriorate further, "significant environmental changes could be expected well into the 21st century." Britain disputes that view, arguing that the pollution is limited to "hot spots" on the margins of the sea and that in general the North Sea is "a wholly healthy body of water".

In 1984, some 5,570,000 tonnes of industrial waste were dumped directly into the North Sea, together with 5,100 tonnes of sewage sludge, and 97 million tonnes of dredge material. Other sources of pollutants included wastes discharged via rivers, the atmospheric deposition of wastes from land-based sources, and emmissions from the 93,677 tonnes of waste incinerated at sea.

As a result, the North Sea received 6,020 tonnes of copper; 31,070 of zinc; 7,840 of lead; 2,830 of chromium; 840 of cadmium; 3,740 of nickel; and some 45 tonnes of mercury. In addition, some 25,000 tonnes of hydrocarbons were either dumped or discharged into the sea, principally from ships and oil rigs; between 744kg (1,600lb) and two and a half tonnes of chlorinated hydrocarbons –

PCBS (*qv*), DDT (*qv*), dieldrin (*qv*) and the like – entered the sea from various sources; and 250,000 tonnes of inorganic nitrogen entered from atmospheric sources, mainly car exhausts and fossil-fuel power stations. Radioactive discharges from the Sellafield reprocessing plant on Britain's Cumbrian coast also added to the total pollution load, the radionuclides making their way into the North Sea via the north coast of Scotland. The estimated input of caesium-137 in 1984 amounted to 2,000 terabecquerels (2,000 million million becquerels).

Britain is the only country in Europe still dumping sewage sludge at sea. Britain also contributes the highest inputs of radioactive isotopes, atmospheric nitrogen and hydrocarbons. Though only trace quantities of the most toxic heavy metals are permitted in sewage sludge, the 5,180,000 tonnes dumped by Britain in 1984 contained half a tonne of mercury and three tonnes of cadmium.

The effects of pollution are most evident along the eastern coast of the North Sea. There, PCB pollution is responsible for the high rates of infertility and miscarriages that have decimated local seal populations and for mass poisonings among eider, spoonbill, herring gull, stern and tern.

Dutch surveys have revealed increased rates of disease amongst flounders, dab and plaice caught off the eastern North Sea – 11 per cent showing evidence of liver cancer, ulcers and finrot. A marked increase in skin disease has also been observed amongst dab caught near a site used for the dumping of titanium dioxide (*qv*) wastes. So too, surveys in the Thames estuary have shown physiological damage (notably finrot and ulcers) amongst 5.4 per cent of fish sampled. A more recent study has revealed that 40 per cent of flatfish in certain parts of the North Sea have cancer. Britain, however, denies that pollution is necessarily the cause of the problem.

Controversy also surrounds the cause of the plankton blooms – the so-called "red tides" – which increasingly affect coastal waters. Denmark, Norway, Belgium, and Germany all blame the blooms on the eutrophication (*qv*) of coastal waters as a result of the increasing quantities of nitrogen and other nutrients entering the sea via sewage, agricultural run-off and atmospheric depositions. Once again, Britain is the odd man out, claiming that a clear-cut relationship between nutrient enrichment of the sea and the plankton blooms has yet to be established.

Britain has consistently opposed stricter controls on the dumping of wastes at sea. In 1984, British officials effectively stymied a West German proposal to halt the dumping of all persistent and bio-accumulative waste into the North Sea. More recently, in August 1986, the British Government rejected a proposed EEC directive which, if implemented, would have halved the amount of waste dumped at sea by 1995 and banned the incineration of waste at sea by 1990.

Whereas Britain appears content to delay action until there is firm evidence of pollution, her neighbours on the other side of the North Sea argue that further delay is irresponsible. As one Dutch scientist puts it, "By the time we have conclusive evidence of pollution, it might be too late to do anything about it. We may have passed the point of no return."

In 1986, environmentalists from West Germany, Britain and Holland launched the "Seas at Risk" campaign. The campaign aims to publicize the threat that pollution poses to the North Sea, the Waddensea and the Irish Sea. Among its goals are a ban on the incineration of wastes at sea and a reduction in the dumping of sewage. (See also: *Fish stocks* Ⓒ Ⓟ, *London Dumping Convention*, *Marine incineration* Ⓟ, *Ocean dumping* Ⓟ, *Oslo Treaty*, *Sub-lethal pollutants* Ⓟ.)

Nuclear accidents

From its inception, the civil nuclear industry has always insisted that nuclear reactors, unlike nuclear weapons, are clean and safe. Indeed, until the early 1980s, the industry refused to acknowledge that radiation from reactors had ever caused a death. In addition to dismissing as remote the chances of a major accident occuring at a reactor, the industry has consistently denied that the health effects of reactor accidents – however serious – could ever compare with the devastation caused by a nuclear bomb.

Minor accidents involving relatively small releases of radioactive material occur all the time at nuclear installations. But what are the risks of a major accident? In 1957, the US Brookhaven National Laboratory put the probability of the most serious accident conceivable occurring in a nuclear plant at less than one per 1,000,000 years of reactor operation. Such an accident would lead to 3,400 deaths, 43,000 injuries and $7,000 million worth of damage to property.

An update of the Brookhaven report, published in 1964, conceded that the effects of an accident at a 1000MW reactor sited within a city could be worse than previously estimated, leading to 55,000 prompt deaths and 70,000 injured. However, the report's estimate of the probability of such an accident occuring remained unchanged.

Ten years later came the WASH 1400 Report, better known as the Rasmussen Report, which gave the chance of a major accidental release of radioactive material from a nuclear reactor as one in a 1,000 million years of reactor operation.

Such calculations are of course highly suspect. As Nobel Laureate Professor Liebe Cavalieri writes, "The case of the Oak Ridge Research Reactor accident is one example of how misleading probability calculations can be. In this accident there were seven sequential failures, each involving redundance of three parallel elements, for a total of twenty-one failures, the absence of any one of which would have prevented the incident. Three of the seven were personnel failures: an experienced operator threw wrong switches in three separate rooms; another operator failed to report finding any of these errors; and so forth. The others were design or installation errors in a reactor with an outstanding performance record. The probability of the event was calculated to be one in 100 billion billion. The event 'was almost unbelievable,' but it happened. Again, in the complex nuclear reactor accident that occurred in 1970 at the Dresden II reactor, the most generous assessment of the probabilities of the separate events could not raise the overall probability above something like one in a billion billion. Yet, here again, it happened."

A number of serious accidents have already occurred to nuclear installations, notably those at Kyshtym (*qv*) in the USSR in the winter of 1957; at Windscale (*qv*) in Britain in 1957; at Three Mile Island (*qv*) in the USA in 1979; and, in 1986, at Chernobyl (*qv*).

These four serious accidents have occurred within the space of 30 years and make nonsense of the Rasmussen Report's official calculations. Indeed, some officials at the US Nuclear Regulatory Commission now estimate the chance of a reactor in the USA suffering a core meltdown before the end of the century at 50:50. This fits with the experience to date. Since the USA possesses about a quarter of the world's reactors, we may see several more serious nuclear accidents before the end of the century.

The possibility of a serious nuclear accident today is indeed only too real. The operation of any reactor is a delicate balance between too little and excessive reactivity. Unleash the normal control mechanisms restraining the nuclear chain reaction and the power can increase phenomenally in a matter of seconds.

At Chernobyl, the power rose to at least 480 times maximum operating power before the reactor exploded. Should the gas circulatory system in a British advanced gas-cooled reactor (*qv*) fail, followed by a failure to scram, an even more devastating accident than Chernobyl could result within seconds.

Similarly, according to calculations by Dr Richard Webb, author of the highly respected book *The Accident Hazards to Nuclear Power Plants*, a super prompt critical accident in a fast reactor could lead to an explosion of up to three kilotons TNT equivalent or more.

Since Chernobyl, it has become increasingly difficult for supporters of nuclear power to maintain the fiction that major nuclear accidents are so improbable as to be not worth worrying about. Instead, there is an acceptance that accidents will occur and the emphasis is now on ways in which their consequences can be mitigated. As the prestigious science journal *Nature* wrote at the time of Chernobyl,"The important question is not so much how accidents like these can be prevented but how we can live with them safely." (See also: *Radiation and health* Ⓛ Ⓝ.)

Nuclear proliferation

The spread of nuclear weapons to countries that do not already have a military nuclear capacity. In the USA, the USSR and the UK, civil nuclear power was a spin-off from the technology developed to produce fissile plutonium for the purpose of making weapons. The risk of nuclear proliferation arises from the ease with which it is possible to do the reverse; to derive fissile plutonium from civil nuclear facilities.

Reprocessing (*qv*) spent nuclear fuel is a necessary step in the development of nuclear weapons. The Non-Proliferation Treaty (NPT) (*qv*) attempts to control the destination of reprocessed plutonium, the aim being to expose any attempts at proliferation while actively encouraging civil nuclear development.

Attempts to limit the "nuclear club" began in earnest after France joined it in 1960 and China in 1964. The NPT was ratified in 1970, and in 1974 India (a non-signatory) became the first state to explode a bomb whose fissile material had been manufactured using imported civil nuclear equipment. The equipment in question was a research reactor supplied to India by Canada. After the test, Canada broke off further co-operation.

Europe's attitude to proliferation has been more cynical than that of the USA or Canada; within two years of the Indian test France had sold a large reprocessing plant to Pakistan and another to South Korea (though neither deal was completed after US pressure on France), and West Germany had concluded an enormous contract with Brazil involving eight reactors, an enrichment plant and a reprocessing plant. Earlier, West Germany and France both provided reactors and technology to South Africa, and France sold nuclear equipment to both Israel and Iraq. By the late 1970s Brazil, Argentina, Taiwan, India, Pakistan, South Africa and Israel (the last two possibly working together) had developed facilities for the separation of plutonium, and Pakistan was also developing technology to build a uranium bomb, using technical data stolen from Holland and equipment bought in Western Europe.

Twenty-nine non-nuclear weapons countries have power reactors and at least 100kg (220lb) of separable plutonium. Twelve also have reprocessing plants; six of these have not signed the NPT. With so many locations holding fissile material, the danger is not only that states will make nuclear weapons, but also that nuclear material can be stolen, and made into "do-it-yourself" bombs using laboratory-scale reprocessing.

Clearly, attempts to limit membership of the nuclear weapons club have failed. A range of smaller countries either have atomic weapons or hold themselves ready to make them at short notice. Many of these countries have refused to submit their facilities to nuclear safeguards (*qv*) laid down by the International Atomic Energy Agency (IAEA) (*qv*) via the NPT.

Members of the EEC also have to submit their civil nuclear facilities to EURATOM safeguards. Until June 1986 the UK government refused EURATOM inspectors access to the Magnox reprocessing line at Sellafield (*qv*), because it is both a military and a civil facility. Even when a EURATOM inspection was permitted, inspectors were not allowed to make their findings public. As a result of its attitude, Britain is being taken to court by the EEC.

Nuclear waste

The nuclear industry worldwide is generating an appalling legacy of radioactive wastes. The properties of those wastes vary enormously, but many are highly radioactive and toxic, and many will remain radioactive for hundreds of thousands of years.

Nuclear waste embraces a great range of material, from lightly contaminated overalls, to uranium (*qv*) mill tailings, to the fiercely radioactive liquid high-level waste that is produced by reprocessing (see illustration). By the turn of the century, the worldwide inventory of high-level wastes is projected to reach 150,000 million curies.

The production of wastes starts at the first stage of the nuclear fuel cycle – with the mining of uranium (*qv*). In the USA, it has been calculated that radon emissions from mill tailings could cause 4,000 cancer deaths a year.

Mill tailings apart, the amount of radioactive wastes produced by a nuclear power programme depends critically on whether reprocessing (*qv*) or long-term storage is practised. A large pressurized water reactor (PWR) (*qv*) produces 11 cubic metres of irradiated fuel per gigawatt per year. If stored, this is the only volume to be dealt with. If reprocessed, 6.25 cubic metres of high-level waste, 40 cubic metres of intermediate-level waste and 600 cubic metres of low-level waste are produced from the original 11 cubic metres of spent fuel.

At the end of the first year after extraction from the reactor, the irradiated fuel still contains, in highly concentrated form, 270,000 times more radioactivity than the ore from which it was derived. Even after 10,000 years, one kilogram would contain, gram for gram, 18 million times more radioactivity than the lambs that Britons were forbidden to eat as a result of Chernobyl (*qv*).

High-level wastes are generally stored in stainless steel tanks, constantly stored and cooled. In the USA, the main storage centre is Hanford (*qv*), a site that has a notorious record for accidents involving waste. Between 1945 and 1973, some 422,000 gallons of liquid waste containing 500,000 curies of radioactivity had leaked out of the tanks. Tritium and ruthenium have been detected in the groundwater; strontium-90 and iodine-131 in the Columbia River; and plutonium-239 in the soil. The levels of plutonium were 5,000 times the permissible level.

Nuclear wastes

Over the next 30 years, the nuclear reactors now operating, or being built, will produce thousands of cubic metres of high level waste. More than 25,000cu m will be produced by the "top ten" countries below.

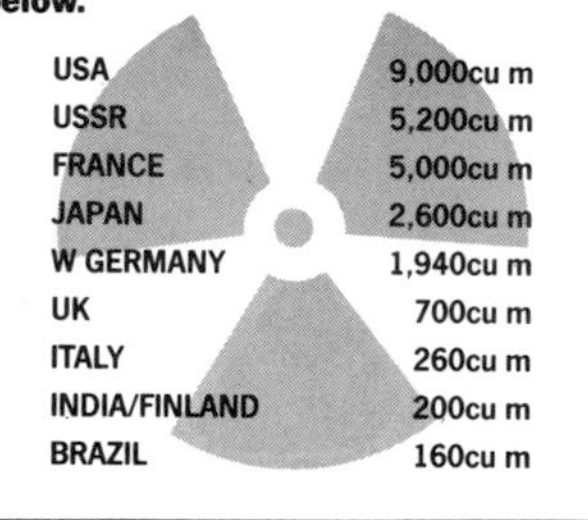

USA	9,000cu m
USSR	5,200cu m
FRANCE	5,000cu m
JAPAN	2,600cu m
W GERMANY	1,940cu m
UK	700cu m
ITALY	260cu m
INDIA/FINLAND	200cu m
BRAZIL	160cu m

In the UK, high level wastes are stored at Sellafield (*qv*). Ultimately, the aim is to solidify the waste in glass through a process known as vitrification. The process is intended to reduce the volume of waste and to make it easier to handle.

A pilot vitrification plant is now in operation at Marcoule in France, but it has only vitrified waste with a radioactivity no higher than 10 curies per cubic metre: once fully operational, it must contend with wastes with a radioactivity up to 300 curies per cubic metre.

Several US and Australian researchers are strongly critical of vitrification. The danger is that the glass blocks will disintegrate at some time in the future.

Even assuming vitrification works, the problem of a long-term repository for high-level wastes remains unresolved – and, many would claim, insoluble. In both Europe and the USA, the nuclear industry talks of burying waste in deep underground sites, but the search for stable geological sites has proved fruitless. Indeed, uncertainties of over geological faults, the inability to predict future geological movement, and the possibility of earthquakes make it impossible to guarantee the integrity of any site for the length of time that the waste must be kept isolated.

In the USA and the UK, intermediate level waste and low level wastes are buried in shallow burial grounds, while in West Germany they are consigned to an abandoned salt mine. All the sites have experienced problems.

In the USA, three out of six sites have been closed down due to contamination problems or breaches in transportation regulations. The West Valley site in New York State was closed after tritium contaminated local groundwater, and the Maxey Flats dump in Kentucky after plutonium was found to have migrated three-quarters of a mile off site within three years of the site opening. At Barnwell, South Carolina – one of the three sites still operating – movement of both cobalt-60 and tritium have been detected.

The only available dump site in Britain for disposing of solid low-level waste is at Drigg, adjacent to Sellafield. The site, which is almost full, is designed to allow its wastes to drain into the Irish Sea, on the grounds that they will be rendered harmless through "dilution and dispersal" (*qv*).

The British Government and the nuclear industry are under considerable pressure to find acceptable disposal routes for intermediate-level waste and low-level waste. Vociferous local opposition at every site chosen for investigation finally caused the government, one month before the June 1987 General Election, to call off the search for a shallow burial site for low-level waste.

Before 1983, most low-level waste produced in Britain was dumped at sea. The practice was halted after a ban was imposed by the National Union of Seamen.

The ban was followed by a international moratorium – consistently opposed by Britain – under the London Dumping Convention (LDC) (*qv*) on all future dumping of nuclear waste of sea. Apart from Britain, the only other countries using ocean disposal were Belgium, the Netherlands and Switzerland. The accumulated radioactivity dumped between 1967 and the ban in 1983 amounted to some one million curies, 90 per cent of it from the UK.

Discharges of low-level nuclear waste into the sea from land (as opposed to dumping from ships) escape the LDC moratorium, and such waste is still discharged by both Sellafield and Cap de la Hague (*qv*).

The failure to find a suitable land-disposal site for intermediate- and low-level wastes in Britain has again focused on the sea, one proposal being to bury the waste under the seabed, with access via tunnels from the land. The use of such a site may be in contravention of the LDC, however, and would provoke both national and international protest.

West Germany's nuclear industry is seeking to dispose of waste in China's Gobi Desert. In July 1987, the country's biggest nuclear company – Kraftwerk Union – announced that a deal was close, the waste being taken in exchange for nuclear technology. (See also: *Khystym* Ⓝ Ⓟ, *Radiation and health* Ⓛ Ⓝ Ⓟ, *Radioactivity* Ⓛ Ⓝ.)

Nuclear winter

A major nuclear exchange could cause rapid cooling over much of the Earth as dark clouds of smoke and dust generated by the explosions and consequent fires spread far beyond the combatant nations. This nuclear winter may last from days to months, possibly even years, causing widespread harvest losses which, when combined with the almost inevitable breakdown in international trade, would threaten the survival of much of the world's population.

A single one megatonne nuclear warhead has the potential to ignite 400sq km (150sq miles) if detonated over an urban or industrial area. A 6,000 megatonne exchange could cause fires over 250,000sq km (96,000sq miles) – about half the area of the 1,000 largest cities in the likely combat zone – and 20,000 to 250,000sq km (7,700 to 155,000sq miles) of rural areas, depending on the season of the exchange. A third to one-half of the smoke generated by these fires would promptly fall out or form "black rain", as happened after the atomic bombing of Hiroshima in 1945. Of the remainder, up to 50 per cent might consist of carbon-based soot less than one-thousandth of a millimetre in diameter. The precise proportion of soot would depend on the locations targeted – city fires would generate more soot than forest fires.

It is the soot component of the smoke that would produce a marked cooling at the Earth's surface. The size, shape and chemical composition of these particles make them unusually strong absorbers of incoming solar radiation while permitting outgoing radiation from the Earth to pass through relatively unhindered. Thus, a cloud of sooty smoke has the potential to warm the atmosphere where the bulk of the cloud lies and to create cold, dark conditions below.

Assuming war takes place in the northern hemisphere, the nuclear winter effect would be greater in the northern summer. At this time of the year, the amount of sunlight normally reaching northern latitudes is at a maximum and the potential for disturbance is greatest.

The acute sensitivity of crops and many natural ecosystems during the growing season means that a spring or summer war could effectively wipe out production even if the temperature drop is at the less extreme end of the range of possibilities. For example, a temperature drop of only 5°C (9°F) over the growing season is sufficient virtually to eliminate grain production throughout the northern hemisphere. Potato production in Poland and Russia would cease and, in England, yields would be cut by half. Livestock and animals would be without their thick winter coats for protection. Coniferous trees in Canada are able to tolerate temperatures of -70°C (-158°F) in the winter but may die if temperatures fall below -10°C (-50°F) during the growing season.

The survivors of the war would face a dark, cold, polluted environment and the likelihood of severe food shortages. Depending on the scale of the population losses during the conflict itself, food stocks might – if they could be distributed – support the survivors through the initial months, but the possible continuance of cooler temperatures into the next growing season combined with a lack of the technology that underpins modern farming practices could mean that food supplies would run short before long.

Following a nuclear war that took place outside the summer months, the initial temperature drop would not be as severe. However, if the cooling persisted into the growing season, mass starvation may only have been postponed.

For Third World nations south of the combat zone, the possibility of nuclear winter adds a new dimension to prospects for survival. In the northern sub-tropics and the tropical zone, the initial temperature reductions are likely to be less severe than those experienced to the north, though they may still be of the order of 5° to 10°C (9° to 18°F) in the case of a summer war. However, the lower natural climatic variability in these regions means that plants and animals are even less well adapted to cope with cooler conditions. During the growing season, a single day of temperatures below 15°C (41°F) would destroy the rice harvest – the staple food for more than 1,000 million people. In Africa and South-East Asia, harvest failure could also result from a weakening of the monsoons due to reduced heating of the Tibetan Plateau. With the possible loss of grain production in North America – currently provider of 90 per cent of the world's food surplus – a food shortfall could not be made up and the affected nations would be thrown back on their own limited reserves. The Sahel is already suffering widespread famine and no reserves would be available. India's food reserves could not support her population for more than a few months.

Cold streams of air penetrating into the tropics and beyond as dense clouds pass overhead would threaten the two-thirds of the world's species of plants, animals

Nuclear winter – a scenario for survivors

The likelihood of a nuclear winter would depend upon the height to which dust and soot were injected into the atmosphere. Up to 12km, the effects would be severe but localized, much of the debris and soot being washed back to Earth as a lethal rain. If the mushroom cloud reached into the upper atmosphere, then light and temperature levels would fall worldwide. The effects would be less severe but prolonged.

The magnitude of the nuclear attack would determine how much dust was generated by the initial explosion. The amount of smoke produced would depend upon the location of the subsequent fires. Short-lived city-centre firestorms would be the worst, producing a light-absorbent, sooty smoke. Dust and smoke combined could lead to a global cooling – a reverse greenhouse effect.

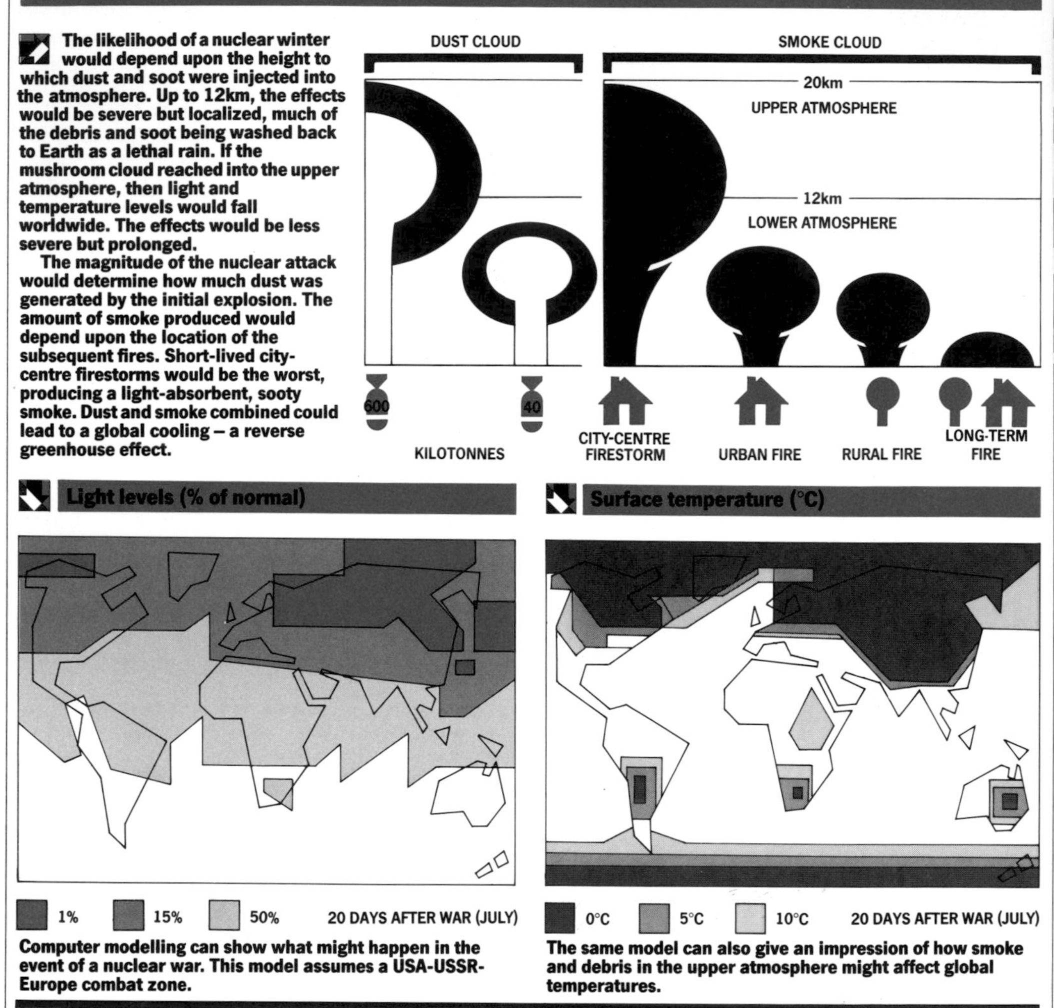

Computer modelling can show what might happen in the event of a nuclear war. This model assumes a USA-USSR-Europe combat zone.

The same model can also give an impression of how smoke and debris in the upper atmosphere might affect global temperatures.

and micro-organisms that live within 25 degrees of the Equator. Since there is no distinct seasonal cycle in the tropics, ecosystems would be vulnerable whatever the season of the war. The destructive potential of cold outbreaks in lower latitudes is illustrated by the impact of the freezing winds that occasionally affect Florida – five such events between 1977 and 1985 caused huge losses of vegetable crops, and fruit and fish stocks. Damaged trees took years to recover. Periodic cool temperatures spreading from the south into the Amazon Basin – "friagems" – are responsible for the deaths of large numbers of birds and fish even in the present day. More than a week of freezing temperatures in rainforest areas could kill all the vegetation above ground and the impact of this would cascade through the food chain, adversely affecting virtually all forms of life.

The 15 per cent of the world's population living south of the Equator would fare better than those in the north. Even at the more extreme range of possibilities, temperature reductions alone are unlikely significantly to affect crop yields over the southern hemisphere.

Many uncertainties surround the predictions of the climatic change that might be induced by nuclear war and it is not possible to estimate the effect except in the most general terms. Nevertheless, it is clear that nuclear war threatens the existence of most of the world's population. Nuclear winter effectively draws all nations into the nuclear debate.

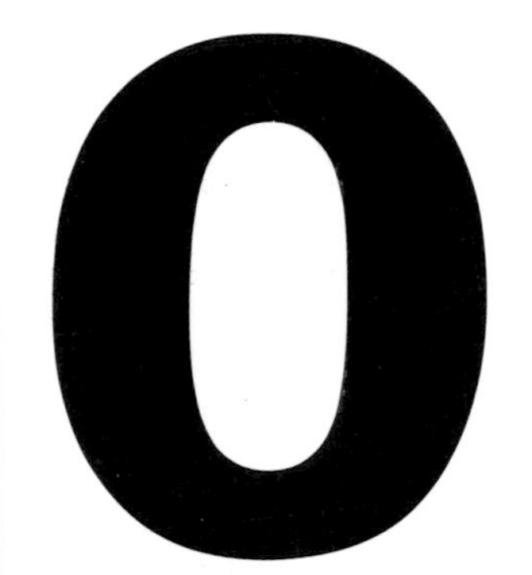

Ocean dumping

Prodigious quantities of dredged spoils, sewage and industrial waste are dumped every year directly from ships into the world's oceans. The major dumping grounds are the Atlantic and the North Sea (*qv*), direct dumping now being banned in most regional seas (*qv*). Proponents of ocean disposal argue that wastes are quickly rendered harmless through dilution and dispersal (*qv*). But many dumping grounds throughout the world are now severely contaminated, with high levels of pollutants in bottom sediments and in fish.

The dumping of heavy metal wastes and persistent or carcinogenic wastes is banned in all seas under the London Dumping Convention (LDC) (*qv*), the international treaty that governs dumping in the world's oceans. The Oslo Treaty (*qv*), which oversees dumping in the North Sea, has similar provisions. Nonetheless, many tonnes of heavy metals and persistent or carcinogenic wastes are still dumped quite legally under provisions which allow for waste consignments to contain "trace quantities" of banned substances. A moratorium on the dumping of nuclear waste (*qv*) at sea has been in force since 1983.

Hundreds of drums of highly toxic and radioactive wastes – dumped before the LDC was introduced – still lie in the world's oceans. Some of that waste was dumped at deep ocean sites on the assumption that the deep ocean was "a tranquil abyss", the sea bed being utterly devoid of currents. Recently, however, researchers at the Woods Hole Oceanographic Institute in the USA have discovered that the deep oceans experience violent underwater storms, which literally scour the sea floor of sediment, transporting large volumes of silt from one area to another. The effects of the drums of waste that have been dumped in the deep oceans is unknown, but even if they survive the battering of the storms, they must eventaully rust, leaking their contents into the sea.

Professor Gunnar Kullenberg, chairman of the main UN expert group on marine pollution, describes current studies of marine pollution as "primitive". "We take each substance separately and then try to estimate what effects it may have. We are not yet able to consider something as complicated as synergistic effects," he told "Siren", a United Nations Environment Programme newsletter, in 1983.

Sea dumping is a prime example of the "out of site, out of mind" approach to waste disposal. It is surrounded by long-term uncertainties and has already caused considerable damage, particularly to fish stocks (*qv*). Alternative methods of disposal for all the wastes currently dumped at sea exist on land, though such alternatives are undoubtedly costly. According to Britain's Water Authorities Association, for example, it would cost $187 million in new equipment and running costs were Britain to agree to a ban on sewage dumping at sea. Industry, too, argues that a ban on sea dumping of its wastes would have a severe recessionary impact, diverting investment from production and undermining competitiveness.

Such arguments, however, ignore both the jobs that would be created through pollution and the savings that would be made through the recycling (*qv*) of materials currently dumped at sea. More important still, they overlook the very real long-term costs that are even now being incurred through dumping.

Those long-term costs are hard to quantify in cash terms, but they are likely to be high. For we are in danger of disrupting an ecosystem whose health is vital to our survival.

Ocean incineration

At present, only four countries – West Germany, Belgium, Great Britain and France – regularly incinerate wastes at sea, burning an average of 100,000 tonnes a year. Though the US Environmental Protection Agency (EPA) has said that it sees ocean incineration as playing an important role in reducing America's burgeoning inventory of hazardous waste, it has yet to sanction burns in US waters.

In anticipation of permission eventually being granted, a US company – Chemical Waste Management Inc, a wholly-owned subsidiary of the giant Waste Management corporation – has built a new incineration vessel, the Vulcanus II. The ship currently operates in the North Sea, where most burns take place.

Proponents of ocean incineration argue that it is an environmentally sound, economic, and safe technology. That view is not shared by the Oslo Commission, the body that oversees the Oslo Treaty (*qv*), and the European Commission, both of which plan to ban the practice, the Nordic countries pushing for a ban by the end of 1991. Within the EEC, only Spain, Britain and Eire wish to retain the option of ocean incineration.

There is pressure, too, within the London Dumping Convention (*qv*) for a ban, though this is being resisted by the USA and several other countries.

Unlike land-based incinerators, incineration ships are not equipped with scrubbers (*qv*) to trap unburned wastes, which are thus emitted directly into the atmosphere. Though research burns show a destruction efficiency of 99.99 per cent, that still leaves one tonne of unburned waste entering the environment for every 10,000 tonnes burned. The danger is that such waste will be concentrated by marine organisms and thus find its way into the food chain. Among the wastes that have been emitted during trials are unburned polychlorinated biphenyls (PCBS) (*qv*) and heavy metals.

The second danger of ocean incineration lies in the possibility of a major accident, followed by a leak. Though Chemical Waste Management put the risk of such an accident at one in 24,000 years, accidents do happen. The results could be catastrophic. In 1983, at the time when the EPA was considering licensing waste burning in US waters, a study warned that a discharge of 500,000 US gallons of liquid PCBS – less than half a ship's cargo – would contaminate the "upper foot of the entire Gulf of Mexico . . . with 2.5 parts per billion of PCB". Since phytoplankton, the very basis of the marine foodchain, live in that upper foot of Gulf water, the resulting pollution would arguably " be sufficient to deteriorate all life in the Gulf of Mexico" – an area of some 1,507,000sq km (582,000sq miles). (See also: *Hazardous waste*Ⓛ Ⓟ, IncinerationⓁ Ⓟ.)

Ogallala Aquifer

The largest source of groundwater in the USA. It stretches from southern South Dakota to northwest Texas and is a classic example of a resource that has been grossly over-exploited.

The 30 years following World War II saw a dramatic expansion in irrigated agriculture in the Great Plains, an expansion that was made possible only through "mining" the groundwaters of the Ogallala. Between 1944 and 1978, according to the Washington-based Worldwatch Institute, the area under irrigation in the six states most dependent on the Ogallala rose from 21,000sq km (8,100sq miles) in 1944 to more than 80,000sq km (30,800sq miles) in 1978. In the Great Plains as a whole, some 95,000sq km (36,700sq miles) of irrigated agriculture – one fifth of the irrigated cropland in the USA – is watered by the Ogallala.

Since 1940, more than 500 cubic km (120 cubic miles) of water have been withdrawn from the Ogallala. The result has been the gradual depletion of the aquifer, with some areas of Kansas, Texas and New Mexico now down to half their reserves. Indeed, a recent study suggests that some of the older irrigated areas of the High Plains of Texas will only be able to continue pumping for a decade or so at the most. At current rates of pumping, some 20,000sq km (7,700sq miles) are expected to be forced out of production by the end of the century. Already, both Oklahoma and Texas have lost 18 per cent of their irrigated farmland. In Colorado, Kansas and Nebraska, almost 6,000sq km (2,300sq miles) have now been taken out of production. The depletion of the acquifer has caused considerable disquiet and even anger in the USA. (See also: *Groundwater*Ⓐ Ⓟ.)

Oil

The most important trend in energy (*qv*) use since the war has been increasing world dependence on oil. By 1973, the world had become dependent on oil for 41 per cent of its energy needs, consuming 56 million barrels a day. After the first energy shock of 1973, a deliberate attempt was made to reduce oil consumption by adopting energy conservation (*qv*) measures and by switching to other sources such as nuclear energy.

As a result, North America succeeded in reducing its oil consumption from 18,600,000 barrels a day to 16,600,000 in 1984; Western Europe from 15,200,000 to 12,300,000 and Japan from 5,500,000 to 4,600,000. During this period, both the Eastern Block and the Third World, however, actually increased their consumption from 17,900,000 to 25,300,000 barrels a day.

Since 1973, total oil dependence has thus only fallen worldwide to 35 per cent of all energy sources. World oil consumption in 1984 was 57 million barrels a day, which is just slightly higher than in 1973.

How long the world can continue to use energy at that rate?

The accepted measure of world oil resources is "proven reserves". In 1950, these stood at 76,000 million barrels. Each year new discoveries are added and "production" subtracted. By 1973, proven reserves had increased to 664,000 million barrels. In the years since, intensive exploration has increased this figure by some five per cent only.

It seems reasonable to suppose that global oil reserves have never been more than 1,600 to 2,400,000 million barrels. Of this, we have already used up some 500,000 million barrels and 700,000 million barrels have been discovered and constitute present proven reserves. This means that there remain between 400 and 1,200,000 million barrels yet to be discovered.

If we continue to make the appropriate effort to discover new reserves, and continue to consume oil at something like the 1985 rate of consumption of 21,000 million barrels a year, then oil resources could last between 50 and 88 years. Increased energy conservation could, of course, make a difference, so could a switch, for some end-use purposes, to renewable sources of energy – but, in spite of the present oil glut, it would be grossly misleading to suppose that our present oil consumption can be maintained for very long.

Organic farming

Farming without the aid of artificial fertilizers (*qv*) or synthetic organic (*qv*) pesticides (*qv*). Such farming has been the norm since agriculture first began. Over the past 40 years, however, all but a small number of farms in the industrial world have abandoned organic agriculture in favour of chemical farming. The move has been encouraged by international organisations such as the UN Food and Agriculture Organisation (FAO) (*qv*), and heavily subsidized by nearly all national governments.

The principles of organic farming lie in the maintenance of soil fertility through careful husbandry, the recycling of agricultural wastes, and the use of natural forms of pest management and weed control.

Soil fertility is affected by the supply of a variety of nutrients in the soil, its structural composition, its water-holding capacity, its aeration, and its capacity to absorb heat.

By using farm wastes in the form of manures, composts and mulches, organic agriculture permits a wide selection of nutrients to be returned to the soil and there is a congruence between the rate of plant nutrient uptake and nutrient release from organic wastes. Bulky organic wastes improve the structure of the soil, while the nutrients they contain increase its fertility. When fallowing is adopted, deep-rooted weeds can bring minerals up from the sub-soil to remineralize the cropland.

Crop rotation, an essential feature of organic agriculture, ensures that different levels of soil are exploited and no particular level overexploited. When green manure crops, such as mustard or comfrey, are included in the rotation and ploughed back into the soil, its organic content is further increased. The introduction of leguminous plants (that is, those with nitrogen-fixing bacteria on their root nodules) such as clover or field beans into the rotation maintains the necessary level of nitrogen in the soil. Rotation also avoids creating a permanent niche for pests, thereby reducing the need for pesticides.

Polyculture, another essential feature of organic farming, also reduces the need for pesticides by reducing the size of the niche available for pests affecting specific crops.

Quite apart from being uncontaminated by chemicals, organically grown food has greater nutritional value than chemically grown food, not least because organic fertilizers do not increase the water content of food to the same degree as artificial fertilizers.

Organic farming has the additional advantage of not polluting surface waters and groundwaters with fertilizers and pesticides, and of being sustainable.

Revenue from organic farms is lower than that from conventional farms – but so are the costs. With non-organic farmers now being hit by the increasing cost of inputs and by falling yields – the result of the inevitable degradation which the excessive use of chemicals inflicts on the soil – organic farming must become increasingly economic.

A recent study conducted by the US Department of Agriculture (USDA) shows that organic farming presents a viable agricultural proposition for the USA. The study found that organic farmers use less energy and more labour per unit of produce than their conventional counterparts. More important, they are also two-and-a-half times more productive per unit of energy consumed than conventional farms. As yet, the US authorities have still to act on the conclusions of the report.

American farmers now owe some $300,000 million to the banks, mainly in loans taken out to buy the chemical inputs and machinery necessary for modern mechanized farming. Unable to service their debts, many are going bankrupt. To those who remain in business, organic farming, with its low costs, is proving increasingly attractive. (See also: *Artificial fertilizers*ⒶⓅ, *Pesticides*ⒶⓅ.)

Organophosphates

Organophosphorus compounds – sometimes loosely known as organophosphates – are powerful insecticides and acaricides. They were discovered during research into nerve gases.

Organophosphates exert their toxic effects by inhibiting acetylcholinesterase, a substance that is involved in the transmission of signals along the nerves. As a result, the nervous system of a poisoned animal suffers massive disruption. Nausea, vomiting, twitching, stomach cramps and diarrhoea lead to convulsions, coma and death in badly poisoned humans. Organophosphorus compounds are the most common cause of insecticide poisoning and can enter the body by mouth, through the skin or as a vapour.

Chemicals that exert this type of effect are known as cholinesterase inhibitors and some people are particularly sensitive to them. Thus very small doses, to which most people would not react, produce uncomfortable and sometimes dangerous symptoms.

Organophosphorus compounds have produced devastating effects on wildlife owing to their high toxicity and low selectivity. However, they are not as persistent in the environment as the chlorinated hydrocarbon compounds, such as parathion (*qv*), fenitrothion (*qv*), malathion (*qv*) and dichlorvos (*qv*). (See also: *Circle of poison*ⓁⓅ③, *Pesticides*ⒶⓁⓅ.)

Oslo Treaty

Signed in 1972 by the major North Sea states, the Oslo Convention governs the dumping of wastes in the North Sea and the North-East Atlantic. The treaty came into effect in 1974.

Like the London Dumping Convention (*qv*), the Oslo Convention has a black-list of substances which may be dumped only in trace quantities, and a grey-list of substances that can be dumped only with prior permission. Black-listed substances include organohalogen compounds; orgaonosilicon compounds; substances "agreed between the contracting parties to be carcinogenic under the conditions of disposal"; mercury (*qv*) and mercury compounds (*qv*); cadmium (*qv*) and cadmium compounds; and persistent plastics.

Britain's interpretation of the "trace quantities" clause for black-listed substances has caused considerable ire among many of her co-signatories to the treaty. For even when dumped in trace quantities, black-list substances build up in the environment.

Thus, in 1982, the dredge spoils dumped by Britain accounted for more than 35 per cent of the mercury entering the Oslo Treaty area – 13.6 tonnes out of a total of 35 tonnes. Indeed, according to a 1986 House of Lords report, ten times more heavy metals enter the North Sea in dredge spoils than in sewage sludge or industrial waste. But because they appear in "trace quantities", they are not regulated. (See also: *North Sea* Ⓟ, *Ocean dumping* Ⓟ, *Ocean incineration* Ⓟ.)

Overfishing

See: *Fish stocks*

Overgrazing

The overstocking of land with grazing animals – cattle, goats or sheep, for example – can all but eliminate edible plant cover, particularly when the animals are clustered in small areas, such as around a waterhole, or kept on the same land for a long period. Eventually, the land is laid bare. The soil is then vulnerable to severe erosion (*qv*), excessive compaction and ultimately, desertification (*qv*).

Traditional pastoral nomads avoided the problem of overgrazing by ranging extensively over large areas. Overstocking, however, has become a problem as nomads have been encouraged to build up their herds as part of livestock development schemes. At the same time, many rangelands have been converted to plantations in order to grow crops for export, thus reducing the extent of available pastures. Well-digging programmes, too, have encouraged the harmful concentration of animals.

Nomads are well aware of the forces that lead to overgrazing. Indeed, in West Africa, one group, the Illabakan Tuareg of Niger, went so far as to request that the Niger Government close down a new well that had been built in their area. Another group, the Wodaabe of Bernou, are reported to be opposed to all new well construction programmes.

Overgrazing is now fairly general throughout the dry tropics, especially where traditional pastoralism is being replaced by intensive-livestock rearing schemes

Ozone layer – weakening our ultraviolet screen

Ozone

The ozone layer lies in the stratosphere, 20-25km above the Earth's surface. It filters out all forms of incoming ultraviolet (UV) radiation, in particular providing a protective screen against the harmful UV-B radiation.

UV-B radiation

UV-B radiation increases the incidence of certain types of skin cancer. The US Environment Protection Agency has estimated that over the next 90 years ozone depletion may be responsible for 800,000 additional cancer deaths. UV-B is also a major cause of cataracts.

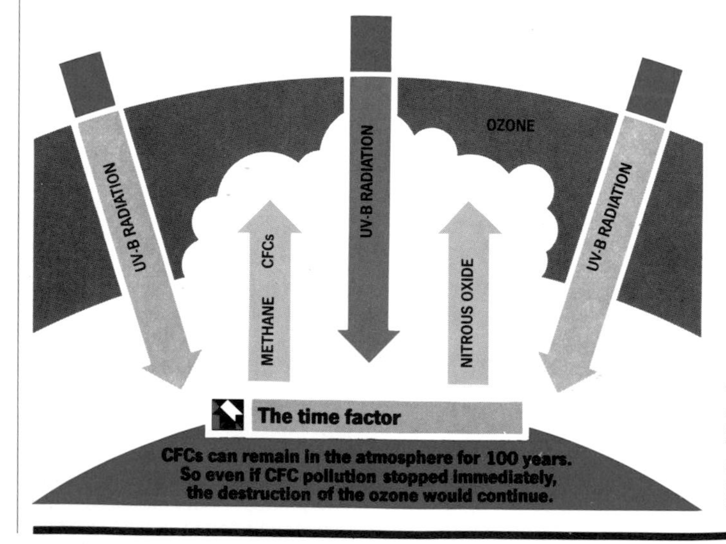

The time factor

CFCs can remain in the atmosphere for 100 years. So even if CFC pollution stopped immediately, the destruction of the ozone would continue.

Ozone destruction

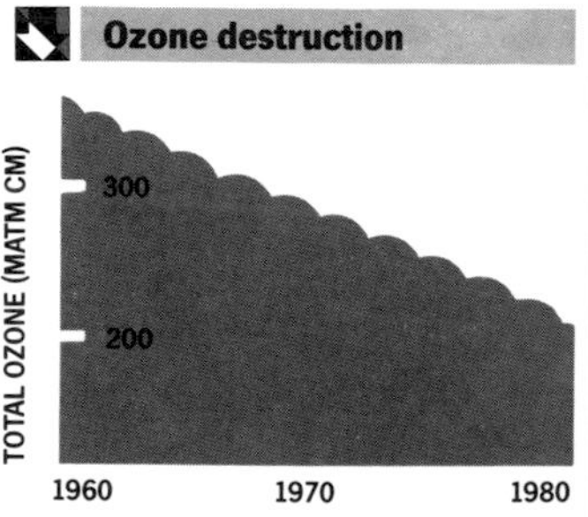

READINGS TAKEN IN OCTOBER AT HALLEY BAY, ANTARCTICA

The prime cause for ozone depletion is the increasing presence of chloroflourocarbons (CFCs) in the atmosphere. Used as aerosol propellants, among other things, they release chlorine when they are broken down by the strong UV-B radiation in the upper atmosphere. Each chlorine atom may destroy 100,000 ozone molecules. Nitrous oxide and methane, both agricultural by-products, also disrupt the ozone layer.

Ozone protection

Though the USA banned the use of CFCs as aerosol propellants in the 1970s, it still permits their use in other products. The UN Environment Programme has called for a ban on their production and use. So far there are few controls over methane and nitrous oxide.

geared to the export market, as in Botswana for instance. There is now a growing consensus that only a reduction in livestock numbers, together with a return to traditional pastoralism can prevent widespread desertification. (See also: *Carrying capacity* Ⓑ Ⓒ, *Desertification* ③ Ⓒ, *Hamburger connection* ③ Ⓐ Ⓒ.)

Ozone layer
The ozone layer is being disturbed by long-lived pollutants. Chlorofluorocarbons (CFCs) and the nitrogen oxides are responsible for ozone depletion, whereas methane tends to increase ozone levels (see illustration).

Since the 1960s, ozone levels over parts of Antarctica have dropped by almost 40 per cent during some months and a "hole" is clearly visible in satellite observations of ozone concentrations over the polar cap. There are, however, a number of conflicting explanations for this trend. Some scientists believe that it may be the result of natural variations in the temperature or wind fields of the upper atmosphere: others claim that change in solar activity is responsible. But rising levels of pollution are the most plausible hypothesis.

Despite residual scientific uncertainty, action has been taken to control CFC releases. At present it seems likely that CFC production will not rise substantially over coming years, limiting the extent of the damage to the ozone layer, though significant cuts in production levels may not occur for some time. Effective control of nitrous oxide and methane releases – both intrinsically related to agricultural and industrial expansion – will be more difficult to establish and it is likely that the threat to the ozone layer will remain a major environmental problem.

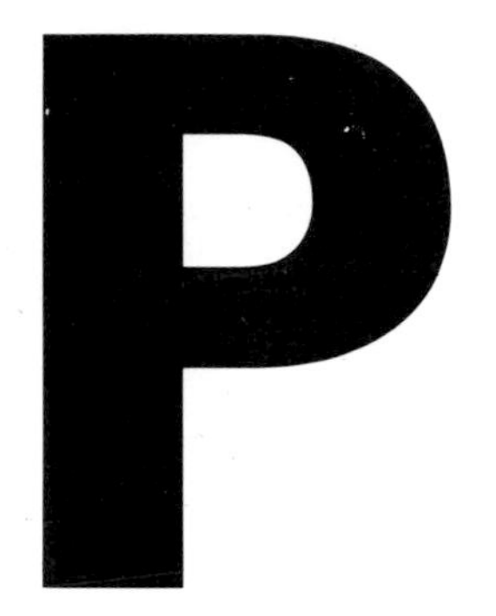

Paraquat
A contact herbicide that is widely used in agriculture, horticulture and gardening around the world. It differs from most other herbicides in that it is highly toxic to mammals. It has been responsible for more poisonings, many of which have been fatal, than any other weedkiller. Less than a teaspoonful of the concentrated material is likely to prove lethal if swallowed, and paraquat is also absorbed through the skin.

In the environment, paraquat is often regarded as fairly harmless since, once sprayed, it adheres so tightly to soil particles that it is not free to poison other plants. New crops can thus be sown soon after paraquat has been sprayed. But this is not invariably true: some soils do not hold on to paraquat so tightly and it can drain into rivers where it is extemely toxic to fish. Paraquat residues on vegetation may also be hazardous. Hares in England have frequently been poisoned by eating sprayed plants.

In some countries spraying with paraquat is replacing the plough as a means of soil preparation. Stubble and weeds are sprayed and new seed is placed in the soil almost immediately by direct drilling. This does not disturb the soil structure, and is economical, but the use of large quantities of paraquat in this way is giving rise to concern about the effects on the environment.

In Britain, concentrated paraquat is listed as a poison and may be bought only by professionals, though a dilute preparation is on general sale. The USA restricts its use to licensed operators and it is severely restricted in Sweden, Finland, Denmark, the Phillipines, New Zealand, Turkey and Israel. It is widely used in the Third World, where it is also said to be the preferred method of suicide. (See also: *Herbicides* Ⓟ.)

Parathion
One of the most toxic insecticides in general use. It is an organophosphate (*qv*) and has been involved in many cases of human poisoning and damage to wildlife.

Because recovery from the effects of parathion is slow, repeated exposure to small amounts can have a cumulative effect. Parathion is restricted in the UK and the USA on environmental and health grounds, but continues to be used in Third World countries, where it has often caused serious, sometimes fatal, poisoning.

Parathion is not as persistent in the environment as the chlorinated hydrocarbon (*qv*) insecticides, but dangerous residues can remain for some time. Serious toxic effects have occurred when workers have handled oranges sprayed with parathion nearly three weeks previously. (See also: *Pesticides* Ⓐ Ⓟ.)

Partial Test Ban Treaty
See: *Limited Test Ban Treaty.*

Patented seeds
In 1964, the British Parliament passed the Plant Varieties and Seeds Act. Further acts were passed in the following years: namely the European Communities Act in 1972 and the Seeds National Lists of Varieties Act 1973. In the US the Plant Protection Act and the Senate Bill S.S.23 were passed in 1970.

The aim of such legislation is to give plant breeders rights over the plant varieties that they develop and enable them to collect royalties on the sale of the seeds for as long as 30 years. In the UK, the Acts also provide for the establishment of a list of approved seeds, and make it a criminal offence to sell seeds that do not appear on the list. It is relatively expensive to register a seed, since it costs about £500 ($800) to prove that it is a distinct variety and hence eligible for patenting. It then costs a further £800 ($1,280) a year to keep it on the list and, in addition, a plot must be sown with the variety and inspected every year by a representative of the Ministry of Agriculture, for which a further payment is required. If the variety is to stay on the list, then at least 5,000 packets must be sold every year. For small family seed companies, such payments can prove crippling.

The net effect of the legislation has been to condemn traditional, often local, varieties of fruit and vegetables to extinction (*qv*), only those varieties produced on a large scale by large companies being allowed to survive.

A driving force behind these new developments is the International Union for the Protection of Plant Varieties (UPOV) in Geneva. Financed by multinational companies that have recently bought up the traditional seed companies, UPOV has close relations with various European Ministries of Agriculture and is backed by the United Nations.

The UPOV has persuaded the EEC that a large number of non-patented traditional varieties should be banned on the grounds that they are not really distinct varieties, but simply synonyms that are very nearly alike. Lawrence Hills of the Henry Doubleday Research Association in England has shown, however, that many of these varieties are genetically distinct and have different advantages such as resistance against different diseases.

For a variety to be listed, it must be distinct in shape and form, uniform in its genetic structure, and it must be stable – that is, it must breed true. However, as Lawrence Hills points out, if these criteria were adopted in the Third World practically "all the local varieties would become illegal." The reason is that "each one is a gene pool in itself, a centre of genetic diversity which is as unexplored as the botanic treasures of the Amazon."

The seeds that are replacing traditional varieties require large amounts of artificial fertilizer and pesticides, products that are manufactured by the companies now involved in patenting seeds. The new varieties tend to be F1 and F2 hybrids, whose main feature is that they cannot be regrown: farmers cannot keep the seeds from one year to plant the next, but must buy new seeds.

An effect of the new legislation will be a reduction in the quality of life. Traditional varieties – bred for their flavour, their storage qualities and nutritional value, and adaptability to local climatic conditions – are rapidly being replaced by new varieties that are often tasteless, of low nutritive content, and not particularly well adapted to local conditions. The massive genetic loss, moreover, will make humanity correspondingly more vulnerable to plant epidemics, climate changes and other agricultural disadvantages. (See also: *Green Revolution* Ⓐ ③, *Pesticide-seed connection* Ⓐ.)

Paving over
Worldwide, vast areas of highly productive land are lost to roads, factories, housing development, shopping centres, car parks and other forms of urban development – a process known as "paving over".

In the USA, 31sq km (12sq miles) of farmland are lost to concrete every day and 11,315sq km (4,360sq miles) go out of production every year, one-third of which is prime agricultural land.

In Britain, some 5,000sq km (1,900sq miles) of land were lost between 1933 and 1963. Much of that land was of the highest quality. Since the 1960s, the rate of farmland loss has accelerated.

In the Third World, too, the loss of prime agricultural land to urban sprawl is a growing problem. In Egypt, some 4,000sq km (1,500sq miles) of fertile land in the Nile Valley were lost to urban expansion between 1955 and 1975 – more, that is, than the 3,700sq km (1,400 sq miles) brought into production through reclamation projects and irrigation schemes.

PCB
See: *Polychlorinated biphenyls.*

PCDF
See: *Polychlorinated dibenzofurans.*

Pentachlorophenol
A general biocide (*qv*) that has been used against fungi, algae, molluscs and other organisms throughout the world. It is a chlorinated hydrocarbon (*qv*) and persists in the environment, where it can accumulate in food webs. Pentachlorophenol is carcinogenic and teratogenic. It is restricted in Canada, New Zealand and the USA but is still available in the UK to people untrained in its use.

Pesticides

A term used more or less as a synonym for biocides. Included in this category are herbicides (*qv*), insecticides (*qv*), fungicides (*qv*), ascaricides (which kill spiders and mites), nematocides (which kill nemotode worms), molluscicides, and rodenticides.

World production of pesticides in 1986 was about 2,300,000 tonnes a year. In the USA, more than 450,000 tonnes of pesticides are used every year or 1.8kg (4lb) for every American. In the UK, about 4,540 million litres (1,200 million US gallons) of liquid containing pesticides are sprayed onto crops every year. Pesticide use continues to increase; it has doubled in the US since 1964, and is increasing worldwide at a rate of 12.5 per cent a year.

Unfortunately, pesticides cause widespread pollution of the environment, seeping into our rivers, killing off fish life, and contaminating groundwater, drinking water, and food, most of which now contains pesticide residues.

Though levels of pesticides in the general environment are usually low, pesticides tend to concentrate as they move up the food chain, a phenomenon known as bioconcentration (*qv*). The consequences for wildlife and human health are severe. Our body fats also now contain pesticides, as do ova and spermatazoa. A World Health Organization (WHO) study in 1987 warned that human breast milk is now seriously contaminated with pesticides and other chemicals that could have serious effects on the health of our children. The WHO experts even recommend that, in certain areas, breast feeding should be avoided.

Most modern pesticides are synthetic organic (*qv*) chemicals, a category that includes many known or suspected carcinogens (*qv*), mutagens (*qv*) and teratogens (*qv*). A 1985 report by the London Food Commission suggested that 49 pesticides in current use in the UK were possible carcinogens, 61 suspected mutagens and 90 possible allergens. In the UK, where controls are particularly lax, 38 pesticides in current use are banned or severely restricted in other countries.

Unfortunately, the literature on the health effects of pesticides is scanty. To date, the US Environmental Protection Agency (EPA) has been able to establish the harmlessness of only 37 of the more than 600 active ingredients used in the 45,000 different formulations at present marketed in the USA. This means that precise and reliable information is available on little more than five per cent of all the pesticides used in the USA.

In 1972, a WHO expert committee on insecticides concluded that there were about 500,000 cases of pesticide poisonings a year, causing 9,000 deaths. Since then pesticide use has increased by 50 per cent and it is now estimated that there are more than a million cases of poisonings a year, leading to about 20,000 deaths. This figure does not take into account long-term health effects.

At least one-quarter of the pesticides exported from the USA to Third World countries have been banned or severely restricted, or have never been registered for use in the USA. There is little if any control on the use of these poisons in the Third World – or, as the Bhopal (*qv*) tragedy demonstrated, on their production.

The pesticide industry argues that pesticides are essential for feeding the world. The standard claim, originally made by Dr Norman Borlaug, the father of the Green Revolution (*qv*), is that without pesticides, world food production would fall by 50 per cent, causing starvation.

This seems unlikely. Indeed, since the introduction of modern pesticides in the 1940s, crop losses in the USA have increased rather than decreased, from an estimated 32 per cent to 37 per cent. This is probably largely due to changing farm practices, with much bigger farms, ever-greater stretches of monoculture (*qv*), the cultivation of high-yielding varieties (HYVs) (*qv*) that are particularly vulnerable to pests, and the greatly reduced farm labour force. It can also be partly attributed, however, to problems inherent to the use of pesticides: notably, their tendency to increase pest infestations by killing off predators that previously controlled target species, and the latters' tendency to develop resistance to pesticides. This problem is particularly acute in the tropics where conditions are most favourable for pest infestation.

Such factors have given rise to what has been termed the "pesticide treadmill" with farmers being forced to use ever greater amounts just to keep crop losses at a given level. What, then, would be the real cost of dispensing with pesticides?

A study undertaken at Cornell University concluded in 1978 that without pesticides financial losses would be about nine per cent higher, while food crop losses, in terms of food energy, would be about four per cent higher. This is a long way from the pesticide industry's 50 per cent claim. In any case, pesticides are mainly used in the Third World for the cultivation of cash crops.

The use of pesticides should not be considered on its own, but as part of the modern farming system, which also requires HYVs, fertilizer (*qv*), elaborate farm machinery and irrigation (*qv*) water – a system which, because of its prohibitive cost and its adverse social and environmental effects, cannot be maintained for very long. (See also: *Desertification* Ⓐ, *Erosion* Ⓐ, *Salinization* Ⓐ, *Tropical agriculture* Ⓐ.)

Phenoxy herbicides

Among the first of the modern herbicides (*qv*) to be developed in the 1940s. They interfere with plant growth causing the plant to overgrow and die.

The two main phenoxy herbicides are 2,4-D (*qv*) and 2,4,5-T (*qv*), the former being used against general weeds and the latter against woody weeds.

For many years it was thought that these chemicals were virtually harmless to humans, since large doses were required to produce immediate symptoms of poisoning. But in the late 1960s concern was expressed that long-term and more subtle harmful effects may result from exposure, particularly to 2,4,5-T.

There has been considerable controversy over whether it is the herbicides that are harmful or the dioxins (*qv*) that contaminate some of them during manufacture. A variety of different chemicals is involved and it seems that the herbicides themselves, even if dioxins are absent, may be carcinogenic. Recent research has suggested that this may well be so with 2,4-D, which does not contain dioxins.

Despite concern about their environmental effects, most phenoxy herbicides remain in unrestricted use. (See also: *Pesticides*Ⓛ Ⓟ.)

Phosphates

Phosphates are essential nutrients for plants but, like many other such substances, they can cause serious pollution if present in excessive quantities. The principal effect is eutrophication (*qv*), which occurs when phosphates, together with nitrates (*qv*), enter lakes or slow-moving rivers in large quantities.

There are three main sources of phosphate pollution. Some runs off agricultural land, since phosphates are major constituents of fertilizers. Naturally occurring phosphates that enter sewage also contribute, but the main source is phosphate water softeners used in synthetic detergents. These enter waterways in sewage discharges and have caused problems in many parts of the world. As a result, attempts have been made to restrict their use. Unfortunately, however, some of the substitutes suggested have proved to be more hazardous than the material they were intended to replace.

In eastern England, where the Norfolk Broads suffer serious phosphate pollution, a plant has been built to remove phosphates from sewage before it is discharged into the local river and this may be the first of many. (See also: *Artificial fertilizers*Ⓐ Ⓟ, *Eutrophication*Ⓒ Ⓟ.)

Photo-voltaic cell

A device for generating electricity directly from sunlight.

At present, photovoltaic cells are still too expensive for general use, but their price may fall to around 35 cents per peak watt, a level competitive with conventional generating systems, before the end of the century. (See also: *Energy*Ⓔ, *Energy conservation*Ⓔ, *Soft energy paths*Ⓔ.)

Plant breeders' rights

See: *Patented seeds*.

Plutonium

The first, and most significant, man-made or "trans-uranic" element to be created. The first tiny sample was produced at Berkeley University in February 1941. Four years and $2,000 million later, a few kilogrammes had been made, enough to fuel two atomic bombs. The first of these was tested over the desert in New Mexico, the second over Nagasaki.

Even before plutonium actually existed, it was known that it would be capable of fission (*qv*) with somewhat better characteristics as a bomb material than uranium-235.

Plutonium was, and is, made in nuclear reactors, by bombarding uranium-238 with neutrons. The uranium-238 transmutes to plutonium-239 by absorbing a neutron into its nucleus, and is extracted from the reactor fuel by a chemical process called reprocessing (*qv*).

Plutonium can itself be used as fuel, or part of the fuel, for conventional, or thermal, nuclear power reactors, or alternatively as fuel in fast breeder reactors (*qv*). However, because of high costs and concerns over safety, fast reactors are few in number.

Countries that reprocess civil reactor fuel have growing stockpiles of plutonium, which present obvious proliferation and pollution risks. The British programme of Magnox power reactors has produced some 47 tonnes of plutonium, of which 30 tonnes is separated and stockpiled. At present, the Central Electricity Generating Board's stockpile is the largest in the non-communist world, and will quadruple by 2030 if the CEGB builds all the reactors it presently envisages, and if reprocessing continues. (See also: *Nuclear waste*Ⓝ, *Nuclear proliferation*Ⓝ, *Radiation and health*Ⓝ, *Radioactivity*Ⓝ.)

Polychlorinated biphenyls (PCB)

First synthesized in 1881, polychlorinated biphenyls have been used commercially since the 1930s, their low flammability, high heat-resisting capacity and low electrical conductivity making them valuable compounds in a wide range of products – from fluorescent light bulbs to hydraulic fluid and, most important of all, electric transformers and capacitators. Though the toxic effects of PCBS were first documented as early as 1936, it was not until the late 1960s – following a major pollution incident in Japan – that their dangers became widely appreciated.

PCBS are immensely stable compounds and can be destroyed only through incineration at temperatures above 1,200°C (2,200°F). Incomplete incineration can lead to the formation of new, and often more toxic, compounds, notably PCDFS and dioxins (*qv*).

A dramatic illustration of the problem occurred in 1981 when an explosion in the basement of an office block in Binghampton, New York, caused an electrical transformer containing PCBS to crack. In the intense heat of the explosion and the resulting fire, the PCBS began to break down, forming a fine ash which was thoroughly contaminated with a hotchpotch of PCBS, dioxins and PCDFS. The polluted ash was transported through the office block by the building's air conditioning system. Four years later, the building was still considered too toxic to enter without protective clothing. In 1985, it was estimated that complete decontamination of the site would cost some $19 million.

PCBS have not been manufactured in Britain since 1977, when Monsanto – the sole UK manufacturer – ceased production. In 1980, the British Government implemented a 1976 EEC directive that banned the use of PCBS except in sealed equipment. As a result of that ban, little new equipment containing PCBS now comes on to the market.

But while the EEC directive has helped curtail the production and use of PCBS, next to nothing has been done by European governments to combat the threat posed by PCBS in old and often dilapidated equipment. It is estimated that 90 per cent of Britain's 20,000 transformers are in such a poor state of repair that they regularly leak PCB insulating fluid. (See also: *Cancer* Ⓛ Ⓟ, *Carcinogen* Ⓛ Ⓟ, *Incineration* Ⓟ.)

Polychlorinated dibenzo-furans

A common contaminant of commercial polychlorinated biphenyls (PCBs) (*qv*), polychlorinated dibenzo-furans are many times more toxic. Indeed, many researchers now believe that PCDFs are responsible for the most harmful effects of PCBs.

Tests carried out in the late 1960s using three commercial brands of PCB – Clophen A 60, Phenoclor DP 60 and Aroclor 1260 – revealed that Clophen and Phenoclor were much more toxic than Aroclor. Suspicions that Clophen and Phenoclor were contaminated with some chemical that was more toxic than PCB itself were confirmed when further analysis of the Clophen and Phenoclor samples revealed the presence of tetra- and pentachlorodibenzofurans, two of the most toxic furan isomers.

The presence of PCDF contaminants in PCBs is generally thought to result from the PCBs having been heated. (See also: *Dioxin* Ⓟ, *Incineration* Ⓟ.)

Population

Since the beginning of the 19th century, the world's population has increased five-fold (see illustration). Currently, 220,000 new

Population – 10,000 million mouths to feed

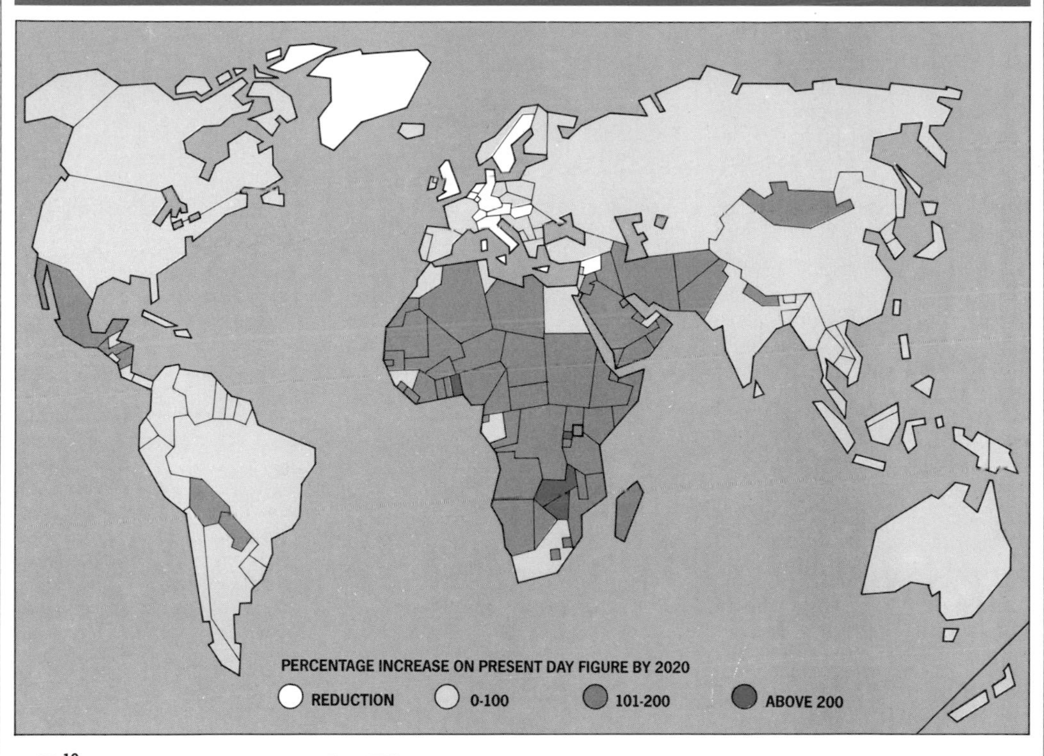

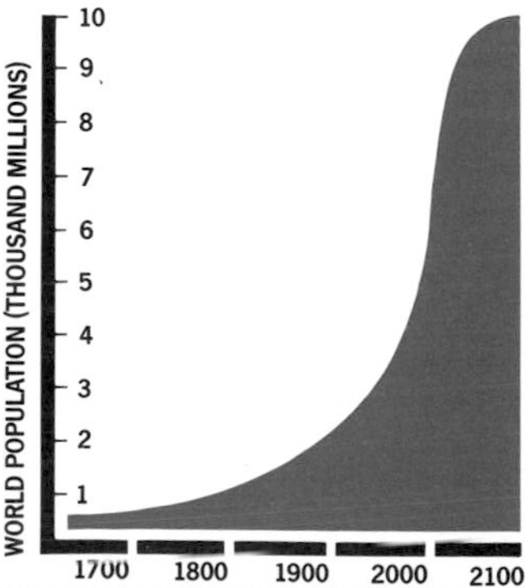

Explanations for the sharp rise in population range from the Malthusian theory that populations inevitably breed up to their food supply to the fact that improved medical treatment and sanitation have greatly reduced mortality. But the explanation that best accords with the data is that the Third World population increase is a direct result of social insecurity bred by development. Development encourages people to invest in children as a way to achieve the security that was formerly provided by their extended family, their lineage, their clan and their community. It also breaks traditional controls on procreation. As a result, population is growing too quickly in the wrong places.

Before the end of the 21st century the world's population will have doubled, reaching 10,000 million people. Ninety per cent of that growth will occur in the Third World: Africa alone is predicted to pass the 1,000 million mark by 2010.

Some European countries have near-zero or even negative growth rates. The population of West Germany, for example, is expected to fall from 61 million in 1987 to 50 million by 2020. Denmark, Austria, Hungary, Sweden, Belgium, Italy, Luxembourg, Switzerland and the UK are also expected to show declines.

The USA's population is expected to rise by almost 53 million and Canada's by more than three million.

babies are born every day – 150 every minute, one every half second.

Ninety per cent of the growth predicted to occur before the end of the 21st century will be in the Third World. Africa alone is predicted to pass the 1,000 million mark by 2010.

In the industrial world, population growth is declining, some countries having zero growth rates. Even so, as the World Commission on Environment and Development (*qv*) (the Brundtland Report) points out, "the population in North America, Europe, the USSR, and Oceania is expected to increase by 230 million by the year 2025, which is as many people as live in the USA today."

But could the Earth support the 10,000 million predicted by UNFPA? Even with 5,000 million people, half the world's population live in a state of grinding poverty, suffering from endemic malnutrition and continually on the brink of starvation.

Under the present economic system, the carrying capacity (*qv*) of many areas of the world has already been exceeded – in many cases, irreversibly so. Indeed, delegates at the Second International Conference on The Ennvironmental Future, held in Reykjavik in 1977, predicted that 1,000 million people will die of starvation before the end of the century as a result of current economic policies. The present famines (*qv*) in Africa and S.E. Asia lend credibility to this view.

But more people could be fed if radical changes were made in the use of land and resources.

For example, millions of square kilometres of the best arable land in the Third World are currently used to grow cash crops (*qv*) for export to the industrialized world. Were that land to be used to grow crops for local consumption, it would not only alleviate hunger but would also reduce the pressure on the environment by relieving displaced peasants of the need to encroach onto marginal lands.

Equally, a less materialistic way of life in the industrialized countries would greatly reduce the environmental impact of rising population levels in both the industrialised world and the Third World. High living standards in the industrialized world ensure that a single western child has about a forty times greater impact on the resources of the world than a child born in the Third World.

In the absence of such reforms, however, there is little chance of a "civilized" solution to the problem of over-population.

The role that can be played by family planning (*qv*) is limited. The notion that, with sufficient economic growth, a demographic transition (*qv*) will occur is largely an act of faith, and the economic growth is, in any case, unlikely to materialize. (See also: *Zero population growth*Ⓛ.)

Population control

See: *Family planning.*

Population explosion

In historical terms, the current population explosion is a very recent phenomenon. For 90 per cent of the time that humans have lived on Earth, the human population remained very low, a generally accepted estimate being around five million people.

During that time, man lived in hunter-gatherer bands, but with the adoption of farming, some 10,000 years ago, the population began to rise. By 500BC, it had reached an estimated 100 million, rising to 500 million by 1300.

The next major spurt in population growth occurred with the coming of the industrial revolution, the population passing the 1,000 million mark in the early 19th century. Since then, as economic and industrial development has spread, the population has increased at a phenomenal rate, moving from 4,000 million in 1975 to 5,000 million in 1987.

Several explanations have been suggested to account for this population explosion. The best known, first put forward by the Reverend Thomas Malthus in the 18th century, is that populations inevitably breed up to their food supply: increasing yields have thus brought increasing numbers.

The theory, though popular, has little basis in fact. Though the output of food has undoubtedly risen, particularly in the last 40 years, the availabilty of food in those countries where population growth has risen fastest has actually declined so far as the poorest sections of society are concerned. The increased production has largely gone for export, and has thus been at the expense of food grown for local people.

In addition, anthropological studies make it clear that traditional societies do not breed up to their available food supply. On the contary, a wide range of strategies – from natural contraceptives to taboos on sex during lactation – are exploited to control population numbers.

A second explanation is that improved medical treatment and sanitation have greatly reduced mortality, thus allowing more people to survive into old age. Undoubtedly this is true, but it is not a sufficient explanation. Disease and ill-health were not the only constraints on population growth prior to the age of modern pharmaceuticals.

A third explanation – the one that best accords with the data – is that the population explosion is a direct result of the social insecurity bred by the development process. In traditional societies, people derive a sense of identity and belonging from their extended family, their lineage, their clan and their community: there is thus less need to invest in children to acquire security.

By destroying those tradional supports, economic development (*qv*) not only ferments the alienation that leads people to seek children, but also breaks the traditional controls on procreation. The inevitable result is an upsurge in population. (See also: *Contraception*Ⓛ, *Demographic transition*③, *Family planning*Ⓛ③ *Population*③.)

Post harvest losses

According to the UN Food and Agriculture Organisation (*qv*), about 30 per cent of the food produced in the Third World is now lost during storage to pests and mould. This is probably an overestimate: nonetheless, high losses do occur and the problem is undoubtedly serious.

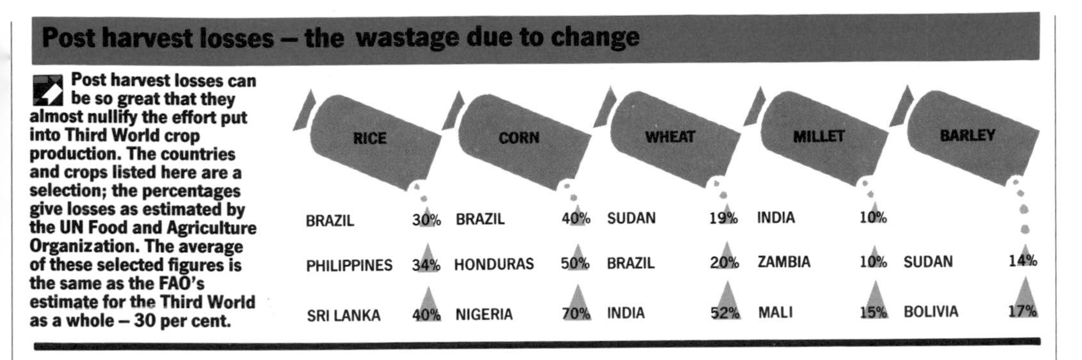

Post harvest losses – the wastage due to change

Post harvest losses can be so great that they almost nullify the effort put into Third World crop production. The countries and crops listed here are a selection; the percentages give losses as estimated by the UN Food and Agriculture Organization. The average of these selected figures is the same as the FAO's estimate for the Third World as a whole – 30 per cent.

Post-harvest losses are officially attributed to primitive food storage methods, the solution apparently lying in the introduction of new capital-intensive technology, much of which involves the increasing use of pesticides.

In fact, traditional food storage techniques are, contrary to what we are told, extremely efficient. Indeed, it now appears likely that the high losses incurred during storage are primarily the result of modern agricultural technology and production processes. This is the view of Dr. Martin Greenley of the Institute of Development Studies at the University of Sussex: "It is the changing structure of production operations associated with increases in land productivity and seasonal changes in cropping patterns that critically affect the level of losses rather than the inefficiency of traditional systems."

In particular, blame has been placed on the high moisture content of modern cereals. This is partly due to the fact that the new varieties frequently mature before the end of the rainy season, partly to lack of artificial drying facilities, and partly to their genetic structure, which makes drying difficult. The high levels of artificial fertilizer required by modern varieties have also been incriminated, since plants whose critical carbon/nitrogen balance has been disrupted by exposure to nitrogen fertilizers are known to react by taking up more water.

For some, the high water content of modern cereals is an advantage. As Dr. Anne Bergeret of France's National Centre for Scientific Research notes, "Dealers have always been tempted to sell a percentage of water or dust for the price of grain." However, if losses during storage are anything like as high as FAO states, the price mankind must pay for such dubious commercial practices is a high one. (See also: *Development* ③, *Green Revolution* Ⓐ③, *High-yielding varieties* Ⓐ③, *Pesticides* Ⓟ.)

Preservatives

A group of food additives (*qv*) used to inhibit the rate at which bacteria will grow and reproduce in food products. While many bacteria are benign, a few, such as botulism and salmonellae, can be fatal. Preservatives can thus protect the health of consumers, which is not true of other types of additives.

In Britain and the EEC, 47 different preservatives are permitted, and they all have E numbers (*qv*) ranging between E200 and E297.

While many preservatives are almost certainly safe, some are suspect. There is some evidence that benzoic acid and the benzoates (E210 to E219) may provoke occasional symptoms of acute intolerance, while sulphur dioxide (E220) and the sulphites (E221 to E226) may well cause adverse effects asthma sufferers.

In the main, however, concern centres on sodium nitrite (E250) and sodium nitrate (E251). These are widely used in many processed and cooked meat products, such as bacon, ham and salami. The nitrate (*qv*) decays into nitrite (qv), and sodium nitrite can combine with food constituents during the cooking process and in the digestive tract to form chemicals called nitrosamines (*qv*) which are notorious carcinogens (*qv*). The official position is the levels at which nitrosamine compounds are formed are too low to pose any significant hazard. Some consumers prefer not to take the risk. (See also: *Cancer* Ⓛ.)

Pressurized water reactor

First developed by the Westinghouse Corporation as a reactor for nuclear submarines, the pressurized water reactor (PWR) is now the most widely used reactor in the world – with 190 in operation, 115 under construction and more than 70 at the planning stage. The first PWR was built at Shippingport, Pennsylvania, and began operating in 1957.

PWR fuel consists of pellets of uranium dioxide enriched to three per cent U-235. The core is contained within a steel pressure vessel filled with ordinary water at up to 150 atmospheres pressure. The water serves as both moderator (*qv*) and coolant, circulating through the core and the steam generators. The high pressure prevents the water from boiling, even though temperatures are in excess of 300°C (570°F).

The PWRs' steam temperature, and thus its efficiency, is low. Nevertheless, the power density is high and a loss of coolant – for instance through a stuck-open valve, as happened in the Three Mile Island (*qv*) accident – can lead to fuel melting and threaten the integrity of the pressure vessel, the heat being provided by fission product decay.

Price-Anderson Act

A serious accident at a nuclear plant could cause damage in the region of billions of dollars. Consequently, US electricity utilities were initially slow to order nuclear reactors because they were unable to find private companies prepared to insure them against an accident. The Federal Government therefore passed the Price-Anderson Act of 1957 which limits corporate liability in a reactor accident to $60 million, and total liability to $560 million. The Act was renewed in 1967 and is still in force. In 1987, however, the Energy and Commerce Committee of the US House of Representatives recommended a new ceiling of $7,000 million. Many democrats argued that companies should be made liable for unlimited damages in the event of negligence. (See also: *Nuclear accidents* Ⓝ.)

Proposition 65

Known technically as the "Safe Drinking Water and Toxic Enforcement Act", Proposition 65 was adopted by Californian voters in a state referendum in November 1986. Despite opposition from Governor George Deukmejian, and from industry (which spent $4,500,000 in an attempt to stop the Act), the proposition was passed by a 2-to-1 majority.

The proposition requires the Governor of California to publish a list of chemicals that are known to cause cancer or birth defects. Under the new legislation, it will become illegal knowingly to allow "significant" amounts of any of the listed chemicals into any source of drinking water. The Act uses a stricter definition of "significant" than that employed in federal food contamination laws.

Citizens will have the right to sue companies they suspect of infringing the new regulations but, in sharp contrast to previous legislation, it will be up to the accused companies to prove their innocence. The burden of proof has thus been shifted dramatically.

The implications of Proposition 65 for the siting of factories and waste disposal facilities are far-reaching. In particular, the new law will force industrial plants away from water sources and hasten the end of landfill as a means of waste disposal. (See also: *Cancer* Ⓛ Ⓟ, *Groundwater pollution* Ⓟ, *Hazardous wastes* Ⓟ.)

QR

Radiation and health

The ability of radiation to damage human health is well-established. At high doses – such as those received by the firemen who died during the Chernobyl accident (*qv*) – radiation causes vomiting, loss of hair, bleeding and death. It is also one of the best-documented environmental carcinogens (*qv*) known to man. But controversy surrounds the effects of low levels of radiation on health.

The living environment is bathed in radiation from a variety of sources, both natural and man-made. Every second of the day, human beings are bombarded with some 100,000 cosmic ray neutrons and 400,000 secondary cosmic rays, about 30,000 radioactive atoms that disintegrate in the lungs, and 200 million gamma rays from soil and building materials. Finally, radiation from the diet leads to about 15 million potassium-40 atoms and about 7,000 natural uranium atoms disintegrating inside the body every hour.

That astronomical number of hits, totalling on average some 60,000 a second, will give a radiation dose of 1,870 microsieverts, or five times smaller than the one rad of X-rays that used to be given up until the 1960s in producing a radiographic picture of bones.

In industrial societies, people are increasingly exposed to radiation additional to background radiation. Airline crew, for instance, can virtually double their annual radiation dose by increasing their exposure to cosmic radiation, which becomes more intense the higher the altitude. Radiation workers in nuclear installations can receive doses during normal operation that are five or even 25 times natural background. People living near nuclear installations may suffer increased radiation. Those, for instance, who consume relatively large quantities of locally-caught fish and shellfish will receive greater-than-normal radiation if the sea is contaminated with radionuclides

Do low levels of natural and man-made radiation constitute a health risk, or can the body cope with the levels currently found in the environment?

The International Commission on Radiological Protection (ICRP) (*qv*) bases its assessment of radiation damage on the premise that radiation is always potentially harmful, however small the dose. The commission also assumes a straight-line relationship between the radiation dose and the probability of cancer induction, getting data for the high-dose end of the graph from what happened to the victims of the Hiroshima and Nagasaki atom bombs.

Some radiation biologists believe that as radiation levels increase, the body's repair mechanisms become overloaded, and, therefore, the risk of damage at high levels is greater than at low levels. But most believe that low doses of radiation are particularly pernicious, since they stimulate repair mechanisms which then operate imperfectly and allow deranged cells, which would otherwise be destroyed by higher doses of radiation, to survive. The deranged cells transform into malignant cells, which proliferate until they have formed a mass sufficiently large to be detected as a tumour.

Historically, acceptable levels have had to be progressively reduced. Thus the level acceptable for people working within the industry was set at 73 rems in 1931, a level which was reduced to 50 rems in 1936, 25 in 1948, 15 in 1954 and five in 1958. Some authorities, such as Professor Radford, President of the BEIR Committee (*qv*) of the US Academy of Sciences, consider that the present limit should be reduced by a further factor of five to ten.

The dose acceptable to the public at large has also been reduced on a number of occasions, the present level being considered by a number of authorities, such as Professor John Goffman of Berkeley University and Dr Alice Stewart of the Cancer Registry, Queen Elizabeth Hospital in Birmingham, to be ten or even 50 times too high.

The deaths to date among radiation workers at the Hanford (*qv*) nuclear establishment in the United States indicate that, in terms of life expectancy, the cancer rate is some 25 per cent higher among those workers who have been most exposed to radiation. Those figures suggest that the ICRP has underestimated by between ten and 20 times the power of low doses of radiation to cause cancers.

Clearly, if these studies are correct, the effects of low doses such as emanated from the fallout from Chernobyl, or have accumulated in the environment around the Sellafield plant in Cumbria, will be considerably worse than estimated.

Nor should we assume that natural background radiation is entirely harmless. According to Dr. Alice Stewart, one of the authors of the Hanford study, more than 70 per cent of childhood cancers are caused by natural background radiation. (See also: *Atomic tests*Ⓝ, *Radioactivity*Ⓝ.)

Radioactivity

Certain atoms are inherently unstable and liable to transform into other elements through the expulsion of particles from the nucleus.

The period of time for half the number of atoms in a given mass of the element to transform is known as the half-life. After 10 half-lives about one-thousandth of the original mass is left, the remainder having transformed into a new element, which itself may or may not be radioactive and unstable.

The half-life of a radioactive substance is individual and specific. Uranium-238, the commonest isotope of uranium found in nature, has a half-life about equal to the present age of the Earth, namely 4,500 million years; uranium-235, the fissile isotope of uranium, has a half-life of 710 million years; polonium-214, one of the decay products of uranium-238, has a half-life of 1,160 micro-seconds.

Fission products resulting from nuclear reactors have a whole range of different half-lives; three important ones, iodine-131, caesium-137 and strontium-90, have half-lives of eight days, 30 years and 28.1 years respectively.

Depending on the element and its isotope, radioactive decay can take different forms. Some isotopes give out alpha radiation (*qv*), while others give out beta radiation (*qv*) or gamma radiation (*qv*). When uranium-238 undergoes spontaneous radioactive decay with the loss of an alpha particle, thorium-234 is created, which is also radioactive and has a half-life of 24 days. The thorium then disintegrates and proactinium-234 is created. After twelve more transformations of one element into another lead-206 is left, which is stable and not radioactive.

The different particles and rays that comprise the radioactivity of elements have different characteristics, depending on their size and energies. Gamma rays require a large slab of lead or concrete to stop them. Alpha particles, on the other hand, are ponderous and have poor penetrating powers. Beta particles are intermediate with regard to their effects and can penetrate through a centimetre or more of living tissue. (See also: *Radiation and health*ⓁⓃ.)

Rainbow Warrior

See: *Direct action.*

Rainforest

See: *Tropical forest.*

Recycling

Every day, thousands of tonnes of materials are thrown away by consumers and industry – New York City alone discards 24,000 tonnes of domestic waste a day. Much of that "waste" could be re-used – either directly or after relatively simple treatment – and as such it represents a potentially valuable resource. Indeed, according to David Morris of the Washington-based Institute for Local Self-reliance, "A city the size of San Francisco disposes of more aluminium than is produced by a small bauxite mine, more copper than a medium copper mine and more paper than a good-sized timber stand."

The ecological benefits of re-using waste materials extend beyond good resource management: recycling also cuts energy consumption and reduces pollution. A 1987 study by the Worldwatch Institute found that the energy equivalent of half a can of petrol is saved every time an aluminium can is recycled. In addition, "One ton of remelted aluminium eliminates the need for four tonnes of bauxite and 700kg (1,540lb) of petroleum coke and pitch, while reducing emissions of air polluting aluminium fluoride by 35kg (77lb)." The report calculates that, by doubling worldwide aluminium recovery rates, "over a million tons of air pollutants – including toxic fluoride – would be eliminated."

Several countries now operate successful recycling projects. In the USA, the amount of paper collected for recycling doubled every year between 1975 and 1980, while in Switzerland more than 40 per cent of glass is recovered, with recycled glass now supplying the material for 60 per cent of glass production.

Recycling has also been adopted by a number of industries. In the electroplating industry, for example, electrolysis has been used to recover gold, silver, tin, copper, zinc, solder alloy and cadmium. Equally, in the motor industry, many companies now recover the polyvinyl chloride in scraps of car-seat fabric by washing the material in a solvent. Previously, the scraps had been incinerated, causing vinyl chloride (*qv*), a potent carcinogen (*qv*) to be released into the atmosphere.

In both Holland and Britain, waste exchange networks have been operating since the 1970s, more than 150 wastes being listed for exchange. In the USA, the recycling of industrial solvents is already a $200 million-a-year business which is expected to reach $1,000 million a year by the end of the decade. (See also: *Hazardous waste* Ⓟ, *Waste reduction* Ⓟ.)

Reforestation

Deforestation (*qv*) has deprived many countries of much of their tree cover, and there is now an urgent worldwide need to plant trees to restore ecological stability and combat erosion (*qv*), desertification (*qv*), groundwater depletion, droughts and floods.

The need for reforestation is clearly greatest in the tropical countries of the Third World and in those areas most prone to desertification. But many temperate zone countries also need to increase their tree cover. Britain, for example, is only nine per cent wooded, relying on imports for 98 per cent of its timber. According to the eminent US ecologist Professor Eugene Odum, the optimum tree cover for countries in temperate areas is 50 per cent. Below that level, ecosystems no longer benefit from the full range of "services" provided by trees – from the absorption of pollutants to soil conservation, flood control and groundwater recharge.

At present, most reforestation projects are little more than commercial plantations under a different name. The World Bank's (*qv*) social forestry (*qv*) programme, currently being promoted throughout the Third World, is a case in point: ostensibly intended to provide villagers with trees for firewood, it has primarily served to supply industry with trees for pulping or rayon manufacture.

The near universal choice of fast-growing trees that yield a high return on investment – notably eucalyptus (*qv*) and conifer – reflects the commercial orientation of most reforestation projects. The result has often been the progressive degradation of soils, with conifers causing an acid pan to build up beneath the surface and eucalypts depleting available nutrients. The problem is not restricted to the Third World: in Europe, the Czechs have learnt from bitter experience that sandy soils used to grow more than seven generations of pines are soon so degraded that they can no longer support commercial forestry.

But successful reforestation schemes do exist. Villagers in many parts of the Third World have mobilized themselves – without the aid of international agencies – to replant degraded lands, with remarkable results. Undoubtedly, one of the keys to their success lies in the fact that the villagers identify with the project, looking after the trees and ensuring their protection. In India, for example, the Chipko (*qv*) movement has inspired numerous reforestation projects throughout India, while similar movements have achieved the same in North and East Africa. (See also: *Fuelwood crisis* Ⓒ③.)

Regional seas

The world's regional seas – most notably the Mediterranean, the Baltic, the Carribean and the North Sea (*qv*) – have all been hard hit by pollution. Being enclosed or semi-enclosed, and often having slow rates of water renewal (it takes between 80 and 100 years for the Mediterranean to renew its waters), regional seas do not have the same cleansing capacity as the open oceans. Until recently, several tottered on the brink of ecological disaster as a result of industrial and muncipal discharges, direct dumping from ships, oil pollution and agricultural run-off. Though a series of international conventions have now been signed in order to clean up the world's regional seas, the pollution threat is by no means over.

In the Baltic, the largest area of brackish water in the world, eutrophication (*qv*), combined with industrial pollution, has so degraded the marine ecosystem that only a few hardy species of worm survive. An estimated 100,000sq km (38,600sq miles) are affected – half of the Baltic's deep waters. Currently, the sea receives some 1,100,000 tonnes of nitrogen and 77,000 tonnes of phosphorous a year, mainly in the form of agricultural run-off. Nonetheless, some progress has been made in reducing pollution. Over the past decade, the levels of mercury in fish have been reduced by 66 per cent in some coastal waters, while a ban on DDT by all the Baltic states has cut levels of DDT in fish substantially. Levels of PCBs have not fallen, however, and have been blamed for reducing the reproductive rate of female gray seals from 80 per cent to 20 per cent.

Because of its high rates of evaporation and slow rate of water renewal, the Mediterranean is particularly vulnerable to pollution, the pollutants tending to accumulate without degrading. In 1976, the United Nations Environment Programme (UNEP) launched an Action Plan to clean up the Mediterranean. The plan was endorsed by 17 nations: but despite international agreement on the need to curb pollution, the Mediterranean remains one of the most polluted seas in the world. Though the direct dumping of wastes from ships is banned, the Mediterannean must still contend with billions of tonnes of waste pumped into its waters every year from land-based sources, an estimated 430,000 million tonnes entering the sea in 1977. Edible shellfish are deemed unsafe for human consumption in all but four per cent of recognized growing areas and levels of pesticides and heavy metals continue to rise in many marine organisms. In 1983, the cost of cleaning up the Mediterranean was put at $10,000 to $15,000 million.

The principal threats to the Carribean lies in pollution from pesticide run-off and industry. Of all the world's regional seas, the Carribean is thought to be the most damaged by pesticides. Many of the pesticides that enter the Carribean via the 16 South and Central American states that border the sea are banned in the USA, which itself discharges prodigious quantities of pesticides and other chemicals via the Mississippi, which drains 41 per cent of the land area of the USA. Industrial pollution, particularly from sugar refineries, mining, offshore oil drilling, and the dumping of dredgings and industrial waste, has also wrought havoc with the fragile ecosystem of the region. Discharges of industrial waste have badly polluted the Gulf of Paria, between Trinidad and Venezuela. Tourism, too, poses a threat to the Carribean, particularly to the regions' coral reefs (*qv*), which are being depleted of their fish and coral for sale to tourists.

Througyh its Regional Seas Programme, UNEP has drawn up action plans for the world's major regional seas. However, their implementation has been patchy and much remains to be done if the ecological health of regional seas is to be ensured. (See also: *Fish stocks*Ⓟ, *Ocean dumping*Ⓟ, *Ocean incineration*Ⓟ.)

Reprocessing

The extraction of plutonium (*qv*) and unused uranium (*qv*) from nuclear fuel after it has been irradiated in a reactor.

Reprocessing is a fairly straightforward chemical process. Fuel rods are stripped of their cladding and then dissolved in a hot concentrated solution of nitric acid and an organic solvent, usually tri-butyl phosphate (TBP). The uranium and plutonium cross over into the TBP, and subsequent chemical operations separate the two elements. They are eventually stored in either liquid (nitrate) or solid (oxide) form.

Problems do not in the main arise from the plutonium and uranium, but from the mass of fission products that remain dissolved in the nitric acid. These fission products are a range of isotopes of middle-order elements, and are intensely radioactivity with varied half-lives. They constitute high level waste (HLW) or, as it is sometimes called, heat generating liquid waste. HLW, by definition, is created solely by reprocessing, and in the UK it is stored in 11 stainless steel tanks at Sellafield (*qv*), each containing some 130 million curies of fission products. The tanks must be constantly cooled and stirred.

Reprocessing to extract plutonium for military purposes is relatively uncomplicated because fuel is resident in the reactor for a relatively short period, which minimizes the general level of radioactivity throughout the reprocessing system. The residence of uranium fuel in power reactors is much longer, because the requirement is for maximum heat extraction per fuel loading. Irradiated fuel from the civil reactors is thus more radioactive, and its reprocessing more difficult. That of fuel from second-generation reactors, which use enriched fuel and achieve much higher burn-ups, is still more difficult and costly.

The assumption that civil fuel should be reprocessed is increasingly being questioned. It has been practised most enthusiastically in the UK and France, for three main reasons. First, fuel for the Magnox reactors was adopted in both countries cannot be stored for long under water; secondly, recovered plutonium and uranium were expected to have commercial value as reactor fuel; and thirdly, reprocessing was thought to minimize waste problems.

But Britain's Central Electricity Generating Board (CEGB) has demonstrated that Magnox (and any other type) fuel can be kept virtually indefinitely in dry stores, and since 1977 the price of uranium has fallen far below the economic cost of recycling it. Fast breeder reactors (*qv*), which can use plutonium as fuel, are still some decades away from commercial exploitation. Finally, re processing creates high level waste, and increases the quantities of other categories of nuclear waste by a factor of 100.

Public disquiet in Europe centres primarily on the discharges from Britain's Sellafield reprocessing plant. The discharges are higher than those from all other European nuclear installations added together, and have turned the Irish Sea into the most radioactive sea in the world. Further, the practice of Magnox reprocessing has made the CEGB the owner of the biggest stockpile of separated plutonium in the West,growing by 2.5 tonnes a year.

The cost of reprocessing has been rising steadily, so that it is now some ten times greater than the cost of long-term dry storage, and it now appears unlikely on cost grounds that the CEGB will contract to have the fuel from its new pressurized water reactor stations reprocessed. Such fuel is more highly radioactive than Magnox fuel and correspondingly more difficult to treat, as has been shown by the French experience at Cap la Hague (*qv*).

The spent fuel from fast breeder reactors is still more radioactive – and hence still more difficult to treat. However, its efficient retreatment is essential for success for any fast breeder programme capable of extending the life of the nuclear industry after the exhaustion of world supplies of uranium-235.

In 1977, the US administration withdrew support for civil reprocessing, and the practice, never technically successful in that country, has not been resumed. The main reason was concern that civil plutonium could be diverted for military or terrorist purposes. Other countries – notably Canada, Sweden and most other European countries – have also renounced the practice, clearly demonstrating that a limited nuclear power programme can be operated without reprocessing, and that the only way to limit international nuclear proliferation (*qv*) is to stop separating plutonium from irradiated reactor fuel. France, the UK, West Germany and Japan are now in the minority. It seems likely that reprocessing in these countries, too, will prove to have been a temporary practice – for economic, if not for ecological and health reasons.

Resettlement programmes

Since World War II, millions of people have been uprooted from their homes and relocated, often forcibly, elsewhere. Some have been moved to make way for highways and roads, some for urban renewal projects, and still others for large dams (*qv*) or agricultural and industrial projects.

In the industrialized world, the numbers involved in resettlement schemes are generally small. By contrast, in Third World countries, resettlement programmes can involve thousands of people. The World Bank (*qv*) alone estimates that between 1979 and 1983, it funded projects that caused the forced resettlement of almost 500,000 people.

Few resettlement schemes in the Third World have proved successful. Where tribal peoples are involved, as often happens, the move from their traditional lands is often enough to provoke the total disintegration of their culture.

The emotional wrench of resettlement is frequently compounded by problems over compensation. In many instances, tribal people have been denied compensation altogether because they lack legal title to land. The same applies to landless squatters, some 10,000 of whom were denied compensation for land lost to Brazil's Tucurui Dam Project.

Even where compensation is paid, it is generally inadequate: the 45,000 people uprooted by Sri Lanka's Victoria Dam, part of the five-dam Mahawelli Project, received just £90 for the loss of their homes and land. Similarly, where land is offered in return for land lost, it is generally available only because it is of inferior quality. In all too many instances, relocees have been forced to exchanged their fertile valley land for little more than scrub.

Clashes between rival ethnic groups have dogged many resettlement programmes and inadequate or inappropriate housing is also a common problem. In Ghana, where polygamy is the rule and men frequently have several wives, the houses provided for those displaced by the Volta Dam only had one bedroom. Instead of having their own room, as in a traditional house, the wives were expected to share the same room – with disastrous results. In Indonesia, those resettled under the country's Transmigration (*qv*) Programme have complained that their new houses lack even a place to cook.

Since 1980, the World Bank has laid down guidelines for resettlement in bank-sponsored projects. However, a 1986 review conducted by the bank reveals that those guidelines are frequently broken. Twenty-five projects were surveyed: in nine of them, the authorities had not even identified the areas where the displaced population were to be resettled.

Many resettlement programmes have been carried out with great brutality, with whole villages being bulldozed and inhabitants being moved at gunpoint. The most notorious example in recent times comes from Ethiopia, where 800,000 people were moved between 1982 and 1986 from the north of the country to the south. Ostensibly, the programme (which was backed by the major Western aid agencies) was to relieve famine in the north. However, critics argue that its real aim was to break the back of local seperatist movements with whom the Ethiopian Government have been fighting a long drawn-out civil war.

Few of those who moved did so voluntarily. Indeed, refugees report that government troops burned their crops and poisoned their land with insecticides to force them into feeding centres, where they were then packed into transport planes for the flight to the resettlement camps. An estimated 100,000 people died as a result of the programme, which was eventually halted after the EEC refused to grant further aid. (See also: *Development*③, *Narmada Valley Project*③.)

Rhine

The largest, and dirtiest, of Europe's rivers.

Even before the 1986 Sandoz fire (*qv*) disaster rendered a 100-200km (60-70 miles) stretch of the river lifeless through a major spill of chemicals, the Rhine was little more than an open chemical sewer – polluted in the main by perfectly legal everyday discharges (see illustration).

With experts predicting that the Sandoz fire has set the clean-up of the Rhine back ten years, the river's future looks bleak. (See also: *North Sea*Ⓟ.)

Routine radioactive discharges

All nuclear installations, including reactors and reprocessing plants, discharge radioactivity into the environment. There is increasing concern that the discharges, though within the limits set by government agencies, such as the Department of the Environment and the Ministry of Agriculture, Fisheries and Food in the UK, have caused extra cancers and disease.

In the early days of nuclear power, radioactive releases were high, as were the permitted levels. For instance, the Dresden boiling water reactor, located 50 miles from Chicago, emitted one quarter of a million curies of radioac-

Rhine – licensing the pollution of drinking water

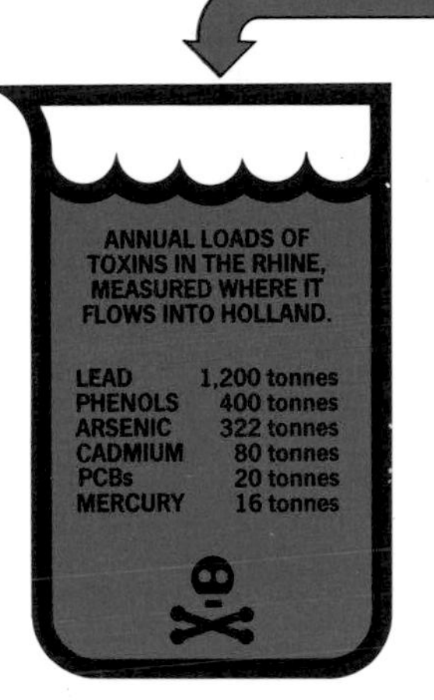

Few experts dispute that the Rhine ecosystem is thoroughly degraded. In 1982 it was reported that of the 11 species of migratory fish that used to frequent the Rhine, nine had disappeared from the river. A 1983 survey found strong evidence of sub-lethal damage to fish – including tumours and abnormal skeletons.

In 1982 the International Water Tribunal (IWT), a group set up by a coalition of Dutch environmentalists, conducted an ecological survey of the entire length of the Rhine. As a result of that survey, the IWT was able to pinpoint a number of discharge pipes that were apparently operating in contravention of their discharge permits. All told, 45 companies were accused by IWT of causing pollution.

Stricter pollution controls have done little to clean up the Rhine. A 1986 study, conducted by the West German Government before the Sandoz fire, revealed that the Rhine is only slightly less polluted than in the 1970s. According to the report, the only pollutant to have declined significantly is DDT. The scientists who compiled the report argue that "if the level of pollution continues unchanged and the cumulative burden thereby continues to rise... irreparable damage to the entire North Sea ecosystem may be inevitable."

The pollution of the river has another disturbing aspect. Some 20 million Europeans still draw their drinking water from it. It is, of course, purified before it is fed into town water supply systems, but even the most advanced technology cannot remove all the pollutants – in fact, in spite of the most strenuous efforts more and more pollutants are seeping into drinking water. In a last-ditch stand, waterworks are using chemical reagents such as chlorine, chlorine dioxide and ozone – but these pose new health problems. Investigations in the USA have revealed increased rates of cancer among people supplied with chemically polluted and chlorinated water.

“20 million Europeans draw their drinking water from the Rhine”

RHINE CATCHMENT AREA

HOLLAND

Principal wastewater discharges into the Rhine (measured as chemical oxygen demand) total about 316,000 tonnes a year, an increase of 58,000 tonnes since 1983. The discharges are licensed by the German authorities: the figures given below are current levels along the river from Holland to Switzerland.

W GERMANY

Krefeld

Dusseldorf

Cologne

Bonn

Koblenz

Factories and towns between the Dutch border and Krefeld are allowed to discharge some 32,000 tonnes of effluent into the Rhine. In this section, more than a third of the discharge comes from the Ruhr Water Authority.

Total effluent legally discharged between Dusseldorf and Mainz is 95,700 tonnes. Dusseldorf contributes more than 32,000 tonnes and Cologne almost 7,000 tonnes, but the greatest single discharge comes from the Bayer works at Burrig – 43,800 tonnes. Included in the total are two works on a tributary, the Wupper, which together contribute 3,000 tonnes.

Mainz

Main

Frankfurt

The Main river, looping south-east of Mainz through Frankfurt, accepts 84,000 tonnes of effluent: the Ashaff and Neckar add 6,000 tonnes.

Mannheim

Neckar

Karlsruhe

FRANCE

Strasbourg

Between Mannheim and the Swiss frontier 98,000 tonnes of effluent flow into the Rhine. More than 40,000 tonnes of that comes from the Waldhof papermill at Mannheim; nearly 33,000 tonnes from BASF at Ludwigshafen; 34,000 tonnes from Hoechst works at Hochst and Griesheim; and 5,300 tonnes from Holzmann at Karlsruhe.

SWITZERLAND

Basel

At Grenzach-Wyhlen, near Basel, the Ciba-Geigy works discharges more than 4,000; other works in the region add more than 6,000 tonnes.

tive gases in 1967, the permissible level, set by the US Atomic Energy Commission, then being 22 million curies. The corrosion of the steel cladding used to encase the fuel was a major factor in the release of gases and equally in rapidly rising levels of radioactive iodine, strontium and caesium, as well as other elements, in the cooling water. The decision was then taken to replace the steel cladding with zircalloy which was much more resistant to corrosion. Meanwhile a study by Professor Ernest Sternglass, a radiological physicist at the University of Pittsburgh Medical School, indicated that infant mortality rates in the counties around the Dresden reactor more than doubled during the period of peak emissions. Equally, a study by the Brement Institute for Biological Safety indicated that leukemia deaths among children shot up more than seven-fold in the decade following the start-up of the Lingen nuclear reactor in West Germany compared to the decade prior to operation. Today, there is increasing evidence of higher cancer rates and premature dalth in the area close to nuclear installations.

Noble gases such as krypton-85, as well as water containing tritium and carbon-14, cannot be easily contained within a reactor or reprocessing plant and tend to escape in their entirety. Tritium, for instance, a weak beta emitter with a half life of 12.35 years, is formed naturally by the action of cosmic radiation on water vapour, some two million curies being formed each year, leading to a global inventory of about 36 million curies. Today, depending on type, a commercial reactor will discharge up to five per cent of its tritium inventory, hence between 1,250 and 2,500 curies per gigawatt per year, in the form of vapour and liquid. Even so, much higher releases occur from ageing reactors, the pressurized water reactor (PWR) at Chooz in France releasing approximately 31,700 and 21,800 curies in 1971 and 1972 respectively, equivalent to as much as 100,000 curies per gigawatt per year. The emissions of tritium from all commercial reactors in use throughout the world amounts to more than one million curies; hence 50 per cent of naturally produced tritium.

Routine discharges are, however, potentially the largest from reprocessing plants. Krypton-85 is discharged in its entirety, approximately 11,000 curies per tonne for PWR fuel, some 30 tonnes per year of spent fuel being generated each year for one gigawatt of electricity. As for tritium, up to 100,000 curies per gigawatt per year are discharged during reprocessing, amounting to 85 per cent of the total tritium in the spent fuel. The two new reprocessing plants to come on line in the 1990s at Cap de la Hague (*qv*) in France, namely the UP 2 800 and UP 3 A, will release up to one million curies of tritium per year, according to current official estimates. Thus the UK and French reprocessing plants alone will release as much tritium as is now generated naturally.

During the years of their operation since World War II the large nuclear complexes at Hanford (*qv*) in Washington State and at the Savannah River Plant in South Carolina have had large discharges of waste containing tritium. The Savannah River Plant, for instance, has five military production reactors, and reprocessing plants. In the mid-1970s, the plant was routinely discharging 300,000 curies per year from the reactors – 75 per cent to the atmosphere – and 400,000 curies per year from the reprocessing facilities – 90 per cent to the atmosphere. The activities then increased, by 1983 the annual tritium release reaching one million curies.

Other radioactive wastes are routinely discharged from nuclear installations, the world's largest discharger being British Nuclear Fuels at the Sellafield (*qv*) site in Cumbria in the UK. Between 1968 and 1979, some 180 kilogrammes of plutonium were discharged into the Irish Sea, the operators of the site, first the UK Atomic Energy Authority, and then British Nuclear Fuels Ltd (BNfl), having had authorizations to discharge up to 6,000 curies of alpha-contaminated waste into the sea each year. During the 1970s BNfl also discharged more than 120,000 curies per year of caesium-137 in addition to other beta-emitting radionuclides. By the early 1980s, BNfl managed to reduce annual discharges of beta activity – including caesium-137 – to just under half the peak value of 250,000 curies discharged into the Irish Sea during the mid-1970s. The aim after a multi-million pound investment programme was to reduce alpha emitters to 20 curies per year and beta to 8,000 curies per year, compared with 67,000 in 1983.

The principle being applies in the UK to bring discharges down, despite large authorizations, is known as ALARP –As Low As Reasonably Practicable – or ALARA – As Low As Reasonably Achievable. Critics, meanwhile, have argued that discharges should be "As Low As Technically Achievable" and, in 1986, the European Parliament's Committee on the Environment, Public Health and Consumer Protection put forward a resolution demanding zero discharges of radioactive wastes into the Irish Sea and that gaseous emissions of radioactive substances, including krypton-85, carbon-14 and tritium be controlled in accordance with available technology. Public exposure to radioactive substances around Sellafield has consistently been ten to 100 times greater compared with French, German and US plants. (See also: *Radiation and health* Ⓛ Ⓝ, *Radio activity* Ⓛ Ⓝ.)

Saccharin
A synthetic sweetener discovered in 1879. A patent for its commercial manufacture was granted in June 1885, and it was then introduced into the market, initially in Europe, but in the USA, too, by the turn of the century.

Saccharin can be found in a wide range of products marketed as "low calorie", "diet" or "diabetic" foods and drinks. These include diet soft drinks, ice creams and fruit preserves, and grape juice concentrates used in making wine. Saccharin is also used as a table-top sweetener.

Saccharin was first suspected of being a health hazard in the 1880s, the French banning its manufacture and importation in 1890. In 1898, the German government restricted its use, expressly banning it from food and drink. In 1912, its use was banned in food and soft drinks made in the USA, but permitted in chewing tobacco, and under prescription.

During World War I and II, sugar shortages prompted many governments to relax their restrictions on saccharin, which they then failed to reimpose when peace was declared.

By the mid-1970s there was clear and consistent evidence from at least four animal studies that saccharin causes bladder cancer. In April 1977, the US Food and Drugs Administration (FDA) proposed a ban on the use of saccharin in all processed food, in soft drinks, and as a table-top sweetener. However, the ban has never been implemented, despite the explicit provisions of the Delaney Amendment (*qv*), which obliges the FDA to ban all known carcinogens from the permitted list of food additives. As a result of intense lobbying by the US food, drink and additives industries, Congress passed a temporary moratorium in the autumn of 1977, preventing the FDA from banning saccharin but requiring all products containing the chemical to carry a health warning. The moratorium has subsequently been renewed at least three times.

The British official attitude has been to accept that saccharin is an animal carcinogen, but to argue that carcinogenicity in humans has yet to be proven. Britain has yet to accept an EEC recommendation that foods containing saccharin should carry a warning of possible dangers, particularly for pregnant women. The use of saccharin in baby foods is, however, restricted.

Saccharin has been banned for all purposes in Canada since 1977, and is banned from food and drinks in France, Greece and Portugal, where it is sold only in tablet form, as a table-top sweetener. (See also: *Additives*Ⓛ, *Artificial sweeteners*Ⓛ *Aspartame*Ⓛ, *Cyclamates*Ⓛ.)

Sahel famine
See: *Cash crops.*

Salinization
When land is too salty to support plant life, it is said to be salinized. Worldwide, an estimated 200,000sq km (73,000sq miles) are thought to be naturally saline –wetland and coastal areas in arid regions being particularly susceptible. But, increasingly, land is being rendered saline through man's activities, the chief culprit being perennial irrigation schemes (see illustration).

All soils contain minute quantities of salt. In areas of regular rainfall, those salts are flushed out of the soil into underlying groundwaters, or carried away to the sea by local streams and rivers. Either way, the salts do not accumulate in the soil. In arid areas of the world, however, where rainfall is minimal and evaporation rates high, soil tends to have a naturally high salt content, with salt making up 12 per cent of the soil profile in badly affected areas. Groundwaters, too, tend to be more saline in arid and semi-arid areas.

Salinization occurs when the delicate salt balance of the soil is upset, thus allowing salts to build up in the root zone of crops or, worse still, to form a saline crust on the surface. Where the water contains sodium or sodium bicarbonate salts – of volcanic origin – the process of salinization goes one step further and the lands become "alkalanized". The soil loses its structure and eventually turns hard as rock, rendering it barren and useless for agriculture. Some 900sq km (350sq miles) are affected by alkalinization in the Soviet Union, whilst Australia tops the world league with 3,000sq km (1,200sq miles) affected.

Perennial irrigation schemes – where land is irrigated year after year without ever being left fallow – are a major cause of salinization in arid areas. Unless the land is well drained, irrigation inevitably causes the water-table to rise, bringing the salt in the soil to the surface. Worldwide, according to the United Nations Conference on Desertification, 220,000sq km (85,000 sq miles) of irrigated land – one-tenth of the world total – are waterlogged.

The evapo-transpiration of irrigation waters also adds directly to the salt load of soils. Studies in the USA suggest that as much as three-quarters of the irrigation water applied each year in America is lost to evapo-transpiration. The result is a fourfold concentration of the salts in the remaining water. Such water often contains more than 2,000ppm of salt. Not surprisingly, the salinity of irrigation water is now itself a factor in causing the salinization of lands.

According to the United Nations Environment Programme (UNEP), between 30 and 80 per cent of the 2,200 million sq km (850 million sq miles) of irrigated land in the world suffer to some degree from salinization, alkalinization and waterlogging. Countries that are badly affected include the following:

Pakistan 100,000sq km (38,600sq miles) out of the 150,000sq km (58,000 sq miles) under irrigation are salinized, waterlogged or both. All told, an estimated 100,000 acres are lost annually to waterlogging and salinization – more than 250 acres a day.

Egypt The problems of salinisation and waterlogging have been described as "grave". A 1976 survey, undertaken by USAID, found that more than 17,000sq km (6,560sq miles) were undergoing slight to severe salinization. According the FAO, 90 per cent of Egypt's farmland suffers from waterlogging,with 35 per cent ex-

Salinization – the negative face of irrigation

Too salty to support plant life: that is the state of more than 660 million sq km of irrigated land today. Poorly drained irrigation schemes are to blame, with as much land now being taken out of production because of salinization and waterlogging as is being brought into production through irrigation.

Most soils contain some salts which are usually flushed out by rainwater. In arid areas low rainfall and high evaporation combine to produce a naturally saline soil; it is in these areas that perennial irrigation schemes may cause the greatest damage.

Unless they are carefully controlled, irrigation schemes may easily concentrate the salt in the environment, and then kill the crops that they were initially set up to sustain.

From fresh water to salt crust

Many irrigation schemes draw their water downstream from dams. Often the salinity of the fresh water has already been increased through evaporation from the dam's reservoir; the water leaving the Aswan Dam is ten per cent more saline than the water entering Lake Nasser.

Evaporation from irrigation channels further increases the salinity of the water, which then often seeps out into the soil, raising the water table. In the Indus Valley the groundwater rose by 30m, taking 14,000sq km out of production through saline waterlogging.

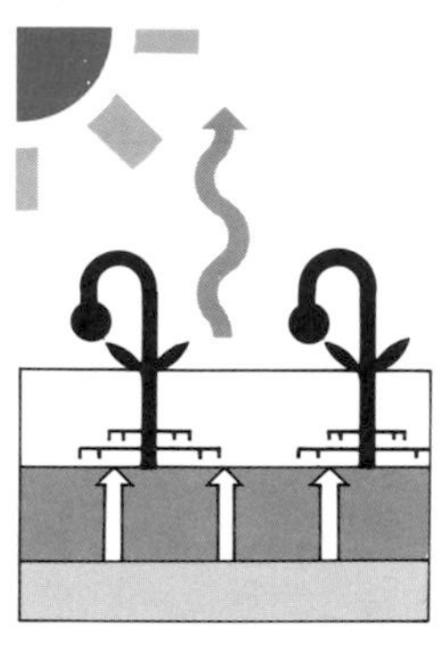

As the groundwater rises, dissolving out salts from the soil, it becomes even more salty. Upon reaching the roots it impairs their functioning, steadily reducing crop yields. In Shaanxi Province, China, wheat yields fell by 80 per cent and cotton yields by a half.

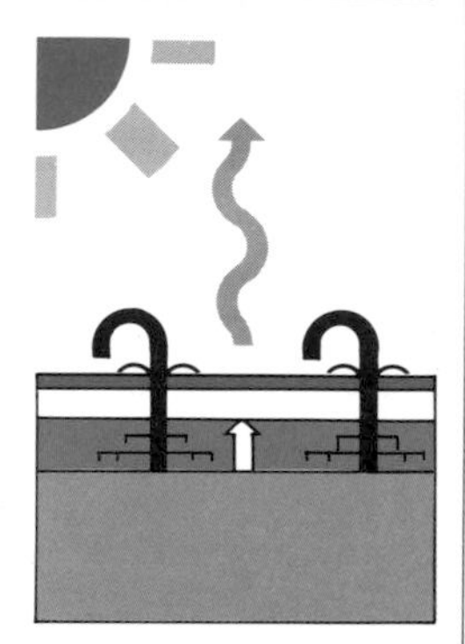

As the groundwater continues to rise through the root zone, it progressively kills the plants. Once close to the surface the water is drawn rapidly up, through capillary action, by the hot air and evaporates, leaving the salts behind on the surface as a hard lifeless crust.

The salty problem

Percentage of irrigated land affected to some degree

● Egypt	90%
● India	15-25%
● Iraq	50%
● Pakistan	68%
● Peru	38%
● USA	25-30%

Prevention...

Often local prevention such as lining irrigation channels, installing drainage to remove the irrigation water before the salts accumulate and digging wells to lower the water table, simply "exports" the salt downstream. The only real solution is desalinization.

...but it costs

Preventing the problem is an expensive business. The US built a $300 million desalinization plant at Yuma in the Arizona desert, near to the Mexican border, to remove the salt from the Colorado River before it crossed the frontier. While, according to the FAO, it costs between $200 and $1,000 to drain just 100sq m of irrigated land.

periencing salinity problems.

Iraq More than 50 per cent of the 36,000sq km (13,900sq miles) under irrigation suffers from salinization and waterlogging. Vast areas in South Iraq are reported by the environmental writer Erik Eckholm to "glisten like fields of freshly fallen snow".

India The amount of land degraded by salinization and waterlogging has variously been estimated at between 60,000 and 100,000sq km (23,000 to 37,000sq miles). If the higher figure is correct, then a quarter of the 430,000sq km (166,000sq miles) under irrigation is affected.

Argentina 20,000sq km (7,700sq miles) of irrigated land are affected by waterlogging, salinization or alkanisation.

Peru According to the Global 2000 Report, 3,000 out of the 8,000sq km (3,100sq miles) of coastal lands under irrigation are salinized, waterlogged or both.

USA According to the US Salinity Laboratory, 25 to 35 per cent of all the irrigated land in the country suffers from salinity problems – and the problem is getting worse. If no remedial measures are taken, the highly productive San Joaquin Valley could lose more than a million acres of farmland in the next hundred years.

It is rare to find perennial irrigation schemes which have not experienced some degree of salinization or waterlogging.

Certain measures can be taken to minimise salinization at the local level. However, such measures are extremely expensive. By and large, their aim is to ensure that saline water is flushed out of the soil before it has time to accumulate. But the flushed salts must go somewhere. Generally, they end up in the nearest river, thus increasing its salt content. For farmers downstream of irrigation schemes, the problem is obvious. They are forced to irrigate their fields with increasingly saline water. In many arid areas, the problem is now acute. In South Australia, for example, the salinity of the Murray River (*qv*) is twice the average for Australian rivers. Other rivers affected include the Euphrates, the Indus, the Rio Grande and the Colorado.

Sandoz fire

In the early hours of November 1 1986, a fire broke out in a warehouse at the Sandoz chemical factory near Basel, Switzerland. As firemen played their hoses on the warehouse, chemicals released by exploding drums were washed into the River Rhine. More than 30 tonnes of pesticides, fungicides and chemical dyes entered the river rendering it lifeless for 100-200km (60-120 miles).

Among the 30 different agricultural chemicals stored in the warehouse were 25 tonnes of the insecticide parathion (*qv*), 12 tonnes of the mercury-based fungicide Tillex, 10 tonnes of the insecticide fenithrothion (*qv*), and 323 tonnes of the insecticide disulfoton (judged to be twice as toxic to rats as potassium cyanide).

Experts are agreed that the accident could have been even worse. Had water sprayed onto an adjacent building come into contact with the sodium stored there – due to a ruptured drum, for example – the result would have been an almighty explosion that could have blown apart nearby storage tanks containing the nerve gas phosgene. If that had happened, then, in the words of one local MP, the result would have been a chemical catastrophe that "would have put Bhopal in the shade".

No sooner had the blaze at Sandoz been put out than scientists monitoring the progress of the slick as it passed through West Germany announced that they had detected high concentrations of the herbicide atrazine – a chemical that was not listed as stored in the Sandoz warehouse. Only later did the chemical giant Ciba-Geigy admit that it, too, had suffered a major spill – purportedly the day before the Sandoz accident – which, according to the company, resulted in the release of 100 gallons of atrazine into the Rhine. West German officials estimated the release to have been 15 times greater than the company claimed.

In the weeks following the accident, company after company was found to have exceeded its discharge permits. Within a month of the Sandoz fire, 12 major pollution incidents had been reported. First there was Ciba-Geigy; then a 1,100kg (2,425lb) of the herbicide 2,4-D (*qv*) were spilled from a BASF plant in Ludwigshafen; then a Hoechst plant on the Main, a tributary of the Rhine, owned up to a major leak of the solvent chlorobenzene; then Lonza admitted that 4,500l (1,200 US gallons) of PVC-contaminated liquor had escaped from its factory near Waldshut. Indeed, accidents are apparently so commonplace on the Rhine that the Basel authorities dismissed the Ciba-Geigy incident as "a bagatelle", adding that "the emission of substances used for agrochemical production into the Rhine happens frequently". According to Greenpeace, the Ciba-Geigy plant had been discharging high levels of atrazine into the Rhine for more than a year.

In just two hours, the Sandoz fire released more pollutants into the Rhine than the river normally receives within a year. The effect of the spill on aquatic life was (and will remain) cataclysmic. Aquatic chemists and biologists fear that it will be at least ten years before the river recovers – and that some stretches may never do so. It is estimated that some 500,000 fish were killed outright by the spill: many of those that survived are likely to be heavily contaminated by mercury, 200kg (440lb) of which were released by the accident. Even normally hardy species, such as eels, were killed in their hundreds of thousands, *Newsweek* reporting that "the few found alive had bulging eyes and ugly sores".

Nor were fish and eels the only species to be affected: for a long stretch downstream of Basel, everything from worms to snails, shrimps, water fleas, larvae and plankton were decimated, thus drastically reducing the food supply available to those few organisms that survived. The impact of the disaster on bird life is likely to be severe. It is feared that the spill will wreck havoc with wildlife in the Wadden Sea, itself already badly polluted by PCBs and other chemicals.

For those living in the Rhine watershed, the accident brought more than polluted fish. The chief threat lay in polluted drinking water – a threat that has not gone away with the passing of the toxic slick. Though water supplies have now been restored to those villages and towns cut off for public health reasons during the disaster, the danger of pollution persists. Scientists fear that the chemicals may have permeated through the river bed of the Rhine into underlying groundwaters. If that is so, then water supplies in the Rhine Basin may be polluted for years to come. (See also: *Groundwater*Ⓟ, *North Sea*Ⓟ, *Rhine*Ⓟ, *Water quality*Ⓟ.)

Schistosomiasis

A disease caused by parasitic flatworms, known as "schistosomes". Three common species infect man: *S. Haematobium*, *S. Mansoni* and *S. Japoni*.

In 1947, an estimated 114 million people suffered from schistosomiasis, the symptoms of which are fever, cough, muscle pains and skin irritation. Today, 200 million people are affected – the equivalent of the entire population of the USA.

The larvae of the schistosomes develop within the bodies of freshwater snails. When people swim or wade in water contaminated by infected snails, the larvae bore through their skin and enter their bloodstream. From there, they move to the liver, where they mature in a few weeks and mate. The resulting eggs leave the human body via urine or faeces.

The dramatic spread of schistosomiasis over the past 35 years is largely the result of large-scale water development projects. Such projects provide ideal habitats for both fresh water snails and the schistosome parasite.

The connection between schistosomiasis and water projects is well established. Not only is the snail vector's habitat greatly extended by water development projects but the conditions are also created for much longer breeding periods. In Kenya, schistosomiasis now affects almost 100 per cent of those children living in irrigated areas near Lake Victoria. In the Sudan, the massive Gezira irrigation scheme had a general infection rate of 60-70 per cent in 1979, with the rate amongst school-children reaching over 90 per cent. All in all, 1,400,000 people were affected. After the building of the Aswan High Dam (*qv*) in Egypt, the infection rate rose to 100 per cent in some communities.

Few doubt that the disease is on the increase. Dr. Letitia Obeng of the United Nations Environment Programme warns that the current incidence of schistosomiasis is "only the thin end of the wedge." (See also: *Dams* Ⓛ ③, *Waterborne diseases* Ⓛ ③.)

Scrubber

A piece of equipment designed to remove toxic gases from chimney emissions. The term scrubbing is mainly used to describe processes that use a liquid to absorb the gases though there are other techniques, involving the use of materials that adsorb gases onto solid particles, which are sometimes referred to as dry scrubbing.

Battersea and Bankside power stations on London's River Thames were among the first to scrub their stack gases to remove sulphur dioxide. The resulting gypsum slurry was easily disposed of in the Thames, which was highly polluted at the time. This option is no longer available and such wastes are now often landfilled, which may cause pollution problems in the long term.

Another problem with wet scrubbing is that in some weather conditions wet, smelly mists are created. This problem can be dealt with by re-heating the stack gases, but this is expensive in terms of energy.

Scrubbing is an important technique for controlling air pollution by gases. Other methods, such as electrostatic precipitators and bag filters, can remove particulate matter such as grit and smoke, but will not affect gaseous emissions. (See also: *Acid rain* Ⓟ, *Coal* Ⓔ Ⓟ, *Incineration* Ⓟ, *Landfill* Ⓟ.)

Sedimentation

According to one estimate, as much as 24,000 million tonnes of sediment are carried each year to the sea by the world's rivers. Of that total, only 9,300 million tonnes are thought to be of natural origin: the rest – some 14,700 million tonnes – results from man's activities. In China the three major rivers of Sichuan carry an annual silt load of 250 million tonnes – enough to cover an area of 1,600sq km (620sq miles) with a good 5cm (2in) of topsoil.

Some of the material carried downstream is deposited on flood plains, some on the beds of rivers and streams, and still more ends up on the ocean floor. Where rivers have been dammed, sedimentation can cause major problems. Prevented from travelling freely downstream, the silt simply accumulates in the reservoirs behind the dams, thus reducing their capacity and, hence, their effective lives.

In temperate areas, the sedimentation of reservoirs is usually a slow process and the average annual loss of capacity not serious. In the tropics, however, the story is very different. The problem lies in the fragile nature of tropical soils – and in the sheer scale of deforestation in tropical areas. Where the catchment area of a dam is forested, the rate of soil erosion (*qv*) is very low: even steep slopes are protected.

Where the forest has been destroyed, however, the rate of soil erosion is dramatic: India alone loses an estimated 12,000 million tonnes of soil each year to erosion. In the Himalayas, the foothills of Nepal, once covered in rhododendron, oak and pine, have been stripped of 40 per cent of their forest cover since 1952. Each acre (4,000sq m) of the now denuded hills loses an estimated 10 tonnes of soil a year. Such is the volume of silt carried downstream by the Ganges and the Brahmaputra that whole new islands have been formed in the Bay of Bengal.

Predictably, dams (*qv*) in the tropics have experienced high rates of sedimentation:

In India, the expected siltation rate at the Nizamsagar Dam in Andhra Pradesh was 530 acre-feet a year. The actual rate was closer to 8,700 acre-feet a year.

In Haiti, the Peligre Dam on the Artibonite River was built to last 50 years. Excessive sedimentation threatens the dam with decommissioning after just 30 years in service.

In China, the Sanmexia Dam, commissioned in 1964, had to be decommissioned in 1964, due to premature sedimentation. Worse still, the reservoir of the Laoying Dam actually silted up before the dam was fully operational.

Once silted up, a dam is next to useless. Even when the last waters of the decommissioned reservoir have eventually drained away, the land beneath will not be suitable for agriculture. Only a small strip of land close to the dam, where the coarser – and therefore less compacted – particles of silt accumulate, will be cultivable.

Excessive sedimentation can also cause floods. Ironically, the problem is often most acute in those areas where embankments have been built to control them. Under normal circumstances, the silt carried by rivers is deposited on the river's flood plains. But where embankments have been built, this is no longer possible. The silt therefore accumulates on the river bed, raising its height until –eventually – it is higher than the surrounding land. Where China's massive Yellow River crosses the Yellow Plains, the bed of the river is now five metres (16ft) above ground level – and, in some places, ten metres higher. Raising the height of the embankments is only a partial solution: the higher the embankments, the greater the volume of water carried through them, and, consequently, the more severe the damage from flooding should a breech occur. (See also: *Deforestation* ③, *Drought flood cycle* ③, *Flood control* ③, *Hydro-electricity* Ⓔ, *Tehri Dam* ③, *Three Gorges Dam* ③.)

Seed-pesticide connection

Recently, the world's major chemical companies have started buying up seed companies on a major scale. Imperial Chemical Industries ICI now owns Societe Europeene de Semences, one of Europe's leading seed-breeding companies, the Garst Seed Company, the biggest hybrid maize breeder in the USA, and Sinclair McGill, one of Britain's leading seed-marketing and cereal-breeding companies.

Meanwhile, in August 1987, Unilever outbid more than 40 chemical companies to buy Britain's newly privatized Plant Breeding Institute and National Seed Development Corporation.

Worldwide, a total of $10,000 million has been spent by chemical companies making seed company acquisitions or participating in research with them. Indeed, in the last few years the old established seed companies have been systematically bought up by chemical companies (see illustration). One reason for this is that the seeds business has become much more commercially attractive as a result of new legislation that permits the patenting of seeds. Also the industry has now hit on the idea of making high yielding varieties (HYVS) infertile, which means that farmers are forced to buy new seeds every year, thereby massively increasing sales. HYVS are also of interest to chemical companies because they are specifically designed to respond to fertilizer applications, which correspondingly increases fertilizer sales. They also tend to be particularly vulnerable to pests, thereby increasing pesticide sales.

Pesticides and seeds are complementary businesses, and chemical companies have learned to cash in on the pesticide/seed market in a number of ways. The first is by coating seeds with chemical protectants, forcing farmers and horticulturalists to buy the chemicals and seeds as a package.

The second method is by designing advertising packages that provide particular combinations of pesticides and seeds.

The third is by gearing research towards the development of seed pesticide combinations that are designed to be used together.

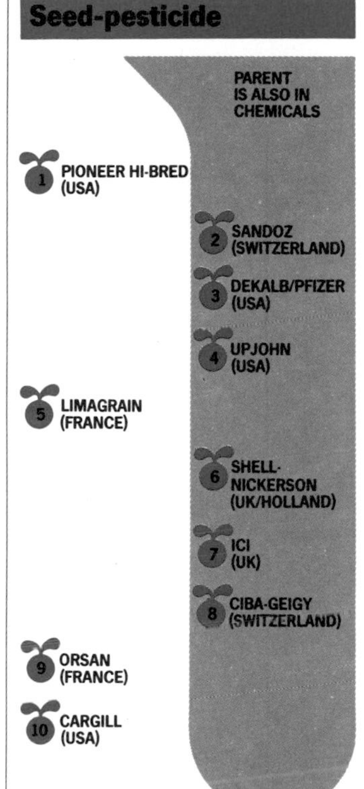

Listing the world's top ten seed companies demonstrates that six of them have parent companies that are chemical producers. The link is designed to outflank the genetic engineers, who may produce plants that don't need pesticides. The nationality of the parent is given in parentheses.

The seed pellet business is being increasingly refined. Growth regulators have been added to the coating, and fluorescent materials have also been included, allowing farmers to plant at night. More and more chemicals are added to the coating so that, in the words of Patrick Mooney, author of *Seeds of the Earth*, the seed is "wrapped in a swarm of special chemicals protecting it from everything from rot to weeds."

Ciba-Geigy includes three chemicals in the coating of its sorghum grains. Two are to protect the seed from encroaching grasses, but the third is there to protect it from Ciba-Geigy's own herbicide "dual", which is normally toxic to grain sorghum. As a result, Ciba-Geigy can now sell much more of that herbicide.

A 1982 EEC report notes that breeders had three goals: to maximize yields, to make the crops easy to harvest by machine and easy to process, and to make them disease-resistant. It then says that breeders were willing to "sacrifice disease control if the pesticides effective against the disease were available." As Patrick Mooney notes "if the breeder is also the pesticide manufacturer, that sacrifice hardly hurts at all."

One consequence of these trends is that the new varieties are designed less and less for their nutritional qualities. A survey in the USA in the early 1970s concluded that only one or two breeders even possessed the facilities for determining the nutritional quality of their new varieties.

A further consequence of these trends is that small farmers must spend more and more money on chemicals. This puts many of them into debt and eventually forces them out of business. In the Third World many become landless labourers and slum dwellers. (See also: *Green Revolution*, *HYVs*, *Patented seeds*, *Pesticides*.)

Sellafield

Major nuclear complex, including Britain's principal reprocessing (*qv*) plant.

The complex is located on the coast of Cumbria, 15 miles (24km) south of Whitehaven. Construction began in September 1947, and the first plutonium-production pile, or reactor, began operating in October 1950. A second pile started up eight months later, and in January 1952 the first reprocessing plant, B204, was commissioned. In March, the first metallic plutonium was made at Windscale, and in October the first British atomic bomb was tested at Monte Bello.

The original piles were closed following the Windscale fire (*qv*) in 1957, but by that time their successors, the four Calder Hall reactors, were supplying the increased demand for plutonium. These, with four more reactors at Chapelcross in Dumfriesshire, were the prototypes of the Magnox reactors (*qv*) that were used for the first British civil nuclear power programme.

Nine twin-reactor Magnox power stations had been built by 1972; and a new reprocessing plant went operational in 1964 to cope with the irradiated fuel from these and the eight military reactors. Management of Windscale was transferred from the United Kingdom Atomic Energy Authority (AEA) to British Nuclear Fuels Ltd (BNFL) in 1971, and in 1976 the company applied for permission to build a further reprocessing plant – the Thermal Oxide Reprocessing Plant or THORP – to handle uranium oxide fuel from the UK's advanced gas-cooled reactors (AGR) (*qv*) and foreign pressurized water reactors (PWR) (*qv*). After the Windscale Inquiry of 1977, consent was granted, but construction of THORP did not begin until 1984.

Since 1950, more than 270 reportable accidents have occurred at Sellafield; incidents involving the contamination of workers have increased as the Magnox reprocessing plant has aged. To date, ten workers have been paid compensation by BNFL (though without the company formally admitting liability), and a further 70 claims are outstanding. Abnormal discharges to sea have also occurred regularly; BNFL was found guilty on four counts, and fined £10,000 ($16,000), in connection with one such discharge in November 1983. The spill was discovered when Greenpeace divers tried to plug the main discharge pipe. Another attempt to plug the pipe was made by Greenpeace in 1987. A quarter to half a tonne of plutonium has been discharged into the Irish sea, and there is a higher-than-normal incidence of childhood cancer in the area surrounding Sellafield.

New, more stringent, discharge authorizations were imposed on the Sellafield plant in January 1985, with annual limits of 600 curies of alpha emitters, and 200,000 curies of beta. But the authorizations for the French reprocessing plant at Cap de la Hague (*qv*), comparable in age and throughput, have been 90 curies of alpha and 45,000 of beta emitters since 1980. The actual discharge of alpha emitters from Sellafield between 1967 and 1981 totalled 31,583 curies, and from la Hague 141.5 curies. Doses to workers per unit of Magnox electricity generated are more than double those at la Hague, despite the lower discharges to sea.

In December 1983, a £150 million ($240 million) programme was announced, to reduce discharges to levels comparable with those from Cap de la Hague. This programme was largely complete by 1986, but Magnox reprocessing will continue to provide the bulk of the discharges and doses at Sellafield until those reactors cease operation and the last of their fuel is reprocessed in 2000 or soon afterwards.

Sellafield is also the site of a prototype 30MW AGR which operated for 15 years and which is now being decommissioned as an exercise in "proving" the economic feasibility of dismantling old reactor hulks. (See also: *Decommissioning* Ⓝ, *Leukemia clusters* Ⓝ, *Nuclear accidents* Ⓝ, *Routine discharges*.)

Seveso
Site of a major chemical accident in 1976. The accident occurred when a vat used in the manufacture of the bacteriocide hexachlorophene exploded at a chemical plant owned by Givaudan, a subsidiary of the Swiss pharmaceuticals giant Hoffman-La Roche.

A cloud of dioxin (*qv*) was released in the explosion, contaminating 18sq km (7sq miles) of surrounding countryside and necessitating the evacuation of more than 900 people. In the year following the accident, the rate of birth defects increased by more than 40 per cent.

The soil in the most heavily contaminated area was so polluted that the top 20cm (8in) had to be removed and buried in a massive plastic-lined pit. The clean-up has proved only a partial success, however. Recent soil samples from the "decontaminated" land have proved "dioxin-positive".

The factory itself was dismantled after the accident, some two tonnes of dioxin-contaminated waste being removed for disposal.

In 1984, a major scandal broke when 41 barrels of dioxin waste went missing in France. The barrels were eventually found in an abandoned abattoir at Anguilcourt-Le-Sart in Northern France. They were subsequently incinerated at a Hoffman-La Roche plant in Basel.

Shifting cultivators
Farmers who use slash and burn techniques to clear forested land for cultivation and who regularly move to new lands as soon as the fertility of their plots begins to decline. (See also: *Deforestation* ③, *Encroaching cultivators* Ⓐ ③, *Swidden agriculture* Ⓐ ③.)

Sievert
The unit of radiation dose absorbed by tissue; being the dose in grays (*qv*) multiplied by a quality factor to account for differences in damage to tissue caused by different kinds of radiation, such as alpha and gamma radiation (*qv*). The sievert (Sv) is replacing the rem: 1 Sv = 100 rem.

Sizewell
Located on England's Suffolk coast immediately to the south of a world-famous water bird reserve at Minsmere, Sizewell Beach is an officially-designated area of outstanding natural beauty. Nevertheless, a Magnox (*qv*) nuclear power station, Sizewell A, began operating there early in 1966, the sixth in the Central Electricity Generating Board's (CEGB) programme of eight such stations. The CEGB owns about 2.5sq km (600 acres) of Sizewell Beach.

The British government, advised by the UK Atomic Energy Authority, has traditionally dictated its choice of reactor type to the CEGB. In the early 1960s, the CEGB wanted no nuclear stations at all, considering them to be uneconomic, but the Magnox programme was imposed. The government's choice to follow the Magnox was the AGR, which had been developed by the UKAEA, but by 1975 the CEGB had experienced severe difficulties in building these reactors. It decided that the best course was to abandon the AGR and buy the American pressurized water reactor (PWR) (*qv*) instead. But such was the prestige invested in the AGR project that the CEGB was not allowed to apply for permission to build its first PWR until 1980.

The public inquiry into whether the CEGB should be granted planning permission for a PWR at Sizewell B sat for 340 days, the longest public inquiry so far held in the UK, and took detailed evidence on the need for the proposed station, its economics, its design and operational safety, and local environmental issues.

The Inspector, Sir Frank Layfield, recommended that planning permission should be granted. After a brief parliamentary debate, permission was granted in March 1987, despite the fact that in the meantime coal prices had fallen dramatically and the Chernobyl (*qv*) accident had taken place.

Critics argue that the inquiry was nothing but a formality, rubber stamping a foregone conclusion. They point out that construction work on the PWR was already under way – with more than £200 million ($320 million) already spent – even before the inquiry opened. It has also emerged that much of the inspector's final report was written by officials from the Department of Energy, one of the parties promoting the Sizewell B reactor. (See also: *Nuclear power*Ⓝ.)

Slash and burn agriculture
See: *Swidden agriculture.*

Social forestry
Many of the reforestion (*qv*) programmes being promoted by the World Bank (*qv*) and other development agencies are intended to encourage "social forestry" – described by the Forest Department of Gujarat, India, as "the creation of forests for the benefit of the community through the active involvement and participation of the community." The aims are both to increase village self-sufficiency in firewood and other forest products, thus reducing the pressure on existing forests, and to reduce rural migration and unemployment.

But contrary to the aims of the programme, few social forestry projects are owned communally: instead most are in the hands of relatively large landowners – a recent study in Karnataka, India, showing that 80 per cent of the plantations are on private land.

The same study revealed that 92 per cent of the trees planted were eucalyptus (*qv*), a tree which in dry areas has a disastrous impact on both the soil and water supplies. In addition, it is of little use to villagers, since it provides no fodder for animals, no green leaves for manure, and wood that is unsuitable for many traditional uses.

The choice of eucalyptus reflects the commercial nature of most social forestry projects. In India, where social forestry is most widely practised, little of the wood grown under the programme is even available to villagers. In one district of Karnataka, 80 per cent of the timber grown went to local pulp mills or to rayon manufacturers.

To make matters worse, much of the land used for social forestry was previously used to grow food crops for local consumption. By 1988, an estimated 2,200sq km (850sq miles) of farmland in Karnataka alone will have been lost to social forestry projects,greatly exacerbating the problem of malnutrition.

Rural unemployment has also increased: unlike farming, growing trees requires little labour once the inititial planting has been undertaken. Many of the displaced workers have no option but to strip trees for sale to the cities as fuelwood (*qv*), thus increasing the pressure on remaining forests.

Despite the failure of social forestry to achieve its stated aims, the programme is still heavily funded by the World Bank. See also: *Deforestation*Ⓒ③, *Reforestation*Ⓒ③, *Tropical forests*Ⓒ③.)

Soft energy paths
A term coined by the US energy consultant Amory B. Lovins and used as the title of his influential book published in 1977. It refers to a complex of industrial, commercial and political strategies designed to reduce the dependence of industrial societies on conventional sources of energy. These call for an intensive energy conservation (*qv*) programme and for a shift from nuclear power and fossil fuel combustion to renewable sources of energy.

Lovins' argument is that conventional forms of energy production provide electricity but are uneconomic in providing heat for cooking and space-heating, for which much of the energy generated in our society is required. Such "end use services", Lovins argues, "are best provided by a sophisticated energy conservation programme, which would include draught-proofing, thermal insulation, window shutters and shades and coatings, greenhouses, heat-exchangers and the like."

It is only when all energy conservation measures have been taken that a country should even consider generating more power. In such an event, the cheapest and most appropriate methods of generation would be combined-heat-and-power (*qv*) stations, low-temperature heat engines run by industrial waste heat or by solar ponds, modern wind machines and small-scale hydro in good sites, and solar cells. (See also: *Alternative technology*Ⓔ.)

Solar collector
A device for the extraction of useful heat from the warmth of sunshine, to heat water or provide space heating.

In arid climates solar collectors are used to power stills for evaporating saline water, the vapour being condensed and collected as fresh water. Focusing mirrors have been used in low latitudes to intensify the heat falling on the receiving surface, allowing solar heat to be used directly to boil water or cook food and, on a much larger scale, to produce steam for power generation.

At the present time, however, solar collectors make a limited contribution to our energy requirements, portly because there has been little government support for research. Unquestionably, however, they must have an important role to play if a soft energy path (*qv*) is adopted.

Stabilizers
Compounds added to food products to stabilize emulsions that might otherwise separate out into a water-based and an oil-based segment. They are, therefore, invariably used in combination with emulsifiers (*qv*), to which they are often chemically similar. Frequently they work by thickening one of the segments and so making the mixture less likely to separate. (See also: *Additives* Ⓛ.)

Stockholm Conference
Conference held on June 15-16 1972 in Stockholm and attended by representatives of 113 nations. It is considered to be a landmark in the development of global environmental concern. It resulted in the setting up of the United Nations Environment Programme, but the programme's achievements have been limited by its inability to make pronouncements, or take actions, that are likely to offend the governments that fund it.

Sub-lethal pollution
Even at levels too low to kill outright, many toxic pollutants cause severe biological damage. Indeed, in some cases, exposure to "sub-lethal" levels of pollution over a long period can be as damaging, if not more damaging, than exposure to high levels over a brief period.

The clinical effects of sub-lethal pollution are many and varied. For example, very low levels of organophosphate (*qv*) pesticides (*qv*) have been shown to inhibit enzyme activity in humans and to impair hormone functions. Exposure to low levels of oil can cause serious chromosal damage to fish eggs, while DDT (*qv*) disrupts the ability of some fish species to return to their home streams, thus breaking a critical phase in their reproductive cycle. Low levels of DDT also cause fish to become hyper-active.

All organisms are weakened by sub-lethal levels of pollution and are therefore more vulnerable to disease. Sub-lethal pollutants have undoubtedly played a key role in the spread of tree diseases (*qv*).

The sub-lethal effects of pollution have generally been played down or ignored when setting safe exposure standards. By the time the damage becomes obvious, remedial action is often too late.

Subsistence agriculture
See: *Traditional agriculture.*

Superfund
The reckless disposal of hazardous wastes has caused the contamination of both land and water at many sites throughout the industrialized world. As a result, massive sums of money are now required to clean-up old dumps.

In the USA, a "superfund" – bankrolled largely by industry – has been set up to pay for the clean-up. Known formally as the Comprehensive Environmental Response, Compensation and Liability Act (CERCLA), the superfund legislation was passed in 1980 – the last bill that President Carter signed.

As originally conceived, the superfund was to raise $4,200 million – but this sum was reduced to $1,600 million under pressure from industry, which is required to provide 88 per cent of the cash. Clauses in the Act that would have enabled private citizens to sue the fund for any damage caused by improper dumping were also scrapped before CERCLA passed through Congress.

In 1986 the original $1,600 million was raised to $8,500, but even that seems nowhere near sufficient to clean up all of America's abandoned "toxic time bombs". At present, 951 old dumps are listed as superfund sites. A 1983 report by the Office of Technology Assessment (OTA), however, estimated that "10,000 sites (or more) may require clean up by Superfund". OTA went on to comment: "With Superfund's existing resources, it is not technically or economically possible to permanently clean up even 2,000 sites in less than several decades." Some of the polluting wastes removed from superfund sites have been found to be contaminating the supposedly "safe" disposal sites where they have now been deposited. (See also: essay on *Water Fit to Drink?*, *Hazardous wastes*Ⓟ.)

Sustainable development
Public disquiet over the ruinous social and ecological consequences of current development (*qv*) policies throughout the Third World has led the development industry, on several occasions, to announce new strategies which, it is claimed, will minimize the destructiveness of the development process. Unfortunately, few of these strategies have ever come to anything.

The current buzz word is "sustainable development", which has been talked about since the early 1970s.

The World Conservation Strategy argues that, for development to be sustainable, it must not interfere with the functioning of ecological processes and life support systems. This means, above all, that food cropland must be used for crops rather than for cattle ranching, that crops must be managed in an ecologically sound way, that watershed forests must be protected, and that genetic diversity must be preserved.

Though the rhetoric has changed, however, development policies have not. Large dams (*qv*) continue to be built, vast plantations and livestock-rearing projects continue to be set up, colonization projects continue to occur in virgin forests – all undertakings which are known to be highly destructive and totally unsustainable in tropical conditions.

Indeed, it is by no means certain that sustainable development is even possible. To begin with, in the competitive world in which we live, only those development projects are undertaken that maximize political advantages and short-term economic returns; and, unfortunately, small, ecologically and socially benign – and hence sustainable – projects rarely satisfy such conditions. In addition, economic development (*qv*) means increased consumption and production. Inevitably, the impact of human activities on ecosystems that are already highly degraded must also increase. Sustainability and development would thus appear contraditions in terms. (See also: *Cash crops* Ⓐ③, *Green Revolution*Ⓐ③, *Tropical agriculture*Ⓐ③.)

Swidden agriculture

Swidden or "shifting" agriculture is an ecologically sound system of farming used in the tropics (see illustration).

Development agencies such as the UN Food and Agriculture Organization and the World Bank charge that swidden agriculture is both "unproductive" and "environmentally destructive". In particular, it is singled out as a major cause of deforestation (*qv*). The charge is unfounded.

Insofar as shifting cultivators are a cause of deforestation, the threat lies not with traditional swidden farmers but with landless peasant colonists – many themselves "environmental refugees" (*qv*) – who are totally unversed in the skills of forest farming.

Many experts in tropical agriculture are now convinced that swiddening represents the only sustainable method of farming a rainforest. For, despite their lush canopy and profusion of plants, tropical rainforests grow on some of the most fragile and infertile soils in the world.

Traditional swidden systems ensure that the soils beneath cleared areas are kept protected at all times. Every plot grows an extraordinary wide variety of edible plants, each maturing at different times and growing to different heights. The effect is to create a multi-layered garden which, in many respects, mimics the natural forest.

Anthropologists report that traditional swidden cultivators are acutely aware of the need to maintain fallow periods, thus allowing the cleared forest to regenerate. Among the Tsembaga of New Guinea, for example, it is estimated that 90 per cent of clearings eventually revert to mature rainforest. The Tsembaga revere sapling trees as "Duk Mi" – the mother of gardens – and assiduous care is taken not to uproot them while weeding.

Anthropologists also attest to the productivity of swidden systems. Indeed, almost all the food consumed and marketed in Tropical Africa is produced by shifting cultivators using swidden techniques. (See also: *Deforestation.*)

Swidden agriculture – farming in the forest

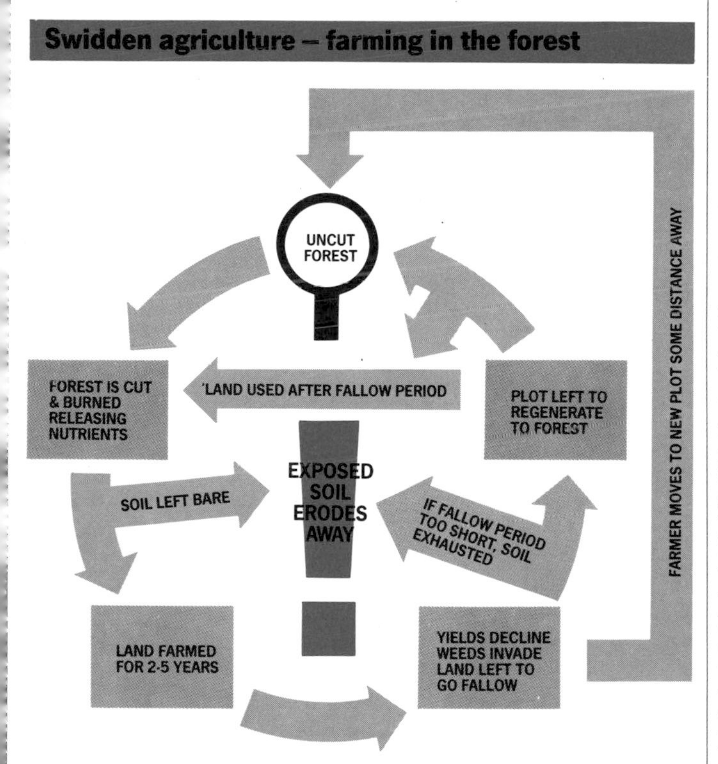

Swidden or "shifting" agriculture is widely practised in tropical forests and through the savannahs of North and East Africa. Small areas of bush or forest are cleared and the smaller trees burned. This unlocks the nutrients in the vegetation and provides the soil with fertilizer in a form that can readily be taken up by plants.

After two or three years of cultivation, the soil is almost exhausted and yields begin to decline. The farmer then abandons the plot and opens a new clearing some distance away. The original plot is left fallow – usually for 15 years or more – so that the jungle can regenerate. In the savannah the cycle is longer, with plots being left fallow for 20 to 30 years.

Traditional swidden systems ensure that the soils of cleared areas are protected at all times. Each plot grows an extraordinarily wide variety of edible plants, which mature at different times and grow to different heights. The effect is to create a multi-layered garden.

Symbiosis

See: *Mutualism*

Synergy

The process whereby two or more chemicals acting in combination produce an effect that is far greater than that achieved by the parent chemicals separately – or which is entirely new and unexpected.

Thus, the solubility in water of the carcinogenic pesticide DDT (*qv*) is increased 10,000 times by the presence of oil, greatly enhancing its dangers to aquatic organisms. Similarly, a study by Professor Irving Selikoff of Mount Sinai Hospital, New York, has shown that whilst asbestos (*qv*) workers who smoke have an eight times higher chance of contracting lung cancer than other smokers, by virtue of their exposure to asbestos, their chances when compared to non-smokers are 92 times higher.

Modern toxicology rarely takes account of synergic effects. Few of the 1,000 new chemical products entering the market every year are tested for their synergic effects, nor are synergistic effects taken into account when setting safety standards. Permissible levels for pesticides in food, for example, are set individually, each pesticide being treated separately. Yet most foodstuffs contain a cocktail of chemicals, the combined effects of which are not even considered by the authorities. (See also: *Additives*Ⓛ, *Cancer*Ⓛ, *Cancer research*Ⓛ, *Carcinogen*Ⓛ.)

Synthetic organic chemicals

The modern chemical industry produces many thousands of synthetic organic chemicals that affect our daily lives in almost every aspect. Fibres, plastics, drugs, pesticides, dyes, food additives and fuels are just a few of the products involved.

All these organic chemicals are compounds of carbon and the skill of the organic chemist lies in shaping the molecules, by heat, pressure and reaction with other chemicals, into the desired end product.

In some cases the compounds produced may be identical with natural products – acetic acid, for example, is produced synthetically but is more familiar as the natural product vinegar – while in other cases – the pesticide DDT (*qv*), for example – the result may be very different from any naturally-occuring substance.

Several characteristics make them dangerous. Many – notably the chlorinated hydrocarbons (*qv*) – tend to accumulate in fatty tissues, with all the attendant problems of bioconcentration (*qv*). Secondly, many have the ability to mimic the characteristics of naturally-occurring substances, with the result that they are readily taken up by living organisms – often, as with pesticides like DDT, with fatal results. In addition, synthetic organic compounds are often resistant to degradation by natural processes, with the result that they persist in the environment and pose major disposal problems.

The majority of synthetic organic chemicals have never been tested for their effects on human health or on the environment. Nonetheless, as the US Environmental Protection Agency notes, synethetic organic account for "a disproportionately large number of recognized carcinogens". (See also: *Additives*Ⓛ, *Cancer*Ⓛ, *Carcinogen*ⒻⓁ, *Hazardous wastes*Ⓟ, *Pesticides*ⒶⓅ.)

T UV

Tartrazine

A bright yellow coal-tar dye (*qv*), also known in the USA as FD&C Yellow No. 5, and throughout the EEC as E102.

It is used unmixed to give a lemon yellow colour, but often mixed with other dyes to provide a variety of shades from cream to orange. Tartrazine has been in use since the late 19th century and remains one of the most widely used coal-tar dyes. It is found, for example, in many soft drinks and confectionery products.

Tartrazine is generally recognized to be responsible for a wide range of allergies and symptoms of intolerance, including child hyperactivity, asthma, migraine, and skin rashes. It has also been suggested that it may cause cancer, though the evidence is inconclusive.

Tartrazine is permitted in the UK, much of the EEC, and the USA, but it is banned in Norway and Finland. There are severe restrictions on its use in Austria, Sweden and France. (See also: *Additives*Ⓛ, *Food dyes*Ⓛ.)

Taste additives

The largest single group of food additives, measured both in terms of the number used and their cost. The European and North American industries are probably using about 3,500 different additives, and that figure does not include the artificial sweeteners (*qv*). The UK industrial market is alone worth more than $88 million a year; the EEC as a whole accounts for about $320 million; and the US market has been valued at $500 million annually.

Though taste additives are the largest single group of additives, they are also the least regulated. Neither Britain, the USA, nor any of the EEC countries has a comprehensive positive list of those permitted. They do not possess E numbers (*qv*) and their identities do not have to be declared to consumers or even to governments. At most, product labels have to list them as part of the ingredients. The food and chemical industries have often argued that regulation is unnecessary and impractical, but consumer groups do not agree. Since they are not regulated, taste additives have hardly been tested at all for their safety or toxicity; they are merely presumed to be safe until evidence to the contrary is provided.

They are sometimes divided into those that are natural, those that are "nature-identical" and those that are synthetic. Those classed as natural occur in, and are extracted from, foods. Those classed as "nature-identical" occur in foods, but, when used commercially, are synthesized in the laboratory. "Nature-identical" additives can thus be incorporated into products that would not naturally contain them. The synthetics are not known to occur in nature.

Some countries, such as West Germany, do have positive lists for the synthetics, but they are untypical. British labelling regulations make a distinction between the meaning of the terms "flavour" and "flavoured". A product that relies on synthetics should be labelled, for example, "strawberry flavour", while a product may be called "strawberry flavoured" as long as it uses either naturals or "nature-identical" flavours. (See also: *Additives* Ⓛ, *Colourings* Ⓛ, *Convenience food* Ⓛ, *E numbers* Ⓛ.)

Tehri Dam

Once commissioned, the Tehri Dam – now being built near the confluence of the Bhagirathi and Bhilangna Rivers in the Himalayas – will flood some 45sq km (17sq miles). The town of Tehri and 23 villages will be completely submerged, and 72 other villages will be partially submerged. More than 70,000 people will be displaced by the dam.

Numerous geologists have warned that the dam is sited in an area of intense seismic activity.

In April 1986, Indian environmentalists took the Government of India and the State of Uttar Pradesh to court in an attempt to cancel the project. One point raised was that high sedimentation rates (*qv*) could render the dam useless within 25 years. Engineers for the hydro-electricity board, however, insist that the dam will last 100 years or more.

The environmentalists were supported in their case by a government-appointed committee charged with assessing the likely environmental effects of the dam.

After six years of deliberation, the committee concluded: "The serious adverse environmental effects of the Tehri reservoir scheme are not commensurate with the potential benefits." The committee recommended that a series of small hydro-electric plants be built instead of a giant dam.

Despite that recomendation, the Government of India is intent on building the Tehri dam. In December 1986, the USSR announced that it would finance the dam – at a cost of 200 million rupees ($15 million). The original cost of the dam was estimated at 19,800,000 rupees ($1,500,000). (See also: *Dams* ③, *Development* ③.)

Teratogen

A substance that causes defects in the offspring of a mother exposed to it. The most notorious teratogen is the drug thalidomide, which deformed the babies of many women who were prescribed it in pregnancy.

Organic mercury (*qv*) compounds have been found to be teratogenic in humans, while dioxins (*qv*) are known teratogens in laboratory animals and suspected of having this effect in people. The herbicide ioxynil has also been found to be teratogenic in animals.

Women are advised to avoid taking drugs in pregancy, but few warnings are given about avoiding exposure to industrial and agricultural chemicals – despite the fact that many substances to which people are frequently exposed have not been tested for teratogenicity.

Third World debt

Between them, the nations of the Third World owe approximately $1,000,000 million to the banks and lending agencies of the industrialized world. The interest on the debts are now so high that the Third World is paying more money to the West in interest payments than it is receiving in aid – a negative transfer which in 1986 amounted to $29,000 million. Some countries are now borrowing money simply to service past debts.

Already several countries, notably Peru and Brazil, have defaulted on their interest payments. Many more look like following suit: at its July 1987 annual meeting, the 50-member Organization for African Unity requested an amnesty on $200,000 million of debts, admitting that "the problem is not one of liquidity but rather of a complete inability to pay."

A large proportion of the debts incurred by Third World governments has been spent on environmentally and socially destructive development programmes, from large dams to massive highway schemes and livestock rearing programmes.

Many of those projects have proved totally uneconomic. A 1987 study by the Asian Development Bank of 40 large irrigation schemes, for example, concluded that financial returns were typically less than 30 per cent of those expected. In India, according to a recent report, every single large dam built since World War II has been an economic disaster: not only have the dams cost more than estimated, but they have failed to provide the expected benefits. In South America, the Guatemalan government is now forced to sell electricity at four times the competitive rate in order to service the debts incurred on its massive hydro-electricity programme. Many farmers and small businesses have simply switched to using diesel and gas.

Other projects have been just as disastrous. In Brazil, it can cost as much as $50 a head to move cattle from the new ranches which have been set up in Amazonia (*qv*) to the nearest railhead, with the animals losing up to ten per cent of their weight en route. Not surprisingly, many livestock projects have proved totally unviable. Other white elephants that have contributed to Brazil's massive debt include its nuclear power programme and its highway development projects.

Corruption has further increased losses, some authorities suggesting that as much as one-third of the money borrowed has been siphoned off by businessmen, bureaucrats and politicians, much of it going into numbered accounts in Switzerland. A study of 18 debtor countries by the Morgan Guaranty Trust Company estimated that some US $200,000 million of loans had been lost to such "capital flight".

To service the loans on debts, Third World countries have been forced to plunder their natural resources for all they are worth, thus adding to the environmental and social costs of the debt crisis. In particular, the amount of land under cash crops (*qv*) has been increased, while the exploitation of forest and mineral resources has accelerated.

Several Western banks, led by Citicorp, have now made provisions against countries defaulting on their debts. Ironically, the acceptance that many loans will never be repaid has opened up the possibility of turning the debt crisis to ecological good. Thus, in July 1987, the US environmental group, Conservation International, negotiated to buy $650,000 worth of Bolivia's debt in return for nearly 16,000sq km (6,000sq miles) of rainforest. The group brought the debt at a discounted rate of $100,000. If national governments could be made to understand the global importance of preserving the world's remaining rainforests, they might be persuaded to write off much, if not all, of the Third World debt in exchange for the preservation of these vital life support systems. Such an arrangement might well provide the only hope for the forests. (See also: *Aid*③, *Dams*③, *Deforestation*Ⓒ③, *Development*③, *Hamburger connection*Ⓐ③, *World Bank*③.)

Threatened species

See: *Extinction*, *Genetic diversity*.

Three Gorges Dam

If it is built, China's Three Gorges Dam on the Chiang Jiang River (formerly the Yangtse) will be the largest dam in the world. Two kilometres (2,166yd) long and 175m (574ft) high, the dam will produce 14,700MW of electricity – the equivalent output of several large nuclear power stations. The final cost is likely to be between $16 and $20,000 million.

Critics of the dam, both within China and abroad, are concerned at its probable ecological and social effects. The dam's reservoir will extend 500km (310 miles) upstream, completely flooding ten cities and partially flooding eight others, in addition to some 440sq km (170sq miles) of fertile farmland. Some 3,300,000 people will require resettlement. Factories, roads and railways will also be flooded: replacing them could cost as much as $3,500 million.

Landslides are common in the area and could threaten the dam. Hou Xueyu of the Chinese Academy of Sciences has argued that the weight of the dam and its reservoir could trigger an earthquake.

Experts are also worried that the dam could have a disastrous effect on fisheries. Eighty species of fish could be adversely affected by the reduced flow of the Chiang Jiang, and by salt water intrusion into the delta region. By trapping silt, the dam will also deprive downstream fisheries of nutrients. Upstream, the accumulated sediment could cause the port of Chongqing to silt up.

In 1987, the Chinese Government announced that the project is to be re-evaluated. (See also: *Dams*③, *Sedimentation*③.)

Three Mile Island

An island in the Shenandoah River near Harrisburg, Pennsylvania, famous as the site of a power station containing two pressurized water reactors (PWRS) (*qv*), one of which almost suffered a catastrophic accident. In the early morning of March 28 1979 several water-coolant feed pumps failed in TMI-2. The reactor closed itself down eight seconds later, but both emergency water supply lines were blocked by closed valves. Though the reactor had closed, the "decay heat" from the nuclear reaction was still causing the core temperature to rise sharply, and no new cooling water was arriving.

To compound the problem, a relief valve in the reactor which opened in the first few seconds of the accident, failed to close, despite a control room indication that it had. This valve bled off much of what little water was in the core. The emergency core-cooling system came into operation, but the operators closed it. The increasing core temperature caused steam bubbles to form in the coolant, accelerating the rate of water loss. Within two hours the water level in the reactor vessel fell below the top of the fuel rods, and their temperature increased drastically. The fuel cladding melted, and reacted with the water to form hydrogen. Later TV pictures indicate that the uranium fuel itself may have started to melt.

The remaining coolant pumps started to shake violently, because they were trying to pump steam; not realising the cause, the operators closed the pumps. Soon afterwards help arrived from the reactor manufacturer, and the core was flooded and brought slowly under control. For several days experts tried to calculate the risk of a hydrogen explosion – in fact, one occurred on the afternoon of the accident – and the amounts of radioactivity released during the accident. In the face of conflicting advice, the Pennsylvania Governor recommended that children and pregnant women within a five-mile radius should be evacuated.

Though the containment (*qv*) held and little radioactivity was released at the time of the accident, the stricken reactor was itself highly contaminated. Six months after the accident, the containment was 2m (6ft) deep in radioactive water, which was leaking in at a rate of 3,800 litres (US 1,000 gallons) a day from the primary circuit. The water contained 250 curies per cubic metre, including a total of 3,000 curies of tritium, a radioactive isotope of hydrogen. Meanwhile, 57 curies of Krypton-85 were trapped in the reactor dome and subsequently, more than a year after the accident, the authorities agreed to it being vented into the atmosphere, despite public protests.

The plant could not be entered for two years, and the first TV pictures, taken a year later, showed a completely collapsed core. The clean-up is not yet complete but, by 1987, its cost was put at more than $1,000 million. At the end of 1986, the plant's owners, General Public Utilities, reported intractable problems in extracting chunks of disintegrated fuel from the reactor. The reactor will be entombed in concrete when adequately decomtaminated. In 1986, its companion reactor TMI-1 was restarted, despite vociferous local opposition.

In 1987, a rise in the death rate among old people living in states neighbouring Three Mile Island was blamed on the accident. Between 1979 and 1982, at least 50,000 (and possibly 130,000) more elderly people died in the affected states than expected, the death rate in Pennsylvania, for example, being 3.6 times higher than the national average. Dr. Jay Gould of Public Data Access, the group that undertook the study, argues that low-level radiation released during and after the Three Mile Island accident may have damaged the immune system of elderly people, thus hastening their deaths from apparently "normal" causes.

The Three Mile Island accident resulted in a re-assessment of safety standards, and the retrofitting of safety devices to many stations. It also brought a disenchantment with nuclear power in the USA, the view among electricity utility executives that nuclear power was too expensive being reinforced by the accident. (See also: *Nuclear accidents*Ⓝ, *Containment*Ⓝ, *Radiation and health*ⓁⓃ.)

Tidal power

The use of sea tides to provide energy to do useful work, most commonly to generate electricity. A large tidal movement is necessary, restricting sites to semi-enclosed bays or estuaries on ocean coasts.

A dam (*qv*) built at a saltwater site, allows tidal water to enter but prevents it from leaving. A lake forms behind the dam, the tide is allowed to ebb, and then the water is released, turning turbines as it flows downstream. In some designs there are two sets of turbines, one operated by the rising tide, the other by the water released at low tide.

The capital cost of tidal power projects is high, but operating costs are low. They can generate large amounts of power, and should enjoy long operating lives provided the bed behind the barrage is dredged regularly to control the accumulation of silt.

Little is known of the long-term ecological effects of tidal barrages. However, by interrupting the flow of water downstream and trapping silt that is disposed of elsewhere, rates and patterns of estuarine sedimentation (*qv*) are inevitably altered. This may affect populations of invertebrates on which wading birds feed, but the lake behind the barrage may increase the total area of open water to the benefit of other aquatic organisms, including waterfowl.

The best-known installation is at La Rance, in Britanny, France, but schemes have also been proposed for the Severn Estuary and other rivers in the west of England. The former would produce 13 terawatt hours per year. An experimental installation has been built at North Uist in the Outer Hebrides. (See also: *Alternative technology*Ⓔ, *Energy*Ⓔ, *Soft energy paths*Ⓔ.)

Tobacco

The adverse medical effects of smoking tobacco are well established. Less well known are the ecological effects of growing tobacco (see illustration).

Worldwide, tobacco accounts for 1.5 per cent of total agricultural exports. For some countries it is the main export crop; Malawi, for example, earns 55 per cent of its foreign exchange from tobacco.

Because tobacco grows best in

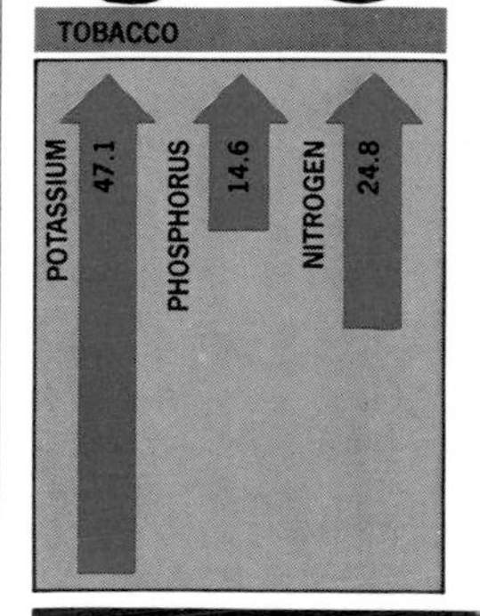

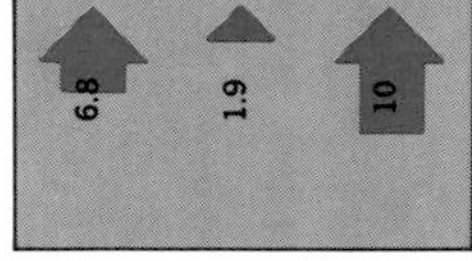

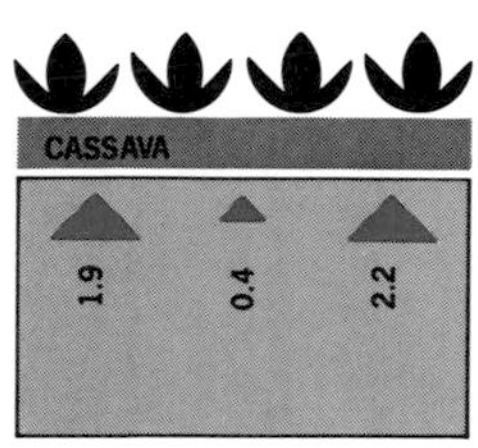

Tobacco is a crop that ruins the soil. It takes out 11 times more nitrogen than a food crop such as cassava, 36 times as much phosphorus, and 24 times as much potassium. It is also responsible for the loss each year of somewhere between 12,000sq km and 50,000sq km (4,600sq miles and 19,000sq miles) of forest, cut down to fuel the curing process. In the semi-arid regions where tobacco thrives, the loss of forest is particularly severe, and encourages desertification. But for more than 100 Third World countries it is a major export crop – and grown on land that could be used for food.

semi-arid areas, the degradation it causes inevitably hastens the spread of the desert.

But the heaviest environmental cost of tobacco production lies in the sheer volume of wood needed to fuel tobacco-curing barns. Every year, the trees from an estimated 12,000sq km (4,600sq miles) are cut down, with 55 cubic metres of cut wood being burned for every tonne of tobacco cured. Some experts put the figure even higher – at 50,000sq km (19,000sq miles).

Tobacco production uses up 0.3 per cent of the world's available arable land – land that could be used to grow food crops.

Traditional agriculture

The systems of agriculture practised by tribal and peasant societies.

Though such systems are frequently portrayed as backward, inefficient and unproductive, the reality is quite different. Indeed, traditional agriculture cannot be judged by the same criteria as are used to judge modern intensive agriculture, since the goals of the two systems are largely antithetical.

Traditional farmers do not aim to produce food for sale but to produce food for consumption by their own familes and to fulfill such social obligations as distributing food to relations and giving feasts. This latter goal is very important since it plays a vital part in maintaining the social cohesion which alone makes possible the social co-operation required for fulfilling many essential traditional practices – such as maintaining irrigation canals, desilting storage ponds, weeding plots and maintaining terracing.

Unlike modern intensive agriculture, the goal of traditional agriculture is not to maximize yields but rather to maximize security by minimizing risks. To quote the anthropologist Dr. James Scott, "Typically, the peasant seeks to avoid the failure that will ruin him rather than attempting a big, but risky, killing. In decision-making parlance, his behaviour is risk-averse. He minimizes the subjected possibility of maximum loss."

To achieve this, traditional farmers exploit a whole range of strategies. In particular, they practice polyculture, as opposed to monoculture (*qv*), growing not only many different crops in the same plot, but also many different ss of the same crop. In Sri Lanka, for instance, farmers would grow up to 400 different varieties of rice, in India, more than 10,000. Each had different qualities: some were useful when the rain was poor, others were resistant to certain pests, others were cultivated because they were highly nutritive and therefore good for pregnant women, or simply for their taste and fragrance. This extraordinary diversity of cultivated crops confers great stability on traditional agriculture.

Great care is also taken to maintain the fertility of the soil. This is achieved by returning to it all the organic matter possible, by cultivating leguminous crops which maintain the soil's necessary nitrogen content, by terracing mountain slopes, and by allowing land to remain fallow for often lengthy periods.

Though high yields are not a primary consideration, they are rarely as low as is made out by the development industry. Indeed, contrary to received wisdom, traditional agriculture is highly productive. According to Susan George, author of *How the Other Half Dies*, "Indigenous cropping systems use labour more efficiently, give more stable yields from year to year and are intrinsically higher yielding than monoculture." She considers that "the subsistence farmer has developed a highly sophisticated system . . . based on good economic sense."

The World Bank also acknowledges the efficiency of traditional farmers. In a report on New Guinea, it notes: "A characteristic of Papua New Guinea's subsistence agriculture is its relative richness. Over much of the country, nature's bounty produces enough to eat with relatively little expenditure of effort."

Yet, despite the productivity of traditional agricultural systems, the development industry is intent on replacing them by intensive farming regimes. One reason is that traditional farmers are unwilling to grow the export crops (*qv*) that provide Third World countries with the foreign exchange to import western goods. Another is that traditional agriculture does not provide a market for the capital-intensive goods and services – tractors, pesticides, dams and so on – that the development industry is geared to selling.

Though a minority of agronomists now argues that the Third World would benefit from a return to traditional techniques, it seems unlikely that this will come about while the development industry remains committed to its present goals. (See also: *Development*③, *Green Revolution*Ⓐ③, *Post harvest losses*Ⓐ③, *Sustainable development*③, *Tropical agriculture*③.)

Transmigration

An Indonesian colonization programme, the largest in the world, which aims to resettle inhabitants from the densely populated central islands of Java, Lombok, Bali and Madura to the sparsely populated outer islands of the Indonesian archipelago, Sumatra, Borneo and West Papua. Some 3,600,000 people had already been moved by 1984 and, if the Government of Indonesia has its way, a further 65 million will be moved in the next 20 years. The programme is heavily funded by the World Bank (*qv*) and other Western multilateral development banks. By June 1985, nearly $800 million had been provided to the Indonesian government, with a further $750 million promised.

The programme has been heavily criticized by environmental and human rights organizations. The colonization programme is being implemented at the expense of the tropical rainforest, which covered 1,480,000sq km (571,000sq miles) of Indonesia in 1950. It is estimated that as much as one-third of the forest cover has been lost since then, and further transmigration will put an intolerable burden on what is left.

Settlement schemes in tropical rainforest areas are rarely capable of supporting sustained agricultural exploitation, due to the poor rainforest soils. According to one study, more than 5,000sq km (1,900sq miles) of lowland development projects in Indonesia were located in forests judged unsuitable for the type of exploitation planned under transmigration. The same study estimated that only about 750sq km (290sq miles) of Central Kalimantan were suitable for development. Indonesian Government targets, however, involve the clearance of nearly 17 times that amount of forest in the period from 1979 to 1989. The figure gives some measure of the extent to which the programme is ill-conceived and environmentally destructive.

One of the reasons for behind transmigration is the need to resettle peasants who have been deprived of their land to acommodate large development projects. It is clear also that transmigration fulfils many of the ambitions of Indonesian nationalism. The military component of the programme is quite evident, especially in sensitive areas such as West Papua and East Timor. The programme also has a part to play in the government's plans to "assimilate" the large tribal population on the outer islands into mainstream Javanese society.

In December 1986, in the face of falling oil prices and widespread opposition to transmigration, the government announced a 65 per cent cut in the programme's budget. The target for the numbers to be relocated in 1987 was reduced from a planned 100,000 people to just 1,000. Hovever, the government is talking of stepping up the quotas for "spontaneous" transmigrants – that is, those who pay their own way. The government has also increased its "housing-settlement" budget, leading to speculation that it is promoting transmigration in disguise.

Meanwhile, the US Senate Appropriations Committee, which allocates US treasury funding to the multilateral development banks, has announced major cuts in its lending in 1987 due to their failure to meet the "fundamental environmental concerns" of the committee. The committee was "especially concerned about the impact of large-scale ... agricultural resettlement schemes in tropical forest regions inhabited by indigenous peoples". Whether the World Bank and the other banks learn their lessons and meet these concerns remains to be seen. (See also: *Deforestation* ③, *Development* ③, *Resettlement*, *Tribal peoples* ③, *Tropical rainforest*.)

Tree diseases

Even before acid rain (*qv*) began to wreak havoc among forests in northern Europe and the east coast of North America, trees were already succumbing in ever-greater numbers to various virulent diseases.

From New England to south-west Australia, various diseases have, since the beginning of the century, attacked chestnuts, elms, oaks, maples, coconut palms and eucalyptus.

In an undisturbed climax ecosystem, though there are indeed tree diseases, they kill off mainly the very young, the very old and the sick. What then has gone wrong?

The most obvious maladjustment is that created by the introduction of alien parasites against which native trees have no defense. Thus the fungus associated with chestnut blight was imported from China, while that associated with Dutch elm disease came from Asia via Europe. The elm bark beetle, without whose help the fungus could not have caused so much damage, also came from Europe.

Unfortunately, modern society favours such ecological invasions by encouraging a high level of mobility and international trade.

Diseases, however, cannot be regarded as being solely caused by a pathogen. The factors that reduce the host's resistance to them must also be considered.

One such factor is the planting of trees in conditions – soil, climatic and biotic – to which they have not been adapted by their evolution. Another is pollution. Damage by smoke, for instance, has been shown to increase susceptibility to pests such as bark beetles, weevils and fir lice. Smoke also tends to reduce resistance to frost. Pollution by cement dust has been found to reduce photosynthesis on lime and elm trees. Sulphur dioxide from blast furnaces, coke plants, fertilizers, soda cellulose, sulphuric acid plants and thermal power stations, has, for a long time, been known to increase susceptibility to pests, as has fluorine from aluminium foundries, and fertilizer factories.

What makes matters still more complicated is that the various factors leading to ecological maladjustment, and hence to the reduced resistance of trees to disease, rarely occur independently. Very often trees are affected by various combinations of these factors which have, at best, an additive effect, at worst, a synergic one (*qv*).

This, makes tree diseases particularly difficult to treat. To spray affected forests with pesticides is merely to treat the symptoms, and in a very crude way. The only real solution is to correct whatever maladjustments are involved. Ideally, evolution would do the work, but that takes time. To speed up the process of adaptation, we can plant soil-improving trees such as the Rowan or the silver birch. However, their timber is of little value, so this is not generally regarded as being economically viable.

Planting different species from the appropriate strains to replace a single monoculture (*qv*) must also reduce vulnerability, but unfortunately it may also reduce immediate financial returns, which may not be regarded as acceptable. When it comes to dealing with a maladjustment caused by a destructive imported parasite, more adventurous remedies such as biological control (*qv*) are clearly required, even though this means taking certain risks.

When pollution is an important factor, then there can be no satisfactory solution but to reduce emissions of the offending pollutants. Because of present priorities, however, this is difficult. It is also difficult to cut down on the mobility and the trade in wood products that leads to the importation of alien parasites.

Tribal peoples

According to the human rights group, Survival International, there are some 200 million people worldwide who can be described as "tribal". Such people live in small cohesive communities that are economically self-sufficient and are themselves usually part of larger, culturally distinct groups, which may or may not be capable of acting as political units.

Tribal peoples, among whom are numbered hunter and gatherers, shifting cultivators, herders, subsistence farmers and fisherfolk, can also be defined by the absence of some of the characteristics common to state-based societies. They may not use money. They may have no systems of writing. Leadership patterns and decision-making processes are complex and power is diffuse. They may have deep-rooted traditions to ensure material equality, though not necessarily social equality, and they have strong ties to their traditional lands. Unfortunately, such lands often lie in territory that can be profitably logged, mined, farmed, ranched, dammed or otherwise "developed". Consequently, tribal lands are increasingly being expropriated for development projects, with numberous tribal groups being forcibly dispossessed in the "national interest". This is often rationalized on the grounds that tribal people are poor, malnourished and disease-ridden, and that it is the duty of governments and international agencies to assure that for their own benefit they integrated into mainstream society.

In Indonesian-occupied West Papua, for example, the Indonesian Government has embarked on a project entitled "Total Development of Indonesian People" and aimed at re-educating West Papuans, whom it described as "still living in a stone-age-like era." To bring these peoples "up to par with the rest of the country", it was decided that the children would be "separated from their parents to keep them from settling into their parents' lifestyles."

Tribal people are also accused of environmentally destructive behaviour. Because they are so "poor," we are told, they cannot afford to take long-term considerations such as resource conservation into account. In particular, the slash and burn agriculture (*qv*) that many forest peoples practise is portrayed as destructive, as are many other time-honoured agricultural practices.

The reality is very different. Tribal peoples are poor only if poverty is identified with material deprivation. Certainly, many have few material goods, but then these are of little value to people who often lead a nomadic way of life. In fact, they do very well without them and there is no evidence that they feel in any way materially deprived.

On the other hand, their biological, social, spiritual and aesthetic needs – which could be regarded as much more important than their material needs – are well satisfied. The diet of tribal peoples, consisting of a rich variety of fresh and uncontaminated foods, is generally excellent; their health status also seems very good; and their extended families and cohesive communities admirably satisfy their basic social needs. As for their agricultural practices, time has proved these to be not only environmentally benign, but also sustainable.

If tribal peoples are to be protected against cultural imperialism, the key to survival lies in the recognition of their rights; their rights to the inalienable collective

Tropical forests – the key to conserving a vulnerable environment

NUTRIENT RECYCLING

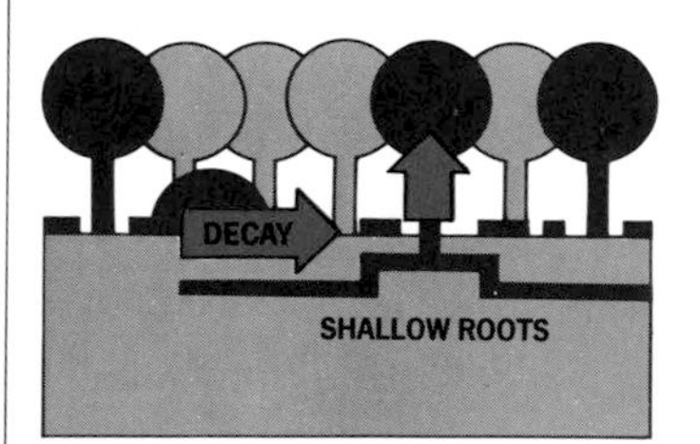

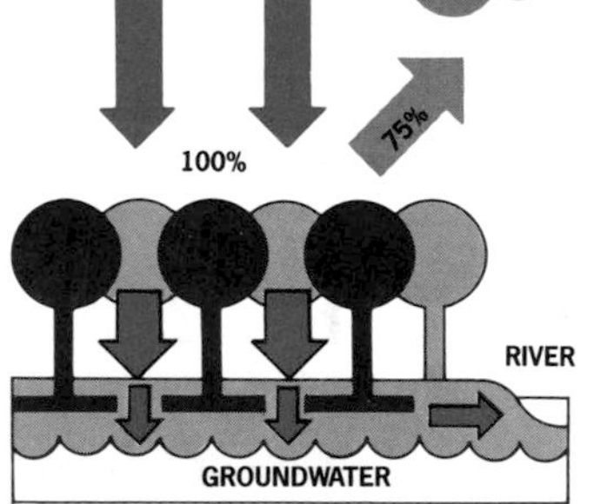

EVAPOTRANSPIRATION
EQUATOR

Tropical forests generally grow on extremely poor soils with most of the essential nutrients locked away within the plants themselves. Nutrient recycling is crucial; the trees have developed roots, which make up 60 per cent of their mass, that absorb any free nutrients within the thick layer of decaying debris on the forest floor before they are leached away.

Up to three-quarters of the rain that falls on tropical forests is lost through evapotranspiration. The rest is held by the forest and slowly released into the ground and rivers. This system protects the fragile soils from the potentially devasting effects of tropical storms and also means that rivers are supplied with water even during the dry season.

The 75 per cent of rain that evapotranspires from the forest plays an important role in maintaining a tolerable climate around the Earth. The heat required to power the evapotranspiration effectively cools down equatorial regions; the clouds formed by the vapour help still further by reflecting solar radiation back into space. Those clouds are then funneled

ownership of their traditional territories; their right to self-determination; and their right to an acceptance of the legitimacy of their own institutions and systems of decision making. (See also: *Environmental refugees* ③, *Resettlement schemes* ③, *Subsistence agriculture* Ⓐ ③.)

Tributyl tin (TBT)
Organic compounds of tin that are toxic to fungi, other plants and molluscs. They are used as wood preservatives, in anti-fouling paints for boats, and socks are impregnated with them to prevent foot odour.

The use of tributyl tin anti-fouling paints has caused serious pollution in estuaries and bays, since the material is toxic at extremely low concentrations. Oysters living in waters frequented by many boats have died or been deformed because of contamination by tributyl tin. Whelks have undergone sex changes, and salmon have been contaminated when these compounds have been used to prevent invertebrate fouling of cages and nets near hatcheries.

The irritant effects of TBT are well known, but more recent evidence, collected by the US Environmental Protection Agency, suggests that it may also attack the human immune system and possibly cause birth defects as well. The use of TBT is currently under review in a number of countries. A ban on certain uses has been introduced in Britain. (See also: *Wood preservatives* Ⓟ.)

Tropical agriculture
A consequence of Third World development has been the introduction of agricultural techniques developed in temperate areas into the tropics – where, for many reasons, they are totally inappropriate. In particular, they have proved unsuited to the local climate and soils.

The heat in the tropics is both intense and continuous. Winter, which elsewhere restrains the growth of weeds, insects, fungi, etc., is absent. The result is an ideal and permanent habitat for potential pests, which often prove almost impossible to control.

Livestock, too, is afflicted by a wide variety of internal parasites that seriously reduce yields, typically to as little as a quarter of the yield in temperate areas.

The intense heat also kills off soil micro-organisms. The consequent low organic content of the soil reduces its water-retaining capacity and makes it particularly vulnerable to erosion, since all the rain falls during the annual monsoons, while during the rest of the year the soil can be dust-dry.

This means that agriculture in the tropics has to be carried out sensitively, and on a small scale, by people whose goal is to provide food for themselves and their families rather than for export. It cannot be carried out on a massive scale for very long.

Not surprisingly, attempts to intensify agricultural practices in the tropics have led to widespread environmental degradation and declining yields. Nonetheless, intensive cropping techniques are still being systematically encouraged by the World Bank (*qv*), bilateral aid agencies, and the specialized agencies of the United Nations. Unless they are abandoned and replaced by others that better satisfy the particular exigencies of tropical soils and climate, environmental degradation and famine can only incrase. (See also: *Cash crops* Ⓐ ③, *Development* ③, *Green Revolution* Ⓐ ③.)

Tropical forests
Running like a girdle around the equator, tropical forests encom-

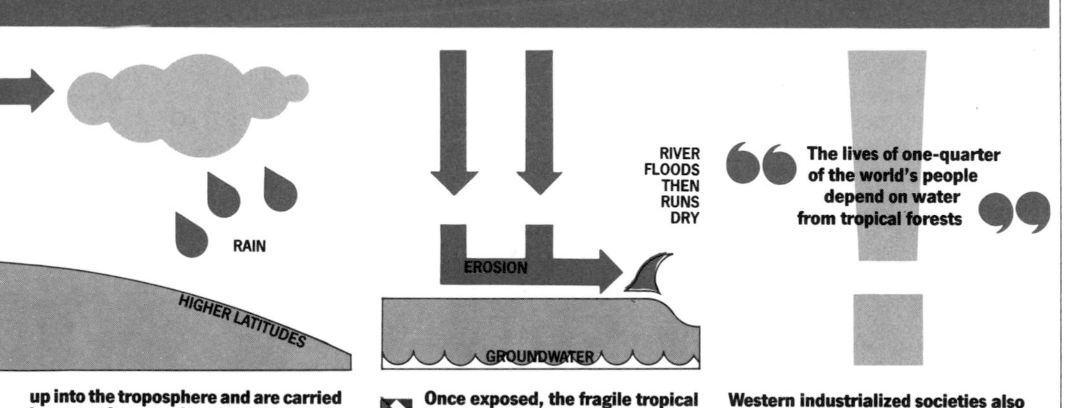

up into the troposphere and are carried by natural convection circulation toward higher latitudes where the vapour is released as warm rain, which raises the temperature of cooler parts of the globe. Without the forests, the temperature difference between the tropics and temperate zones would be far greater.

Once exposed, the fragile tropical soils soon erode away, silting up rivers and disrupting the downstream ecology. Rainwater is no longer retained for long enough to replenish groundwater supplies. Rivers flood with the heavy rains but quickly dry up afterwards. Without the forest's evapotranspiration there are local as well as global climate changes.

Western industrialized societies also depend upon the tropical forests of the world. Many of our modern medicines originated from the forests along with foods, timber, oils and other products. But above all our knowledge of the forest is limited that the larger part of the loss can only be imagined.

pass some 9,500,000 sq km (3,670,000sq miles), the greatest forested area by far being in South America, particularly in the Brazilian part of the Amazon, but stretching into Peru, Ecuador, Colombia, Venezuela and French Guyana. The remaining tropical forests, some five million sq km (two million sq miles) are found in 16 countries of tropical Asia and in Central Africa. The tropical forests of the Ivory Coast and of Nigeria are now virtually gone, cut down over the past 40 years.

The United Nations Food and Agriculture Organization estimates that 100,000 sq km (62,000sq miles) of rain forests are now being lost each year, but it does not take into account forests that have been felled, and have regrown into poor, secondary and degraded forests. On the other hand, the United States National Academy of Sciences puts the total figure of felled primary tropical forest as 200,000sq km (124,000sq miles) each year. At current rates of deforestation, nearly all tropical forests will have been eradicated from the planet within 50 years.

The tropical rain forest, with its extraordinary diversity and interaction of species, grows on very poor soils, primarily through the evolution of symbiotic associations between trees and microflora that enable a remarkably efficient recycling of minerals (see illustration).

In spite of the instability of the soil, many governments in tropical countries encourage colonization of the forests – Indonesia's Transmigration Programme (see *Transmigration*) and Brazil's Polonoroeste project being cases in point.

In 1983, the World Resources Institute, the World Bank (*qv*) and the United Nations Development Programme convened an international taskforce with the brief of drawing up "a priority action programme to address deforestation issues on a broad front."

The taskforce's report, *Tropical Forests: A Call to Action*, was published in 1985. It outlined a five-year action programme – to cost $8,000 million – aimed at providing fuelwood, promoting agroforestry, reforesting upland watersheds, conserving tropical forest ecosystems, and strengthening institutions for research, training and extension.

The plan has been criticized on a number of grounds. First, it makes no attempt to curtail, let alone halt, such major causes of tropical deforestation as large dam projects (*qv*), plantation programmes and livestock-rearing projects. Yet, together, such projects have been directly or indirectly responsible for much of the deforestation which has occurred over the last 40 years.

Second, the plan is less concerned with the preservation of forests than with the setting up of commercial plantations of fast growing species such as eucalyptus (*qv*), which not only have a serious adverse impact on the environment but which also make little contribution to the local economy. Indeed, much of the wood is likely to be sold to commercial concerns in the cities.

Thirdly, the plan does not even mention the rights of those indigenous peoples who inhabit the world's tropical forests and who depend on them for their livelihood.

Several environmental groups have now called on the World Bank to reconsider its financial backing for the Action Plan. Some environmentalists have suggested that the world's tropical forests should be exchanged for the Third World Debt (*qv*) and held in trust in perpetuity by an international body. A limited step in that direction was made in July 1987, when a large tract of Bolivian rainforest was exchanged for part of Bolivia'a national debt. (See also: *Deforestation*③, *Eucalyptus*©, *Greenhouse effect*Ⓕ, *Reforestation*© ③, *Social forestry*③, *Third World debt*③, *World Bank*③.)

2,4,5-T

A phenoxy herbicide (*qv*) which first gained notoriety as one of the ingredients in a defoliant – codenamed Agent Orange (*qv*) – used widely in the Vietnam War. During its manufacture, 2,4,5-T becomes contaminated by small amounts of dioxins (*qv*) which are highly toxic and known to cause birth defects and cancers in laboratory animals.

In many areas other than Vietnam 2,4,5-T has been blamed for birth defects, cancers and a variety of other illnesses. Proving that 2,4,5-T – or its dioxin contaminants – caused the damage has been difficult and many experts disagree over the evidence. Some countries have taken action, however. Sweden has banned 2,4,5-T and, as an environmental protection measure, the USA has greatly restricted its use. Commenting on the evidence against 2,4,5-T, the US Environmental Protection Agency (EPA) stated, "The quality, quantity and variety of data demonstrating that the continued use of 2,4,5-T contaminated with dioxin presents risks to human health is unprecedented and overwhelming." 2,4,5-T is also banned in India, Norway and the USSR.

In Britain, 2,4,5-T is still in widespread use since the official government view is that it is not harmful so long as it is used according to the instructions and provided that the dioxin content remains below 0.01 parts per million. That level, however, is twice the limit recommended by the EEC. Moreover, union officials argue that "the instructions" take little account of the realities of spraying under field conditions. One hundred local authorities and several major trades unions have banned its use.

2,4,5-T, however, is still sprayed in large quantities in the Third World with no restrictions.

The manufacture of 2,4,5-T has now ceased in every country of the world except New Zealand. Previous manufacturing plants were plagued by accidents: the New Zealand plant suffered a serious accident in 1987 and was closed temporarily. (See also: *Phenoxy herbicides*Ⓟ, *2,4-D*Ⓟ.)

2,4-D

A phenoxy herbicide (*qv*) used as a general weedkiller. It was originally thought to be fairly harmless to humans. However, in 1969, a US commission concluded that 2,4-D was likely to cause birth defects and recommended that it "should immediately be restricted to prevent the risk of human exposure". That recommendation was contested by the US manufacturers, Dow Chemicals, and today 2,4-D remains one of the most widely used herbicides in the industrial world. Recent research suggests that 2,4-D may also be a carcinogen (*qv*). (See also: *Agent Orange*Ⓛ Ⓟ, *Biocides*Ⓛ Ⓟ, *Cancer*Ⓛ Ⓟ, *Dioxin*Ⓟ, *Pesticides*Ⓛ Ⓟ, *Phenoxy herbicides*Ⓟ, *Teratogens*Ⓛ.)

Underground nuclear tests
Since the signing of the Partial Test Ban Treaty (PTBT) (*qv*) in 1963, the atomic tests of the USSR, the UK and the USA have been conducted underground. Bowing to international diplomatic and legal pressure, France, a non-signatory of the PTBT, followed suit in 1975. China, also a non-signatory, has conducted only underground tests since 1980. The sole Indian test was conducted underground in 1974.

Underground tests are intended to limit the amounts of radioactivity released into the environment – unlike the previous, much more damaging, atmospheric tests. However, numerous leaks of radioactivity from underground tests have been documented at the US Nevada and French Muroroa (*qv*) test sites. In 1970, for example, a test codenamed "Banebury" resulted in a cloud of radioactive dust being vented into the atmosphere. (See also: *Atomic tests*Ⓝ Ⓛ.)

Uranium
The heaviest naturally-occuring element. Most uranium (99.3 per cent) consists of the 238 isotope. The rare (0.7 per cent) 235 isotope is more unstable and liable to fission (*qv*), either spontaneously or under neutron bombardment. All nuclear reactors generate heat by uranium fission.

The 238 isotope is fertile, that is, it can absorb a neutron and transmute to plutonium 239 (*qv*), which is fissile. It is possible to run a nuclear reactor using as fuel uranium containing the natural proportions of the two isotopes – the Magnox (*qv*) is an example – but most modern reactors use fuel in which the proportion of U-235 has been increased to two or three per cent through uranium enrichment (*qv*).

Uranium enrichment
Techniques that increase the proportion of fissile uranium-235 in the natural metal. The natural proportion of fissile uranium-235 to non-fissile uranium-238 found in the natural metal is 0.7: 99.3. Highly-enriched uranium contains 90 per cent or more U-235, and is required for uranium fission bombs, and as fuel for the small pressurized water reactors (PWRS) that power nuclear submarines. Uranium enriched to 2.5-3 per cent U-235 is used as fuel in many power reactors including advanced gas-cooled reactors (*qv*) and commercial PWRS (*qv*). British highly-enriched uranium is obtained from the USA in exchange for plutonium from the UK military stockpile, under the 1958 US/UK Mutual Defence Agreement (*qv*).

A British low-separation plant began operation at Capenhurst in 1953, initially to enrich the depleted uranium obtained from reprocessing (*qv*) to levels suitable for re-use in Magnox reactors (*qv*). Enrichment is difficult to achieve on a commercial scale, as different isotopes of the same element are chemically identical; separation depends on the tiny difference in the atomic weights between them.

A gaseous diffusion technique was developed by the wartime Manhattan Project, and is still used in the USA, but this method consumes huge amounts of electricity. A consortium of British, West German and Dutch interests formed Urenco in 1970 to develop enrichment using a centrifuge technique; production began at Capenhurst in the mid-1970s, and Urenco now competes successfully on world markets.

A new enrichment technique is under development in the USA and Europe, based on laser isotope separation. This promises to be a much more efficient method, providing low-cost fissile material at any required purity. It may provide a means of extracting plutonium and uranium from irradiated reactor fuel without conventional reprocessing, and to any required degree of purity. (See also: *Uranium*Ⓝ.)

Uranium mining
There is no doubt that uranium miners receive far higher exposures to radiation than any other nuclear workers.

Uranium ore contains 0.1-0.2 per cent uranium metal. The tailings, 99.8 per cent of the mass and 85 per cent of the radioactivity, are left at the mine location. Tailings continue to emit radon for many hundreds of thousand of years. Nonetheless, in the USA, thousands of houses have been built on or near tailing sites. Since the early 1970s the practice has been to cover the tailing heaps with either clay or water, a procedure of limited effectiveness. In Colorado, a centre of uranium mining, over 40 per cent of houses have radon levels above safe limits.

Better mine ventilation and improved health care facilities are beginning to cut the high cancer rate among uranium miners. Other factors are the sharp reduction in demand for uranium, which has brought employment in mines down by 30 per cent of 1978 levels.

Lung cancer is the most prevalent occupational disease, with an incidence among American and Canadian miners of 4.8 times that found in the general population, with those who smoke suffering the highest rates. A US Nuclear Regulatory Commission (NRC) report showed that uranium mining in the USA produces 4,000 extra lung cancers every year; this translates to between 15,000 and 30,000 lung cancer deaths caused by supplying uranium for a large modern nuclear station, such as a pressurized water reactor (*qv*), in Britain, over its 35-year life.

Several early US mines were located in Navajo reservations, and the incidence of lung cancer among Navajo miners (who do not smoke) has been found to be 14 times greater than normal. Navajo miners have also suffered from a higher-than-normal level of birth defects and adverse pregnancies because of the presence of the tailing heaps close to residential areas. Other statistics, from the 1930s, show that half of those employed as uranium miners eventually died of lung cancer.

Canada is now the West's largest producer of uranium, followed by the USA. The other producers are South Africa, Namibia, France and Australia. No statistics are available for South Africa and Namibia. The UK has not obtained uranium from Namibia since 1984. (See also: *Radiation and health* Ⓝ Ⓛ, *Uranium* Ⓝ.)

Urbanization

One of the most critical developments of the last decades, as critical as the population explosion, is the vast shift of population from the countryside to the cities. The urban population of the Third World was about 100 million in 1920; it is 1,000 million today. Between 1985 and the year 2000, Third World cities are likely to grow by another 750 million people. Almost half the world's population will then live in cities.

The most obvious implication of this change is that there must be a massive increase in the physical infrastructure in the world's cities to accommodate all these new city dwellers. No country can conceivably afford this. The richest countries of the world cannot even afford to maintain their existing urban infrastructure, which in both the USA and the UK is disintegrating fast. In the UK more than £27,000 million ($43,000 million) is required to rebuild the disintegrating sewage system, and more than £5,000 million ($8,000 million) to renew the gas mains, while billions of pounds are needed to modernize the country's crumbling stock of local government houses. None of this money is likely to be made available.

In the Third World the situation is very much worse. Half the population of Delhi already lives in slums and the Delhi Planning Authority admits that by the end of the century the figure will be closer to 85 per cent. In the main Third World cities, slums attain a degree of squalor that it is impossible for people in the West to imagine.

The second implication is that food production has to be reorganized. In particular, small farms that are not geared to achieving surpluses must be replaced by big capital-intensive farms, just to produce the amount of food required for city dwellers. A massive transport and distribution system is clearly also required. Facilities for providing clean water must also be correspondingly expanded – this is a massive problem that few cities have overcome. There is also an almost insoluble problem of waste disposal, particularly in Third World countries.

A further implication is health. If we take into account pollution, the tremendous density of population, the lack of facilities, the contamination of drinking water and so on, conditions in the massive Third World slums could not be less propitious.

Officially, migration occurs because of poverty and overpopulation in the countryside, and the search for jobs and the amenities of city life. But the main reason for the growth of the slums (other than natural urban population growth) is the displacement of peasants from their lands as a natural consequence of the modernization of agriculture and the setting up of large-scale development schemes. For instance, if India goes ahead with its ambitious dam building plans, the programme must inevitably lead to the displacement of tens of millions of rural people from their land. The Narmada Valley Project alone will lead to the displacement of a million people. (See also: *Green Revolution* Ⓐ ③, *Resettlement programmes*, *World Bank* ③.)

Vinyl chloride

The raw material for the manufacture of the plastic PVC (polyvinylchloride), sometimes referred to as vinyl chloride monomer or VCM. It is an inflammable gas that has a narcotic effect if inhaled in quantity, an effect which was well known for many years. Less well known was its ability to cause a rare form of liver cancer called angiosarcoma. This was found to occur in workers regularly exposed to vinyl chloride at a much higher frequency than in the general population, and industrial safety standards have had to be revised downwards sharply to protect these employees.

Britain has adopted an exposure standard for the workplace of two parts per million (ppm) – twice that allowed in new factories in West Germany but lower than the 3ppm permitted in old German factories. A 1ppm standard has been in force in the USA since 1974. The US authorities have banned the use of vinyl chloride in aerosols, and its disposal is tightly controlled.

PVC may contain traces of vinyl chloride monomer, and this may leak into materials packaged in this type of plastic. The main environmental hazard occurs, however, when PVC is burned. Corrosive hydrogen chloride gas is released, but of greater significance in the long term are the dioxins (*qv*) formed when PVC burns at too low a temperature. (See also: *Cancer* Ⓛ, *Carcinogen* Ⓛ.)

Wakersdorf

West Germany's first reprocessing (*qv*) plant, at Karlsruhe. It began operating in 1969, and is capable of a throughput of about 40 tonnes of irradiated fuel a year. Under German law, arrangements have to be made for irradiated fuel management and disposal before a reactor is permitted to begin operation. Plans were therefore drawn up in 1969 for a huge federal reprocessing plant and vitrification and disposal facility, to handle fuel from all the country's reactors. However, in 1974 the consortium formed to build the plant decided abruptly to withdraw on commercial grounds.

The federal government decided to press ahead, and chose a site in Lower Saxony called Gorleben for the reprocessing complex. Though very near the East German border, the site was chosen because it is a salt mine and thus an extremely stable geological feature, considered suitable for the disposal of vitrified waste. After vociferous local opposition, hearings into the proposal were arranged, which opened at almost the same instant as the feedwater pumps failed at Three Mile Island (*qv*) on March 28 1979. Two months later the state premier announced that Gorleben would not go ahead.

The idea of a vast central facility was abandoned, and Wakersdorf was chosen as the site for a relatively modest (350 tonnes a year) reprocessing plant. Construction began in May 1986 and is expected to be complete by 1992. (See also: *Cap la Hague* Ⓝ, *Fast breeder reactor* Ⓝ, *Sellafield* Ⓝ.)

Waste reduction

The best solution to the problem of hazardous wastes (*qv*) lies in reducing (or better still, eliminating) their production. Radical as that proposal might sound, it can often be achieved relatively simply through a variety of strategies, including minor changes in manufacturing processes, or by substituting polluting products for non-polluting products.

By 1986, INFORM, a New-York based research organisation, had identified 44 waste reduction practices. Of these, "11 achieved reductions of individual waste streams that ranged from 80 per cent to 100 per cent."

Already several companies have cut considerably the amount of waste they generate. The most celebrated example involves the Minnesota Mining and Manufacturing Corporation (3M) which began its "Pollution Prevention Pays" programme in 1975. By 1984, the company had eliminated 10,000 tonnes of water pollutants, 90,000 tonnes of air pollutants and a 140,000 tonnes of sludge. The rewards were not only ecological but also financial: according to 3M, the programme saved $192 million in less than ten years.

The scope for waste reduction is enormous but it has still to be widely practised. A survey by INFORM of 29 major US companies found that only 12 had waste production programmes – and that these were being applied only to a minute fraction of the wastes generated. Significantly, nine out of the 12 companies had adopted waste reduction as a last resort, generally "only after regulatory and/or operational pressures forced management to focus on waste reduction opportunities." Once in practice, however, the waste reduction measures saved money at five out of those nine plants. (See also: *Hazardous waste* Ⓟ, *Recycling* Ⓟ.)

Water quality

See: essay on *Water Fit to Drink?*

Water transfer projects

Already many regions of the world – including China, India, Canada and the USUSA – are having difficulties meeting even their present-day requirements for water. Forecasts predict severe shortages (with attendant economic loss) by the year 2000. The conventional answer to regional water shortages has been to build storage reservoirs. But there are only a limited number of sites where reservoirs can be built – and many of these have already been exploited or are under development. Several countries are now looking to large-scale water transfer projects to provide their future water requirements, the idea being to transport water from water-rich areas to drought-prone regions.

Some of the proposed projects are mind-boggling in their sheer scale and complexity. One scheme, the North American Water and Power Alliance (NAWAPA), would involve transferring water from northwest Canada, Alaska and the northwestern United States to Mexico, South-Central Canada, the Great Lakes area and the southwestern United States. In China, another such project (already under construction) will transfer 14,000 million cubic metres of water over 1,150 km (715 miles) from Chang Jiang in the southeast of China to the North China Plain.

In April 1986, the Soviet Union cancelled a proposed water diversion scheme that would have reversed the flow of Russia's northern rivers to channel water to the arid lands of Central Asia. President Gorbachev's personal economic advisor, Professor Abel Aganbegyan, criticized the plan as unecessary and ill-conceived. "The idea of redistribution was based on the strategy of endlessly bringing new land under the plough, and on forecasts of the progressive fall in the water levels of our rivers, leading to a catastrophic fall in the level of the Caspian Sea. These forecasts have proved incorrect." (See also: essay on *Water Fit to Drink?*, *Dams*③.)

Waterborne disease

Diseases transmitted by bacteria, insects and other organisms that live or breed in water. Examples include cholera, schistosomiasis (*qv*), malaria (*qv*), river blindness, diarrhoea, leprosy, yellow fever, trachoma, scabies, polio and elephantiasis, a monstrous swelling of the limbs..

Such diseases kill an estimated 25,000 people a day – millions every year. More than 160 million people suffer from malaria; 200 million from schistosomiasis; and 250 million people – the equivalent of the entire population of the USSR – from elephantiasis. In 1980, according to the Washington-based World Resources Institute, five million children under the age of five died from diarrhoea – "about ten every minute of the day".

In urban areas, lack of pollution controls, particularly from sewage outlets and industrial plants, are a major cause of gastro-intestinal disease in Third World countries, particularly in slum areas. In rural areas, dams and other water projects, notably irrigation (*qv*) programmes, have played a major role in spreading schistosomiasis, malaria and other diseases.

In 1978, the United Nations launched its International Drinking Water Supply and Sanitation Decade 1980-90, the aim being to supply piped water and latrines to more than 1,000 million people at a cost of $140,000 million. Lack of funding (the UN raised less than a quarter of its target figure) and inadequate maintenance have dogged the programme. In many rural areas, pumps corroded or simply broke down, never to be repaired.

Nonetheless, by 1985, the UN claimed to have supplied 345 million people in the Third World, most in rural areas, with a clean supply of water. Some 140 million had received new waste disposal facilities. In 1987, 1,200 million people were still without reasonable access to safe water, with 1,900 million lacking basic sanitation.

Though laudable in its aims, the drinking water decade has itself created many problems. By providing piped water to rural and urban areas, it has increased dramatically the consumption of water, thus depleting water supplies. As a result, according to N.D. Jayal, former adviser to the Planning Commission of India, fewer people in East Asia now have access to clean and adequate water than they did in the 1970s – this despite heavy expenditure on drinking water programmes. In Maharastra, India, groundwater depletion has increased the number of villages lacking a source of water from 112 in the mid-1970s to 23,000 in 1985. Attempts to increase water supplies through dam projects will inevitably exacerbate the problem by increasing the prevalence of waterborne disease. (See also: *Development*③, *Urbanization*③.)

Wave power

The harnessing of the vertical motion of sea waves to generate electricity.

Wave power devices have been tested in Britain and other countries, and one has been built off Bergen, Norway, but in Britain, where much of the early development took place, official support for them has declined. The official view is that they are expensive, and vulnerable to storm damage. Moreover, if they are to produce significant amounts of energy, they are likely to be large and thus possibly hazardous to shipping.

In fact, however, wave power machines do not have to be large. Several apparently viable designs for small installations appear capable of producing useful amounts of energy. An example is the Lanchester Polytechnic/Sea Energy Associates "Clam" design. Another is the oscillating water column device completed by Kvaerner Brug in Norway. The prototype has a peak rating of 500KW, and generates electricity at an estimated 3.4 pence (5.4 cents) a unit. The design is said to be suitable for larger installations. (See also: *Alternative technology*③, *Energy*Ⓔ.)

Wetlands

Wetlands are among the most productive ecosystems in the world. Covering six per cent of the Earth's land surface, these marshlands, swamps and bogs play a vital role in cleansing water of pollutants, regulating floods, and providing a habitat for numerous species of plants and animals (see illustration). Yet, throughout the world, wetlands are menaced by drainage schemes, land reclamation, pollution and the reservoirs of large dams (*qv*).

The major threat to wetlands comes from drainage and development schemes. More than half of the original 870,000sq km (335,000sq miles) of wetlands in the USA have been lost, 87 per cent of them being converted to farmland or rangelands.

In Africa,the giant Okovango swamp in Botswana is threatened with drainage, the plan being to develop the area for mining and

Wetlands – from firewood to water purification

An estimated 60-70 per cent of fish in US coastal waters, and two-thirds of shellfish, rely on wetlands for food, or for spawning grounds and nurseries. In Europe, the coastal wetlands of the Wadden Sea help to support almost 60 per cent of the North Sea's brown shrimp, 50 per cent of the sole, 80 per cent of the plaice and nearly all the herring.

As well as supporting fisheries, wetlands provide man with a wide range of other benefits. Mangroves yield roofing materials, synthetic fibres, fodder, tannin for leather, sugar, fuel and alcohol. Rice is a wetland plant, as are the oil palm, the sago palm, and many other plant foods. Wetlands also provide fuel in the form of peat and firewood.

Wetlands act as a buffer against floods, absorbing flood waters before they reach higher ground. Mangrove swamps also protect against tidal waves and storms.

Wetlands play a significant role in filtering out pollutants. Wetlands plants are able to absorb excess nutrients without succumbing to eutrophication, and the bacteria in many plants are able to transform a nutrient such as nitrogen into a gas, thus dispersing it into the atmosphere. In the USA, wetlands in Florida and elsewhere have been used to treat sewage, with remarkable success.

Wetlands are also excellent scavengers of pesticides, heavy metals and other toxins. Somehow, the plants immobilize these pollutants and prevent them from passing up the food chain. They can remove between 20 and 100 per cent of heavy metals.

Unfortunately, the cleansing powers of wetlands are not inexhaustible and many have succumbed to pollution.

When flood waters recede, wetlands provide rich pasture for livestock. In West Africa, as many as 1,500,000 people rely on floodplains for grazing cattle, sheep and goats.

agriculture. In the Sudan, work is already well-advanced on the Jonglei Canal, which will divert water from the marshlands of the Sudd swamps to Egypt and north Sudan. The project will reduce the grassland floodplains of the Sudd by an estimated 50 per cent, disrupting the traditional migration routes of the local pastoralists and depriving wildlife of vital habitats. Elsewhere, in Pakistan, the Farraka Barrage (*qv*) has wrought havoc with the wetland forests of the Ganges Delta.

Many wetlands have succumbed to pollution, particularly in areas where there is run-off of agricultural chemicals or a concentration of heavy industry. One wetland that has been badly polluted is the Wadden Sea, where many fish and marine mammals are contaminated with high levels of PCBS (*qv*) and chlorinated hydrocarbons.

Since 1971, wetlands have been "protected" by the Convention on Wetlands of International Importance Especially as Waterfowl Habitats. Known less formally as The Ramsar Convention, after the Iranian town where it was signed, the convention is ratified by 43 nations, the most recent signatories being the United States, Ireland, Suriname and Belgium, all of which joined in 1986. To date, the convention has saved some 200,000sq km (77,000sq miles) of wetlands. Sadly, this is only a minute proportion of the 8,500,000sq km (3,280,000sq miles) of wetlands worldwide. (See also: *Development*③, *Pesticides*Ⓐ Ⓟ.)

Whales

The first records of commercial whaling date from the 10th century, when the Basques started hunting in the Bay of Biscay. By the 16th century, they had decimated the Biscayan right whales and had moved north to Greenland, Iceland and Newfoundland in pursuit of the bowhead whale. They were joined by Dutch and English whalers. Together the whalers hunted the bowhead to commercial extinction by 1680. The 18th and 19th centuries marked the period of the great American whalers. The sperm whale was their target, and at the peak of the period, between 1835 and 1846, 600 ships were hunting them in the Atlantic. By the 1920s the stocks were practically exhausted. In the Pacific, the gray whale in Baja California was decimated within 45 years of the first kill.

By the turn of this century, interest had shifted to the Antarctic. There, the pillage was aided by the invention of the explosive harpoon and the development of steam vessels, which enabled the pursuit of faster whales. In 1904, the British established a whaling station on South Georgia, where thousands of blue whales were processed each year. In 1914, the Norwegians, in competition with the British, developed the first factory ship. Within ten years, such ships had stern slipways, allowing a whale to be hauled on board and processed within an hour.

As a result of these technological improvements, species after species has been hunted to the brink of extinction. In the 1930s, as the blue whale was getting rarer, attention was turned to the smaller fin whale. By 1960, the fin whale stocks had been commercially depleted and whalers concentrated on the minke whale, which until then had been left in peace.

The International Convention for the Regulation of Whaling, ratified in 1946, marked the first attempt to regulate the hunting of whales on an international basis. The International Whaling Commission (IWC), set up by the convention, voted in 1982 to impose a five-year moratorium on commercial whaling, starting in October 1985. But the commission has no power to enforce its decisions and, by the end of the 1985-1986 season, 4,969 minke whales had been killed. Several countries, notably the Soviet Union, Japan, South Korea, Norway and Iceland, have exploited loopholes in the IWC rules, which allow whales to be killed for scientific research. In 1987, the IWC passed a US resolution calling on governments not to issue licences for "scientific" whaling. None of the IWC nations has to abide by the decision but any nation that fails to do so will face stringent US sanctions on imports and will be barred from fishing in US waters. Nonetheless, Iceland, South Korea and Japan have all said they intend to continue scientific whaling. If whaling continues unabated, despite the IWC's efforts, there seems little chance that these gentle and intelligent creatures will grace our seas for much longer. (See also: *Krill*Ⓒ, *Overfishing*Ⓒ.)

Wind power

The wind has been used to turn sails and work grain mills and water pumps for centuries, but, since the 1940s, entirely new "windmills" have been developed for generating electricity. With rigid blades rather than sails these bear little resemblance to their picturesque but less efficient ancestors, owing much instead to the introduction of new lightweight materials and advances in aerodynamics, especially in the design of their rotor blades.

Individual models range in size from those designed to produce a few kilowatts for domestic or farm use, to large models intended to serve entire communities or to feed power into a grid network and generating about one megawatt. Still larger generators, of up to three megawatt capacity, have been built.

In the continental USA, several thousand wind generators supply some 50 gigawatts of power to grid systems. Many more supply domestic users directly. In Hawaii, the Kahua Ranch "wind farm" feeds two megawatts of power into the local grid. In other countries, wind power is still largely seen as a way to supply remote communities.

In the USA in all, 600MW of wind turbines have been installed, mainly in California; in Denmark 1,400 grid linked machines are now in operation. In the UK it has been estimated that wind power could provide 54 terawatt hours of electricity – about a fifth of annual demand – more than the nuclear industry provides at present. This would mean a 1,000 large (4MW), and 3,000 medium-sized (500KW) wind turbine generators (WTGS) on land and 2,000 very large (8MW) ones offshore. (See also: *Alternative technology* Ⓔ, *Energy* Ⓔ, *Soft energy paths* Ⓔ.)

Wind power

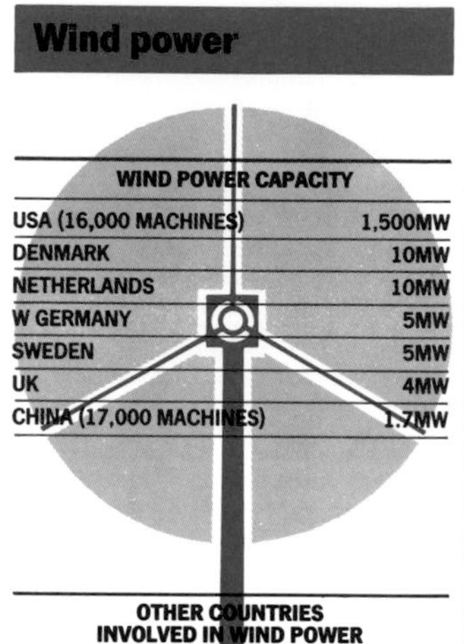

OTHER COUNTRIES INVOLVED IN WIND POWER

AUSTRALIA	INDIA
CANADA	ITALY
FRANCE	JAPAN
GREECE	SPAIN

Wind power has reached a critical point in its development; it is beginning to pass from the research phase into large-scale commercial electricity production. The above list shows estimates of national power capacity, given in megawatts, for the leading countries in the field.

Though there is a growing trend in many countries toward larger machines, China has so far chosen to concentrate on quantity rather than capacity. It has more machines than the USA, but each produces only a single watt.

Windscale

See: *Sellafield.*

Windscale fire

Construction work on the first British plutonium-production piles (or reactors) began in September 1947 at Windscale. The first was operational in October 1950, the second eight months later. Their design did not allow the heat generated to be used for electricity production; instead they were air-cooled, and the heat dispersed by huge electric fans. On the morning of October 8 1957 the physicist operating the No 1 pile was involved in a series of routine procedures that involved raising and lowering the core temperature, but apparently without the Pile Operating Manual available to him. He allowed the core temperature to rise to the point at which fuel began to melt, unaware that the gauges did not take their readings from the hottest part of the core. There was no sign of trouble until, 42 hours later, instruments in the stack filters indicated that radioactivity was being uncontrollably released.

The reactor core was on fire, accelerated by the cooling fans. Flames shot out of the core from the fuel discharge face and, at the height of the accident, 11 tonnes of uranium and the graphite moderator were on fire. Carbon dioxide was used in the first attempt to put out the fire, but served only to intensify it. Despite the real possibility of a hydrogen explosion, fire hoses were coupled directly into a line of fuel channels and water poured onto the blazing core. During the morning of October 11 the fire, which had been out of control for more than 24 hours, finally subsided.

The stack's filters, which had only been incorporated into the pile design as an afterthought, probably prevented a major accident from becoming a disaster. Even so, huge quantities of radioactivity were released, the most dangerous of which were some 20,000 curies of iodine-131. (Only in 1983 was it admitted that polonium was also released, presumably because an earlier admission would have revealed that Britain was manufacturing a type of polonium-triggered hydrogen bomb that was already, by 1957, out of date.) Two million litres (528,000 US gallons) of milk were poured away from farms in the surrounding 500sq km (190sq miles), but the local health effects cannot be known in detail, because the identities of those people near the fire were not recorded. Subsequent epidemiological assessments indicate that 33 people may have died as a result of the accident.

After the fire it was obvious that No 1 pile would never operate again. Both it and the No 2 pile were shut down and filled with concrete – they had anyway been superseded. A year earlier the new Calder Hall reactors, the military precursors of the civil Magnox (*qv*) reactors, came on stream at Windscale and took over the function of supplying military plutonium. They had (and still have) the advantage over the air-cooled piles of producing electricity as a by-product, thus off-setting the costs of plutonium production. (See also: *Nuclear accident*Ⓝ, *Sellafield*Ⓝ.)

Wood preservatives

Mixtures of compounds designed to protect vulnerable timber from attack by insects, such as woodworm and death watch beetle, and fungi, such as wet and dry rot. A wood preservative will therefore often contain an insecticide and a fungicide, though woodworm killers and some anti-rot products can be bought separately.

Several toxic materials have been used as insecticides in these products. Dieldrin (*qv*) was popular but has been phased out for use in Britain and several other countries. Lindane HCH (*qv*) has also been used, but because of its lethal effects on bats it has been replaced in some products by permethrin, which is less toxic but also less long-lasting. Products used to preserve exterior timber, such as fences, do not usually contain a specific insecticide, but some of the chemicals used – notably creosote – are poisonous to insects as well as fungi. Where termite attack is expected, toxic insecticides may be used.

Some timbers are impregnated under pressure or vacuum with a copper/chromium/arsenic mixture that poses risks when the timber is burned or sanded. Tributyl tin (qv) compounds are also used for interior timbers such as joists and rafters. Probably the most harmful compound used in this way is pentachlorophenol (*qv*).

The World Bank

Established in 1944 at the Bretton Woods Conference, attended by 44 nations, the World Bank is now by far the biggest and most influential of the multilateral development banks (MDBs). Not only does it invest some $16,000 million a year in the Third World, but those development projects that it approves or partly funds have little difficulty in obtaining further funds from other MDBS and bilateral aid organizations.

The World Bank works in close association with its sister organization, the International Monetary Fund (IMF) (*qv*). Whereas the bank is responsible for long-term investments, the IMF's role is to provide bridging finance to countries with short-term balance of payments difficulties.

The World Bank is made up of three main bodies: the International Bank for Reconstruction and Development (IBRD), which raises money on international financial markets and lends at commercial rates of interest; the International Development Agency (IDA), which provides loans at concessional rates of interest to the poorest countries and which obtains its funds from member subscriptions, donations and grants from the IBRD; and the International Finance Corporation (IFC), which lends money to the private sector with government guarantees.

The bank is controlled by member countries whose votes are in accordance with the size of their respective donations. This means that the World Bank is largely controlled by the United States, the biggest donor. The World Bank is located in Washington and its director is invariably a US citizen.

The bank's policies reflect Western and, in particular US, foreign policies vis-a-vis the Third World, which is not altogether surprising since it and the IMF were set up largely to assure the continued expansion of the Western industrial economy. The bank is firmly committed to free trade and its policies help to promote the sale of Western manufactured goods to Third World countries, which, in turn, are encouraged to export agricultural produce and other raw materials needed in the West.

Since the bank is a commercial organization, and requires that its loans be repaid, it will lend money only for projects that have some chance of being profitable. A special department of the bank, the Operations Evaluation Department (OED), advises it on the likely economic viability of proposed schemes.

Social and environmental considerations have, up till now, been almost entirely ignored, in spite of continued assurances to the contrary. In 1984, the bank admitted in an internal memorandum, "As a matter of routine, environmental issues are not considered." It was also admitted that it "does not have the capacity to conduct sector work on environmental issues on a routine basis." At the time, the bank's Office of Environmental Affairs had just five members of staff, who were expected to evaluate the environmental impact of more than 300 new projects a year. Out of the 6,000 employees of the bank, there was just one professionally trained ecologist.

Projects funded by the bank that have had adverse effects on the environment and on local people include many large dams and other water development schemes; road construction programmes, particularly in forested areas; and many livestock rearing programmes.

In recent years, three of the bank's schemes have been subject to particularly strong criticisms: the Polonoroeste project in Brazil; the Narmada Valley Project (*qv*) in India; and the Transmigration (*qv*) scheme in Indonesia.

In 1986, the bank's environmental record was severely criticized by the Appropriations Sub-Committee of the US Senate, chaired by Senator Kasten of Wisconsin. Kasten said that if people knew the truth about the projects that the World Bank and other multilateral development banks were financing, they would be "out on the streets" demanding why their money was "being spent on this kind of destruction." Kasten's committee refused to provide further US backing for the Polonoroeste project.

The bank received another blow when Hugh Foster, its alternate US Executive Director, voted against financing a series of hydro-electric projects in Brazil, describing them as "pure folly" and "environmental disasters".

The bank has now promised to mend its ways. Its development committee issued a report entitled "Environment: Growth and Development" in April 1987, in which it set out new environmental guidelines for its activities. Future projects will have to meet three criteria: "economic growth, poverty alleviation and environmental protection."

It remains to be seen whether the bank's new interest in environmental protection is genuine. Environmentalists charge that previous promises to protect the environment have proved hollow, citing a 1971 undertaking "to assure that the projects financed by (the bank) do not have serious ecological consequences."

It is also debatable whether or not the goals set out in the new guidelines are compatible. In particular, it is difficult to see how the pursuit of economic growth can be reconciled with environmental protection. See also: *Development* ③, *Sustainable development* ③, *Third World debt* ③.)

World Commission on Environment and Development
See: Brundtland Report.

X-ray
See: *Radiation and health.*

Zero population growth
State where the number of deaths in a population equals the number of births, with the result that there is no natural increase in the size of the population over time. However, populations which have achieved zero population growth may still increase as a result of immigration.

In the USA, concern over the environmental and social impact of overpopulation has spawned a powerful lobby in favour of zero population growth, including a pressure group of that name.

Despite the ecological imperative of reducing human numbers, however, population growth rates are only part of the problem. Levels of material consumption are also of critical importance. Even if the US population were to stabilize at its present level, for example, current standards of consumption could not be sustained for long without placing an intolerable burden on the environment. (See also: *Demographic transition* ③, *Family planning* Ⓛ③, *Population* ③, *Population explosion* ③.)

Organizations

Australia and New Zealand

Albury-Wodonga Environment Centre, Shop 3, 608 Dean Street, Albury NSW 2640 *Conservation, pollution*
Australian Conservation Foundation, GPO Box 1875, Canberra ACT 2601 *All issues*
Barrier Environment Group, West Post Office, Broken Hill NSW 2880 *Conservation, pollution*
Cairns and Far North Environment Centre, 1st Floor, Andrejic Arcade, 55 Lake Street, Cairns QLD 4870 *Conservation, pollution*
Canberra and South East Region Environment Centre, Childers Street Buildings, Kingsley Street, Acton ACT 2601 *All issues*
Capricorn Conservation Council and Environment Centre, 135 William Street, Rockhampton QLD 4700 *Conservation, pollution*
Central Australia Conservation Council, PO Box 2796, Alice Springs NT 5750 *Conservation, pollution*
Conservation Council of Victoria, 247–251 Flinders Lane, Melbourne VIC 3000 *Conservation, pollution*
Conservation Centre and Conservation Council of SA Inc, 120 Wakefield Street, Adelaide SA 5000 *Conservation, pollution*
Conservation Council of the South East Region and Canberra, GPO Box 1875, Canberra ACT 2601 *Conservation, pollution*
Conservation Council of WA Inc, 1st Floor, 794 Hay Street, Perth WA 6000 *Conservation, pollution*
Environment Centre (Victoria), 247–251 Flinders Lane, Melbourne VIC 3000 *Conservation, pollution*
Friends of the Earth, 366 Smith Street, Collingwood VIC 3066
Friends of the Earth – New Zealand, PO Box 39–065, Auckland West, New Zealand *All issues*
Geelong Environment Council, PO Box 639, Geelong VIC 3320
Greenpeace – New Zealand, Nagel House, 5th Floor, Courthouse Lane, Auckland, New Zealand *All issues*
Illawarra Environment Centre, Suite 11, 157 Crown Street, Wollongong NSW 2500 *Conservation, pollution*

Launceston Environment Centre Inc, 34 Patterson Street, Launcestion TAS 7250 *Conservation, pollution*
Mid North Coast Environment Centre, c/- "Coolenberg", 60 Lake Road, Port Macquarie NSW 2444 *Conservation, pollution*
Nature Conservation Council of NSW, c/- NSW Environment Centre, 176 Cumberland Street, Sydney NSW 2000 *Conservation, pollution*
Newcastle Ecology Centre, Room 6, Trades Hall, Union Street, Newcastle NSW 2300 *All issues*
New Zealand Values Party, Box 116, New Plymouth, New Zealand *Political, all issues*
North Coast Environment Council, c/- J. Tedder, Pavans Road, Grassy Head, via Stuarts Point NSW 2441 *Conservation, pollution*
North Queensland Conservation Council Inc and Townsville Environment Centre, 477 Flinders Street, Townsville QLD 4810 *Conservation, pollution*
Rainforest Information Centre, PO Box 368, Lismore NSW 2480 *Conservation*
South Coast Environment Centre, Shop 5, Sams Market Arcade, Orient Street, Batemans Bay NSW 2500 *Conservation, pollution*
Tasmanian Conservation Trust Inc, 102 Bathurst Street, Hobart TAS 7000 *Conservation, pollution*
Tasmanian Environment Centre Inc, 102 Bathurst Street, Hobart TAS 7000 *Conservation, pollution*
The Big Scrub Environment Centre, 88A Keen Street, Lismore NSW 2840 *Conservation, pollution*
The Environment Centre, Beaman Park, James Street, Yeppoon QLD 4703 *Conservation, pollution*
The Environment Centre (NT), Shop 8, The Track, Darwin NT 5794
The Environment Centre of NSW, 176 Cumberland Street, Sydney NSW 2000 *Conservation, pollution*
The Wilderness Society, 130 Davey Street, Hobart TAS 7000 *Conservation*
The Wilderness Society, PO Box 188, Civic Square ACT 2608 *Conservation*
Total Environment Centre, 18 Argyle Street, Sydney NSW 2000 *All issues*
Tweed Valley Conservation Trust, 35 Tweed Arcade, Queen Street, Murwillumbah NSW 2484 *Conservation, pollution*
Wide Bay/Burnett Conservation Council, 16 Alvie Street, Mayborough QLD 4650 *Conservation, pollution*
World Wildlife Fund – Australia, Level 17, St Martins Tower, 31 Market Street, GPO Box 528, Sydney NSW 2001 *All issues*
World Wildlife Fund – New Zealand, PO Box 6237, Wellington, New Zealand *All issues*

Canada

Alberta Environmental Law Centre, 202 10110–142 St, Edmonton, Alberta T5N 1P6 *All issues*
Alberta Wilderness Association, Box 6398, Station D, Calgary, Alberta T2P 2E1 *Conservation*
Canadian Arctic Resources Committee, 46 Elgin Street, Room 11, Ottawa, Ontario K1P 5K6 *Conservation*
Canadian Coalition on Acid Rain, 112 St Clair Avenue West, Toronto, Ontario M4V 2Y3 *Pollution*
Canadian Coalition for Nuclear Responsibility, CP 236, Saccursale, Snowden, Montreal, Québec H3X 3T4
Canadian Environmental Law Association, 234 Queen Street West, Toronto, Ontario M5V 1Z4 *All issues*
Canadians for Conservation of Tropical Nature (CCTN), Faculty of Environmental Studies, York University, 4700 Keele Street, Toronto, Ontario M3J 1P6 *Third World, conservation*
Conservation Council of New Brunswick, 180 St John Street, Fredericton, NB E3B 4A9 *Wildlife*
Consumers United to Stop Food Irradiation, RR #1, Illderton, Ontario NOM 2AO *Lifescan, pollution, nuclear*
Ecology Action Centre, 1657 Barrington Street, Suite 520, Halifax, Nova Scotia B3J 2A1 *All issues*
Energy Probe, 100 College Street, Toronto, Ontario M5G IL5 *Energy*
Federation of Ontario Naturalists, 355 Lesmill Road, Don Mills, Ontario M3B 2W8 *Conservation*
Friends of The Earth, 53 Queen Street, Room 16, Ottawa, Ontario
Green Party, 831 Commercial Drive, Vancouver, British Colombia, Canada *All issues*
Greenpeace, 427 Bloor St West, Toronto, Ontario M5S 1X7 *All issues*
Institute of Concern for Public Health, Suite 343, 67 Mowat Avenue, Toronto, Ontario M6K 3E3 *All issues*
Manitoba Naturalists Society, #214, 190 Rupert Avenue, Winnipeg, Manitoba R3B 0N2 *Conservation*
Nuclear Awareness Project, 730 Bathurst St, Toronto, Ontario M5S 2R4
Ontario Environment Network, PO Box 125 Station P, Toronto, Ontario M5S 2Z7 *All issues*
Ontario Public Interest Research Group, 229 College Street, Toronto, Ontario K1P 5C5 *All issues*
Operation Clean Niagra, c/- M Howe, 83 Gage Street, Niagra-on-the-Lake, Ontario M5S 2R4 *Pollution*
Pollution Probe, 12 Madison Avenue, Toronto, Ontario M5R 2S1 *Pollution*
Probe International, 100 College, Toronto, Ontario M5G 1L5 *All issues*
Recycling Council of Ontario, PO Box 310 Station P, Toronto, Ontario M8S 2S8 *Pollution, lifescan, conservation*
Société Pour Vaincre la Pollution, CP 65 Place d'Armes, Montreal, Québec H2Y 3E9 *All issues*
SPEC – The Society for the Promotion of The Environment Conservation, 2150 Maple St, Vancouver BC. V6J 3T3 *All issues*
Tomorrow Foundation, 10511 Saskatchewan Drive, Edmonton, Alberta T6E 4S1 *All issues*
West Coast Environmental Law Association, 1001 – 207 West Hastings Street, Vancouver BC V6B 1H7 *Nuclear, pollution, lifescan, conservation*
World Wildlife Fund (CANADA), 60 St Clair Avenue East, Suite 201, Toronto, Ontario M4T 1N5 *All issues*

United Kingdom

Alternate Technology Group, Open University, Milton Keynes, Buckinghamshire *Research on alternate technologies*
Campaign Against Lead in Petrol, 68 Dora Road, London SW19 7HH

Civic Trust, 17 Carlton House Terrace, London SW1Y 5AW *Conservation*
Compassion in World Farming, Lyndham House, Greatham, Petersfield, Hampshire *Agriculture*
Conservation Society, 12a Guilford Street, Chertsey, Surrey *All issues*
Council for the Protection of Rural England, 4 Hobart Place, London SW1N 0HY *Conservation*
Countryside Commission, John Dower House, Crescent Place, Cheltenham, Gloucestershire Gl50 3RA *Conservation*
Earth Resources Research Ltd, 25 Pentonville Road, London N1 9JY *All issues*
Ecoropa (UK), Henbant, Crickwell, Powys NP8 1TA *All issues*
Farm and Food Society, 4 Willifield Way, London NW11 7XT *Agriculture*
Food Additives Campaign Team, 25 Horsell Road, London N5 1XL *Lifescan*
Friends of the Earth, 377 City Road, London EC1 *All issues*
Green Alliance, 60 Chandos Place, London WC2 *All issues*
Green Deserts, High Rougham, Bury St Edmonds, Suffolk *Third World, conservation*
Green Party, 10 Station Parade, Balham High Road, London SW12 9A2 *All issues*
Greenpeace, 30–31 Islington Green, London N1 *All issues*
London Food Commission, PO Box 291, London N5 1DU *Agriculture*
Men of the Trees, Turns Hill Road, Crawley Down, Crawley, West Sussex RH10 4HL *Third World, conservation*
Nature Conservancy Council, 19 Belgrave Square, London SW1X 8PY *Conservation*
National Centre for Alternative Technology, Llwyngwern Quarry, Maclynlleth, Powys *Energy*
National Society for Clean Air, 136 North Street, Brighton, Sussex *Pollution*
Organic Growers Association, Aeron Park, Llangietho, Dyfed *Agriculture*
Oxfam, 274 Banbury Road, Oxford OX2 7D2 *All issues*
Panos Institute, 8 Alfred Place, London WC1E 7EB *Lifescan*
Pesticide Trust, C/o Earth Resources Research, 258 Pentonville Road, London N1 9JY *Agriculture*
Royal Society for Protection of Birds (RSPB), The Lodge, Sandy, Bedfordshire SE1 2DL *All issues*
Soil Association, 86 Colston Street, Bristol BS1 5BB *Agriculture*
Survival International, 310 Edgeware Road, London W2 1DY *Third World*
The Ecologist, Worthyvale Manor, Camelford, Cornwell *All issues*
The International Institute for Environment and Development, 3 Endsleigh Street, London WC1 0DO *All issues*
Urban Centre for Appropriate Technology, 82 Colston Street, Bristol *Energy, conservation*
Vegetarian Society, 53 Marloes Road, London W8 6LA *Agriculture*
Woodland Trust, Autumn Trust, Dysart Road, Gratham, Lincs NG31 6LL *Conservation*
World Food Assembly, 15 Devonshire Terrace, London WC2 3DW *Agriculture, Third World*
World Forest Campaign, 6 Glebe Street, Oxford OX4 1DG *Conservation*
World Wildlife Fund (UK), Panda House, 11 Ockford Road, Goldaming, Surrey *Conservation*

United States

Defenders of Wildlife, 1244 19th Street NW, Washington DC 20036 *Wildlife, conservation*
Earth Island Institute, 300 Broadway, Suite 28, San Francisco, CA 94133–9905 *All issues*
Environmental Defence Fund, 1616 P Street NW, Washington DC 20036 *All issues*
Environmental Law Institute, 1616 P Street NW, Washington DC 20036 *All issues*
Environmental Policy Institute, 1616 P Street, Washington DC 20036 *All issues*
Friends of the Earth, 530 7th Street SE, Washington DC 20003 *All issues*
Friends of Trees, PO Box 1466, Chelan WA 98816, *Conservation*
Food First, 1885 Mission Street, San Francisco, CA 94103–3584 *Third World, agriculture*
Global Tomorrow Coalition, 1325 G Street NW, Suite 1003, Washington DC 2005 *All issues*
Greenpeace, 1611 Connecticut Avenue NW, Washington DC 20009 *All issues*
Intermediate Technology Development Group, 777 United Nations Plaza, New York NY 10017 *Third World, energy*
National Audbon Society, 801 Pennsylvania Avenue SE, Washington DC 20003 *Conservation, pollution*
National Wildlife Federation, 1325 Massachusetts Avenue NW, Washington DC 20005 *Conservation, pollution*
Natural Resources Defence Council, 1350 New York Avenue NW, Washington DC 20003 *Conservation, wildlife*
New Alchemy Institute, 237 Hatchville Road, East Falmouth, MA 02536 *Energy, conservation*
Permaculture Association, PO Box 202, Orange, MA 01364 *Agriculture, conservation, pollution*
Rachel Carson Trust, 8940 Jones Mill Road, Chevy Chase, MA 20815, *Agriculture, pollution, conservation*
Rainforest Action Network, 466 Green Street, Suite 300, San Francisco, CA 94133 *Third World, conservation, agriculture*
Sierra Club, 330 Pennsylvania Avenue NW, Washington DC 20005 *Conservation, wildlife*
Survival International, 2121 Decatur Place, Washington DC 20008 *Third World*
The Oceanic Society, 1536 16th Street NW, Washington DC 20036 *Nuclear, pollution*
The Wilderness Society, 1400 Eye Street NW, Washington DC 20005 *Conservation, wildlife*
World Resources Institute, 1750 New York Avenue NW, Suite 230 Washington DC 20006 *All issues*
Worldwatch Institute, 1776 Massachusetts Avenue NW, Washington DC 20036
World Wildlife Fund (USA), 1250 24th Street NW, Washington DC 20037 *Conservation*

Index

Bold entries and page numbers refer to entries in the A-Z section. Figures in italics refer to illustrations.

Acknowledgments

The General Editors wish to thank the editorial staff of Mitchell Beazley Publishers and the following contributors; Michael Allaby, John Ballantine, Peter Bunyard, Marcus Colchester, Alexander Goldsmith, Bill Hare, Jackie Karas, Mick Kelly, David Lomax, Robert Mann, Erik Millstone, Brian Price, Jonathan Spink, John Valentine; also Hilary Datchens and Helen Lamper for tneir help with the imputing.

Artwork credits
Artwork on the following pages by Grundy and Northedge: 23, 30, 34, 36, 56, 58, 62, 67, 68, 75, 84, 86br, 93, 97, 99, 103, 106, 108, 110, 117, 121, 127, 130, 135, 139, 143, 145, 147, 158, 160, 169, 172, 176, 180, 190, 194, 199, 207, 210, 213, 217, 221, 224, 231. By Paul Drayson and Aziz Khan: 32, 37, 40, 69, 70, 76, 81, 83, 86bl, 114, 123, 124, 151, 155, 162, 168, 188, 201, 232.
34/35 based on information contained in the World Nuclear Industry Handbook 1987, supplement to Nuclear Engineering International, November 1986, published by Reed Business Publishing.
83/86 br/168 based on information in "World Resources 1987"; tables 8.1/8.4/8.6, p. 115/116/118, published by Basic Books, Inc., © World Resources Institute and the International Institute for Environment and Development 1987.
139 based on information supplied by Hilary Bacon drawn from the following sources: ● "Weak Electromagnetic Effects in Biomedical Systems", presented by Dr. C.W. Smith at the All-India Institute of Medical Sciences in March 1985; "High-Sensitivity Bio-Sensors and Weak Environmental Stimili", reprinted in *International Industrial Biotechnology*, April/May 1986; *Electro-Magnetic Man*, by Hilary Bacon and Simon Best, published by J.M. Dent (in press). ■ ▲ Dr. E. Leeper and Dr. N. Wertheimer in *International Journal of Epidemiology*, Vol. 109, No. 3, 1979 and Vol. 2, No 4, 1982, *American Journal of Epidemiology*, Vol. 3, No. 3, 1980; Enander, Hellstrom and Tomenius in *International Symposium on Occupational Health and Safety in Mining and Tunnelling*, June 1982; Barnes, Savitz and Wachtel see *Science News*, Vol. 131, Feb. 1987.
169 based on information supplied by the Australian National Parks and Wildlife Service.
213 based on information in *The Economist*, 15 August 1987, p. 56, © by The Economist.
124 based on information in "World Resources 1986", table 9.7, p. 153, published by Basic Books, Inc., © World Resources Institute and the International Institute for Environment and Development 1986.
169 based on information provided by the Australian Conservation Foundation.
190 based on information in "No Place to Hide: Nuclear Winter and the Third World", Earthscan briefing document No 43.

Research
Hilary Bacon, Michael Clark, Chris Cooper, Maurice Hanssen, Clark Robinson Ltd.

Picture credits
Cover: John Garrett (background and torso); Science Photo Library (Earth); 9: Patrick Piel/Gamma/Frank Spooner Pictures; 12, 13, 15: Nina Leopold Bradley; 16/17: Patrick Piel/Gamma/Frank Spooner Pictures; 18/19: Charlie Waite/Landscape Only; 21: K.N.A. Studio X/Gamma/Frank Spooner Pictures; 25: Rex Features Ltd; 26: Mark Edwards/Panos Pictures; 33: Novosti/Gamma/Frank Spooner Pictures; 38: Svenskt/SIPA Press/Rex Features Ltd; 43: J.L. Atlan/Sygma/The John Hillelson Agency Ltd.; 44: Novosti/Gamma/Frank Spooner Pictures; 48: Alain Nogues/Sygma/The John Hillelson Agency Ltd., 51: ESA/Science Photo Library; 52: Popperfoto; 55: Fay Godwin's Photo Files; 65: Crown Copyright, courtesy Building Research Station; 71: Alan Grant/Life © 1954 Time Inc./Colorific!; 72,73: Regis Bossu/Sygma/John Hillelson Agency Ltd.; 76/7: Richard Baker; 79: Spectrum Colour Library; 80: Mark Edwards/Panos Pictures; 82: Popperfoto; 87: Sipa Press/Rex Features Ltd.